Springer Proceedings in Mathematics & Statistics

Volume 380

This book series features volumes composed of selected contributions from workshops and conferences in all areas of current research in mathematics and statistics, including data science, operations research and optimization. In addition to an overall evaluation of the interest, scientific quality, and timeliness of each proposal at the hands of the publisher, individual contributions are all refereed to the high quality standards of leading journals in the field. Thus, this series provides the research community with well-edited, authoritative reports on developments in the most exciting areas of mathematical and statistical research today.

More information about this series at https://link.springer.com/bookseries/10533

David D. Hanagal · Raosaheb V. Latpate ·
Girish Chandra
Editors

Applied Statistical Methods

ISGES 2020, Pune, India, January 2–4

Springer

Editors
David D. Hanagal
Pune, Maharashtra, India

Girish Chandra
Division of Forestry Statistics
Indian Council of Forestry Research
and Education
Dehradun, Uttarakhand, India

Raosaheb V. Latpate
Department of Statistics
Savitribai Phule Pune University
Pune, Maharashtra, India

ISSN 2194-1009 ISSN 2194-1017 (electronic)
Springer Proceedings in Mathematics & Statistics
ISBN 978-981-16-7931-5 ISBN 978-981-16-7932-2 (eBook)
https://doi.org/10.1007/978-981-16-7932-2

Mathematics Subject Classification: 97K80, 62-XX, 62-08, 62-02, 62-06

This Springer imprint is published by the registered company Springer Nature Singapore Pte Ltd.
The registered company address is: 152 Beach Road, #21-01/04 Gateway East, Singapore 189721, Singapore

Preface

The vast applications of statistics through its various dimensions including data science, data mining, stochastic and reliability modelling, sampling and estimation techniques, decision-making, etc., have been developed in recent times. The applications of statistics are increased as the newly methods are being adopted by the researchers who involved in the multidisciplinary areas, viz. astronomy, forensic studies, clinical trials, agriculture, forestry and environment, epidemiology and finance, and even in the business administration in order to take correct decision about the policy and other interventions. In addition, statistics is emerged as a powerful tool in the successful governmental development and industrial progress through tackling difficult challenges. Applied statistics has become an indispensable discipline.

This book is a collection of recent developments in several areas of statistics in the form of chapters (18 in total) written by eminent statistician in their areas of expertise. We tried our best to invite those authors who could capture new developments in statistical methodology and their possible use in important diversified disciplines. The real applications of a wide range of key topics, including small area estimation techniques, Bayesian models for small areas, ranked set sampling, fuzzy supply chain, probabilistic supply chain models, dynamic Gaussian process models, grey relational analysis, multi-item inventory models, etc., are well presented. The possible use of the other models including generalized Lindley shared frailty models, Benktander Gibrat risk model, decision-consistent randomization method for SMART designs and different reliability models is also discussed. This book includes many detailed practical and worked examples that illustrate the real-life applications of recently developed statistical methods. The titles included in this volume are designed to appeal to applied statisticians, students, research project leaders and practitioners of various marginal disciplines and interdisciplinary research. The relative scarcity of reference material covering statistical applications as compared with the readily available books also enhances the utility of this book. We are sure that the book will benefit researchers and students of different disciplines for improving research through the methodological and practical knowledge of applied statistics.

Chapter "Bayesian Order-Restricted Inference of Multinomial Counts from Small Areas" provides the use of Bayesian paradigm to adaptively pool the data on body mass index cell probabilities over small areas. To estimate the finite population proportion of healthy individuals in each household, a hierarchical Bayesian sub-area beta-binomial models presented in Chapter "A Hierarchical Bayesian Beta-Binomial Model for Sub-areas". Chapter "Hierarchical Bayes Inference from Survey-Weighted Small Domain Proportions" also focuses on the hierarchical Bayes approach of small area estimation for survey-weighted proportions of district level employment. Chapter "Efficiency of Ranked Set Sampling Design in Goodness of Fit Tests for Cauchy Distribution" discusses the use of ranked-set sampling in goodness of fit tests by considering the particular Cauchy distribution. The fuzzy supply chain single period (newsboy) inventory model has been used to obtain optimal order quantity, retailers profit, manufacturers profit and total supply chain profit under decentralized supply chain in Chapter "Fuzzy Supply Chain Newsboy Problem Under Lognormal Distributed Demand for Bakery Products". Chapter "Probabilistic Supply Chain Models with Partial Backlogging for Deteriorating Items" deals with the newly developed inventory models by assuming various probability distributions for demand and deterioration rate under shortages of items. In Chapter "The Evolution of Dynamic Gaussian Process Model with Applications to Malaria Vaccine Coverage Prediction", several popular test function-based computer simulators to illustrate the evolution of dynamic Gaussian process models have been used along with its application to predict the coverage of malaria vaccine worldwide.

Chapter "Grey Relational Analysis for the Selection of Potential Isolates of *Alternaria Alternata of Poplar*" narrates the use of gray relational analysis method for obtaining the best fungal isolates of *Alternaria Alternata* of poplar *(Populus deltoides)* tree. Chapter "Decision Making for Multi-Items Inventory Models" proposes the multi-item inventory models under declining demand with the Weibull distributed deterioration rate for credit period and quantity discount. The Bayesian estimation of generalized Lindley-shared frailty models based on reversed hazard rate for Australian twin data is proposed in Chapter "Modeling Australian Twin Data Using Generalized Lindley Shared Frailty Models". Chapter "Ultimate Ruin Probability for Benktander Gibrat Risk Model" obtains the ultimate ruin probability for Benktander–Gibrat risk model using the Laplace transform, generalized exponential integrals, Meijer G-function and Bromwich integral. Chapter "Test of Homogeneity of Scale Parameters Based on Function of Sample Quasi Ranges" presents a multi-sample test for homogeneity of scale parameters against simple ordered alternative based on function of sample quasi-ranges given censored data, as well as for data contaminated with outliers. To combine the advantages of Q-learning-decision-consistent strategies and response-adaptive designs while controlling for covariate balance, a Bayesian response-adaptive, covariate-balanced and Q-learning-decision-consistent randomization method for SMART designs is proposed in Chapter "A Bayesian Response-Adaptive, Covariate-Balanced and Q-Learning-Decision-Consistent Randomization Method for SMART Designs". The Bayesian inference for finite population characteristics is presented in Chapter "An Introduction to Bayesian Inference for Finite Population Characteristics". Important reliability applications through repairable systems with arrival time of server and stress-strength reliability estimation for multi-component system are given in Chapters "Reliability Measures of Repairable Systems with Arrival Time of Server" and "Stress-strength Reliability Estimation for Multi-component System Based on Upper Record Values Under New Weibull-Pareto Distribution". Chapter "Record Values and Associated Inference on Muth Distribution" describes the record values and associated inference on Muth distribution. The book ends with the statistical linear calibration in data with measurement errors and given in Chapter "Statistical Linear Calibration in Data with Measurement Errors".

Pune, India David D. Hanagal
Pune, India Raosaheb V. Latpate
Dehradun, India Girish Chandra

Acknowledgments

We extend our thanks and appreciations to the authors for their continuous support during finalization of the book. We would like to express our sincere thanks to Shamim Ahmad, Senior Editor, Mathematical Sciences, Springer Nature, for his continuous support and cooperation from planning to the finalization of this volume. We would like to thank the anonymous referees for their valuable comments and suggestions for improvement of this book. The support received from late Dr. Hukum Chandra till the acceptance of this book is incredible.

David D. Hanagal
Raosaheb V. Latpate
Girish Chandra

Obituary

Dr. Hukum Chandra

(7 November 1972 to 26 April 2021)

Dr. Hukum Chandra passed away on 26 April 2021 at the age of 48. He contributed to this book as a resource person. He was an eminent scientist who pioneered the inception and popularization of "Small Area Estimation" technique in the official statistics system of India. He was working as National Fellow and Principal Scientist at the ICAR-Indian Agricultural Statistics Research Institute, New Delhi, India. He did his M.Sc. in statistics from the University of Delhi, Ph.D. from the University of Southampton, UK, and Postdoctoral Research from the University of Wollongong, Australia. He worked on diverse areas of methodological and applied problems in statistics, including survey design and estimation methods; small area estimation; bootstrap methods; disaggregate-level estimation and analysis of agricultural, socio-economic and health indicators; spatial models for survey data; statistical methodology for improvement in agricultural and livestock statistics; energy management in production agriculture; evaluation of agriculture census and survey schemes. He has received number of awards and appreciations for his research contributions such as National Award in Statistics from the Ministry of Statistics and Programme Implementation, Government of India; ICAR National Fellow Award; Cochran-Hansen Award from International Association of Survey Statisticians; Young Researcher/Student Award of the American Statistical Association; Lal Bahadur Shastri Outstanding Young Scientist Award of ICAR; Recognition Award of the National Academy of Agricultural Sciences; Prof. P. V. Sukhatme Gold Medal Award; and Dr. D. N. Lal Memorial Award of Indian Society of Agricultural Statistics. He was a recipient of the Commonwealth Scholarship offered by the Commonwealth Scholarship Commission in the UK. He was Elected Member of International Statistical Institute, The Netherlands; Fellow of National Academy of Agricultural Sciences, India; and Fellow of Indian Society of Agricultural Statistics. He has worked as Council Member of the International Association of Survey Statisticians. As International Consultant of Food and Agricultural Organization of the United Nations, he has worked in Sri Lanka, Ethiopia and Myanmar to strengthen the Agricultural Statistics System. He has published more than 125 research papers in reputed journals of high impact factor. He has published four books, several technical bulletins, project reports, chapters, working papers and training and teaching reference manuals. He has delivered a number of invited talks in many national and international platforms of repute worldwide. He has supervised four Ph.D. and four M.Sc. students. Our heartfelt tribute to Dr. Hukum Chandra.

Contents

Editors and Contributors

About the Editors

David D. Hanagal is Honorary Professor at the Symbiosis Statistical Institute, Symbiosis International University, Pune, India. He was previously a professor at the Department of Statistics, Savitribai Phule Pune University, India. He is an elected fellow of the Royal Statistical Society, UK and also an elected fellow of the Indian Society for Probability and Statistics. He is an editor and on the editorial board of several respected international journals. He has authored five books, six book-chapters and published over 135 research publications in leading journals. He guided 9 Ph.D. students in different areas of Statistics namely, Reliability, Survival analysis, Frailty models, Repair and replacement models, Software reliability, and Quality loss index. He has delivered more than 100 invited talks in many national and international platforms of repute worldwide. He has supervised nine Ph.D. students. He also has worked as a visiting professor at several universities in the USA, Germany, and Mexico, and delivered a number of talks at conferences around the globe. His research interests include statistical inference, selection problems, reliability, survival analysis, frailty models, Bayesian inference, stress–strength models, Monte Carlo methods, MCMC algorithms, bootstrapping, censoring schemes, distribution theory, multivariate models, characterizations, repair and replacement models, software reliability, quality loss index, and nonparametric inference. With more than 40 years of teaching experience and more than 35 years of research experience, he is an expert on writing programs using SAS, R, MATLAB, MINITAB, SPSS, and SPLUS statistical packages.

Raosaheb V. Latpate graduated from the Department of Statistics, Dr. Babasaheb Ambedkar Marathwada University, Aurangabad, in 2005. He had completed his Ph.D. degree from Dr. Babasaheb Ambedkar Marathwada University, Aurangabad. He is working as Assistant Professor in Department of Statistics and Center for Advanced Studies, Savitribai Phule Pune University, Pune, India. He has organized three International conferences and three national conference/workshop/Faculty

Development Programme on statistical methods and applications. Dr. Raosaheb Latpate is a member of number of professional societies and institutions, namely, the International Statistical Institute, International Indian Statistical Association, Society for Statistics and Computer Applications, and the Indian Society for Probability and Statistics. He has published project reports, book chapters. His research interests include genetic algorithm, fuzzy set theory, supply chain management, logistics and transportation problem, simulation and modeling and Sample Survey.

Girish Chandra is presently working as Scientist in the Division of Forestry Statistics, Indian Council of Forestry Research and Education (ICFRE), Dehradun (an autonomous body under the Ministry of Environment, Forest and Climate Change, Govertment of India). Before joining ICFRE (HQ) in 2013, he worked at the Tropical Forest Research Institute, Jabalpur and at Central Agricultural University, Sikkim for about 7 years. Dr. Girish is a recipient of Cochran–Hansen Prize 2017 of International Association of Survey Statisticians, the Netherlands. He is also honoured with ICFRE Outstanding Research Award 2018 besides Young Scientist Award in Mathematical Sciences from the Government of Uttarakhand, India. He has published over 45 research papers in various respected journals and have three books. He has organised two national conferences on Forestry and Environmental Statistics. Dr. Girish is a member of various scientific institutions, including the International Statistical Institute, International Indian Statistical Association, Computational and Methodological Statistics.

Contributors

Priyanka Anjoy ICAR-Indian Agricultural Statistics Research Institute, New Delhi, India

Sangeeta Arora Department of Statistics, Panjab University, Chandigarh, India

Hassan Bakouch Department of Mathematics, Faculty of Science, Tanta University, Tanta, Egypt

M. R. Bhosale Department of Statistics, Shri Shahumandir Mahavidyalaya, Pune, India

Girish Chandra Division of Forest Statistics, Indian Council of Forestry Research and Education, Dehradun, India

Hukum Chandra ICAR-Indian Agricultural Statistics Research Institute, New Delhi, India

Lu Chen National Institute of Statistical Sciences, Washington, DC, USA

Xinyu Chen Department of Mathematical Sciences, Worcester Polytechnic Institute, Worcester, MA, USA

Tianjiao Dai Department of Biostatistics, The University of Texas MD Anderson Cancer Center, Houston, TX, USA

Anil Gaur Department of Statistics, Panjab University, Chandigarh, India

Santosh Gitte Department of Statistics, Mumbai University, Mumbai, India

Pragya Gupta Department of Statistics, Central University of Rajasthan, Rajasthan, India

David Hanagal Department of Statistics, Savitribai Phule Pune University, Pune, India

M. Harshvardhan Indian Institute of Management Indore, Indore, Madhya Pradesh, India

Kanchan Jain Department of Statistics, Panjab University, Chandigarh, India

Shubhashree Joshi Department of Statistics, Bangalore University, Bengaluru, India

Harmanpreet Singh Kapoor Department of Mathematics and Statistics, Central University of Punjab, Bathinda, India

R. U. Khan Faculty of Agricultural Sciences, Department of Plant Protection, Aligarh Muslim University, Aligarh, India

Sandesh Kurade Department of Statistics, MES Abasaheb Garware College, Pune, India

Raosaheb Latpate Department of Statistics, Savitribai Phule Pune University, Pune, India

Kalpana K. Mahajan Department of Statistics, Panjab University, Chandigarh, India

M. Mahdizadeh Department of Statistics, Hakim Sabzevari University, Sabzevar, Iran

S. C. Malik Department of Statistics, M.D. University, Rohtak, India

Balgobin Nandram Department of Mathematical Sciences, Worcester Polytechnic Institute, Worcester, MA, USA

Arvind Pandey Department of Statistics, Central University of Rajasthan, Rajasthan, India

Parameshwar V. Pandit Department of Statistics, Bangalore University, Bengaluru, India

Pritam Ranjan Indian Institute of Management Indore, Indore, Madhya Pradesh, India

Bayesian Order-Restricted Inference of Multinomial Counts from Small Areas

Xinyu Chen and Balgobin Nandram

Abstract Body mass index (BMI) can be a useful indicator of health status, and people can fall in different cells. Estimating BMI cell probabilities for small areas can be difficult, due to a lack of available data from national surveys. We have data from a number of counties in the USA, and it is sensible to assume that BMI may be similar across the counties for each cell. Overall, the cell probabilities for each county follow a unimodal order restriction, and so, it is sensible to assume the same for the individual counties (small areas). Moreover, we assume that the counties are similar with some variations. In this setting, it is convenient to use the Bayesian paradigm to adaptively pool the data over areas. Therefore, we use a hierarchical multinomial Dirichlet model with order restrictions, to model the cell counts and the cell probabilities, thereby permitting a borrowing of strength across areas. We provide efficient Gibbs samplers to make inference about the cell probabilities for multinomial Dirichlet models with and without order restrictions (a model with the same pooling structure). To make inference, we compute the posterior distributions of the cell probabilities for both models. In general for most counties, as expected, the posterior distributions of cell probabilities of the model with order restrictions have significantly less variation, as measured by posterior standard deviations and coefficients of variation, than those of the model without order restrictions.

Keywords Bayesian computation · Body mass index · Multinomial distribution · Monte Carlo methods · Unimodal order restrictions

---

X. Chen (✉) · B. Nandram
Department of Mathematical Sciences, Worcester Polytechnic Institute, 100 Institute Road, Worcester, MA 01609, USA
e-mail: xchen7@wpi.edu

B. Nandram
e-mail: balnan@wpi.edu

D. D. Hanagal et al. (eds.), *Applied Statistical Methods*, Springer Proceedings in Mathematics & Statistics 380, https://doi.org/10.1007/978-981-16-7932-2_1

1 Introduction

In many surveys, questionnaires have items that are categorized into several cells. These items may be filled in by people from different areas or groups, which may be small. Estimates of cell probabilities for individual areas may not be reliable, and a statistician might need to pool data from different small areas (Rao and Molina, 2015). Furthermore, there may be important information over the cells from each area and this information can be incorporated into a model to provide additional improvement. So, our problem is to obtain a methodology to pool information across areas and to incorporate information across the cells in each area. The Bayesian paradigm is attractive for this problem, and we start with the hierarchical Bayesian multinomial Dirichlet model; then, we incorporate the order restrictions over the cell probabilities into this model. We have a specific application on body mass index (BMI), a lifestyle variable in the USA, where BMI can be categorized into five cells, where the order restriction might hold.

Body mass index (BMI) is a person's weight in kilograms divided by the square of height in meters. BMI provides a simple numeric measure of a person's fatness. A person with a higher BMI may have higher chance to get certain diseases (e.g., diabetes). BMI can be used to categorize people's weight that may lead to health problems, but it cannot provide medical diagnostics of the health of an individual. Knowing BMI status well among different areas can help politicians to make better health plans and improve medical care. We use data from the third National Health and Nutrition Examination Health Survey (NHANES III) to provide improved inference for each of 35 largest counties with a population of at least 500,000. But the sample size of small areas such as counties may be too small to generate reliable and accurate estimates. Borrowing strength across small areas to find more accurate estimates is necessary and possible. The hierarchical Bayesian model is straightforward and easy to understand, and Markov chain Monte Carlo method can overcome computational difficulties. Here, it is natural to use the hierarchical Bayesian multinomial Dirichlet model to understand the BMI data. It is important to note that when all counties are combined into a large sample, the order restriction that we use in our model holds, but because of the sparseness of the data within counties the order restrictions might fail.

BMI data are usually categorized into different cells such as underweight (cell 1), normal (cell 2), overweight (cell 3), obese1 (cell 4) and obese2 (cell 5). We assume that there are ℓ areas and the cell counts are denoted by n_{ij}, $i = 1, \ldots, I$, $j = 1, \ldots, K$, where $K = 5$ and $I = 35$ in our application on BMI. We assume that $(n_{i1}, \ldots, n_{i5})$ are multinomial counts with probabilities $(\theta_{i1}, \ldots, \theta_{i5})$ and in a Bayesian model $(\theta_{i1}, \ldots, \theta_{i5})$ follow a Dirichlet distribution with common hyperparameters, which have noninformative priors. This is the hierarchical Bayesian multinomial Dirichlet model; see Nandram (1998). Our new model puts the same order restriction over the θ_{ij} for the ith area. So in the second stage, a Dirichlet distribution with parameters $\boldsymbol{\mu}$ and τ will be an appropriate choice. It can be considered as a baseline, and parameters μ and τ increase the model flexibility. Without any specification or prior knowledge, a vague and flat prior should be used for parameters μ and τ.

Wu et al. (2016) combined domain estimation and the pooled adjacent violator algorithm to construct new design-weighted constrained estimators of wage for U.S. National Compensation Survey. They assumed constrained estimators satisfying the monotonicity. Malinovsky and Rinott (2010) presented predictors with an appropriate amount of shrinkage for the particular problem of ordered parameters in the context of small area estimation. Their performance is close to that of the optimal predictors. Heck and Davis-Stober (2019) provided a comprehensive discussion about linear inequality constraints, such as the set of monotonic order constraints for binary choice probabilities on the parameters of multinomial distributions for psychological theories. They also described a general Gibbs sampler for drawing posterior samples. A suitable order restriction assumption can increase model precision. Li (2008) made a great overview about statistical inference under order restrictions. He also considered the inference of ordered binomial probabilities in frequentist statistics. From Wu, Meyer and Opsomer's research about order restriction to Li's review, they proved that the order constraints should be considered in order to improve efficiency and minimize bias, which can be done in different aspects.

Dunson and Neelon (2003) proposed a general and easy to implement approach for order-restricted inferences on regression parameters in generalized linear models. Their approach is interesting because instead of choosing a prior distribution with the support on the constrained space, which is expected to result in major computational difficulties, they proposed to map draws from an unconstrained posterior density using an isotonic regression transformation. In particular, Gelfand et al. (1992) suggested first choosing a prior density without considering the constraint and then discarding draws inconsistent with the constraint. However, they were not working within the context of small area estimation and their problem is not about order cell probabilities in several multinomial distributions. Therefore, our approach of incorporating order restriction into the prior distributions is natural in our study on BMI.

In the small area context, most of these papers cover order restriction across areas (e.g., Wu et al., 2016). However, in this paper, we are not interested in order restriction across areas, but rather we are interested in order restriction across the cell probabilities within each area. Nandram (1997) provided a good discussion about a hierarchical Bayesian approach for taste-testing experiment and appropriate methods for the model. To select the best population, he studied three criteria based on the distribution of random variables representing values on a hedonic scale using the simple tree order (see also Nandram, 1998).

Nandram, Sedransk and Smith (1997) improved estimation of the age composition of a population of fish with the help of order restrictions. They proposed different order restrictions for different fish length strata. With the help of the Gibbs sampler, they showed that order restrictions provided large gains in precision for estimating the proportion of fish in each age class. The research of Nandram, Sedransk and

Smith (1997) was motivated by Gelfand et al. (1992) and earlier Sedransk et al. (1985).

Since people have similarity that the majority in each county will be in the same level of BMI, it is reasonable to assume that the cell probabilities share a common effect and have the same order restrictions in each county. Actually, it seems that most people will have a third-level BMI, which is overweight, among those counties. So, it is reasonable to believe that the cell probabilities are unimodal in each county and the third level is the mode. With this information, our estimates for each county can be improved using a multinomial Dirichlet model with order restrictions such as $\theta_{i1} \leqslant \theta_{i2} \leqslant \theta_{i3} \geqslant \theta_{i4} \geqslant \theta_{i5}$ for the ith area. One feature of our approach is that Dirichlet distribution with parameters $\boldsymbol{\mu}$ and τ embodies the common effect and the same order restriction. At the second stage of model, parameter $\boldsymbol{\mu}$ has a similar order restriction as cell probabilities $\boldsymbol{\theta}_i$. It has more flexibility without increasing computation difficulty. The work in this paper is a large step forward from Chen and Nandram (2019), which appeared the Proceedings of the American Statistical Association.

The article is organized as follows. In Sect. 2, we present the hierarchical Bayesian multinomial Dirichlet model with order restrictions. In Sect. 3, we present our algorithms; specifically, we describe how to generate samples from posterior distributions and how to handle difficulties caused by order restrictions. In Sect. 4, we show how to analyze the BMI data in our application. Specifically, we show how to run the Gibbs sampler, assess the convergence of the Gibbs sampler and, more importantly, demonstrate how much improvement there is under the order restrictions. In Sect. 5, we also present a Bayesian diagnostic for the model with order restrictions and we discuss difficulties associated with a standard Bayesian diagnostic measure that may not be appropriate. Section 6 has a summary of our work. Also, there is an appendix with technical details and an important table, which shows the improvement that can occur under the order restrictions.

2 Multinomial Dirichlet Models

In this section, we describe the Bayesian methodology for the cell counts over the small areas. First, we give a review of the hierarchical Bayesian multinomial Dirichlet model without the order restriction. Then, we describe the hierarchical Bayesian multinomial Dirichlet model with the order restriction.

It is convenient to give some standard notations here. The multinomial distribution, $\boldsymbol{n} \sim \text{Multinomial}(n_{.}, \boldsymbol{\theta})$, is a discrete distribution over K-dimensional nonnegative integer vectors $\boldsymbol{n}$, where $\sum_{j=1}^{K} n_j = n_{.}$ and $\boldsymbol{\theta} = (\theta_1, \ldots, \theta_K)$. The probability mass function is given as

$$f(\boldsymbol{n}|\boldsymbol{\theta}) = \frac{\Gamma(n_{.} + 1)}{\prod_{j=1}^{K} \Gamma(n_j + 1)} \prod_{j=1}^{K} \theta_j^{n_j}, \qquad \sum_{j=1}^{K} n_j = n_{.}, \qquad n_i \geqslant 0.$$

This is a generalization of the binomial distribution. The Dirichlet distribution, $\boldsymbol{x} \sim$ Dirichlet($\boldsymbol{\alpha}$), is parameterized by positive scalar $\alpha_j > 0$ for $j = 1, 2, \ldots, K$, where $K \geqslant 2$. The probability density of $\boldsymbol{x}$ is

$$f(X|\boldsymbol{\alpha}) = \frac{\Gamma(\sum_{j=1}^{K} \alpha_j)}{\prod_{j=1}^{K} \Gamma(\alpha_j)} \prod_{j=1}^{K} x_j^{\alpha_j - 1}, \sum_{j=1}^{K} x_j = 1, x_j \geq 0, j = 1, \ldots, K.$$

The Dirichlet distribution is multivariate generalization of the univariate beta distribution. It is convenient that the Dirichlet forms a conjugate prior for the multinomial distribution, thereby leading to relatively simpler computations.

2.1 *Model Without Order Restriction (M1)*

Nandram et al. (2019) have a useful discussion of hierarchical Bayesian multinomial Dirichlet model without order restriction and the methodology needed to fit it.

We provide the Bayesian hierarchical multinomial Dirichlet model. Letting n_{ij} be the cell counts, θ_{ij} the corresponding cell probabilities, $i = 1, 2, \ldots, I$, $j = 1, 2, \ldots, K$ and $n_{i.} = \sum_{j=1}^{K} n_{ij}$.

The general hierarchical Bayesian model is

$$\boldsymbol{n}_i|\boldsymbol{\theta}_i \overset{ind}{\sim} \text{Multinomial}(\boldsymbol{n}_{i\cdot}\,, \boldsymbol{\theta}_i),$$

$$n_{ij} \geqslant 0, \qquad \theta_{ij} \geqslant 0, \qquad \sum_{j=1}^{K} \theta_{ij} = 1,$$

$$\boldsymbol{\theta}_i|\boldsymbol{\mu}, \tau \overset{ind}{\sim} \text{Dirichlet}(\boldsymbol{\mu}\tau),$$

$$\pi(\boldsymbol{\mu}, \tau) = \frac{(K-1)!}{(1+\tau)^2},$$

$$\mu_j \geqslant 0, \qquad \sum_{j=1}^{K} \mu_j = 1, \qquad \tau > 0,$$

where, without any prior information, we take $\boldsymbol{\mu}$ and τ to be independent. Also, $E(\theta_{ij}) = \mu_j$, $\sum_{j=1}^{K} \mu_j = 1$ and $\boldsymbol{\mu}$ are cell means and τ is a prior sample size. Nandram, Sedransk and Smith (1997) had a similar model for stratified random sampling, not small areas, and they set the hyperparameters to be fixed. Therefore, our computations for order- restricted inference are much more difficult.

2.2 Model with Order Restrictions (M2)

We incorporate the order restriction into the hierarchical Bayesian Dirichlet multinomial model. We use a grid method in Gibbs sampler. This is more efficient than the method by Nandram (1998). Letting n_{ij} be the cell counts, θ_{ij} the corresponding cell probabilities, $i = 1, 2, \ldots, I$, $j = 1, 2, \ldots, K$,$\boldsymbol{n}_{i.} = \sum_{j=1}^{K} n_{ij}$ and we believe the mode of $\boldsymbol{\theta}_i s$ is θ_{im}, $1 \leqslant m \leqslant K$.

Specifically, we take

$$\boldsymbol{n}_i|\boldsymbol{\theta}_i \overset{ind}{\sim} \text{Multinomial}(\boldsymbol{n}_{i.}, \boldsymbol{\theta}_i), \qquad \boldsymbol{\theta}_i \in C \quad i = 1, \ldots, I,$$

where $C = \{\boldsymbol{\theta}_i : \theta_{i1} \leqslant \ldots \leqslant \theta_{im} \geqslant \ldots \geqslant \theta_{iK}, i = 1, \ldots, I\}$, and assume C is known. As mentioned above, in our BMI study, $C = \{\boldsymbol{\theta}_i : \theta_{i1} \leqslant \theta_{i2} \leqslant \theta_{i3} \geqslant \theta_{i4} \geqslant \theta_{i5}, i = 1, 2, \ldots, 35\}$.

At the second stage, we take

$$\boldsymbol{\theta}_i|\boldsymbol{\mu}, \tau \overset{ind}{\sim} \text{Dirichlet}(\boldsymbol{\mu}\tau), i = 1, \ldots, I,$$

$$\pi(\boldsymbol{\mu}, \tau) = \frac{K(m-1)!(K-m)!}{(1+\tau)^2}, \qquad \mu_j > 0, \quad \sum_{j=1}^{K} \mu_j = 1, \quad \boldsymbol{\mu} \in C_{\mu}.$$

Since $E(\theta_{ij}) = \mu_j$, $\boldsymbol{\mu}$ should have the same order restriction as $\boldsymbol{\theta}_i$, which is $\boldsymbol{\mu} \in C_{\mu}$,

$$C_{\mu} = \{\boldsymbol{\mu} : \mu_1 \leqslant \ldots \leqslant \mu_m \geqslant \ldots \geqslant \mu_K\}.$$

Using Bayes' theorem, the joint posterior distribution of all variables is

$$\begin{aligned}\pi(\boldsymbol{\theta}, \boldsymbol{\mu}, \tau|\boldsymbol{n}) &\propto \prod_{i=1}^{I}\{\prod_{j=1}^{K} \theta_{ij}^{n_{ij}} \frac{\prod_{j=1}^{K} \theta_{ij}^{\mu_j\tau-1} I_C I_{C_{\mu}}}{D(\boldsymbol{\mu}\tau)C(\boldsymbol{\mu}\tau)}\}\frac{1}{(1+\tau)^2} \\ &\propto \prod_{i=1}^{I}\{\frac{\prod_{j=1}^{K} \theta_{ij}^{n_{ij}+\mu_j\tau-1} I_C I_{C_{\mu}}}{D(\boldsymbol{\mu}\tau)C(\boldsymbol{\mu}\tau)}\}\frac{1}{(1+\tau)^2},\end{aligned}$$

where I_C and $I_{C_{\mu}}$ are the indicator functions under those order restrictions, and

$$C(\boldsymbol{\mu}\tau) \overset{denote}{=} \int_{\boldsymbol{\theta}_i \in C} \frac{\Gamma(\sum_{j=1}^{K} \mu_j\tau)}{\prod_{j=1}^{K} \Gamma(\mu_j\tau)} \prod_{j=1}^{K} \theta_{ij}^{\mu_j\tau-1} d\boldsymbol{\theta}_i,$$

$$D(\boldsymbol{\mu}\tau) = \frac{\prod_{j=1}^{K} \Gamma(\mu_j \tau)}{\Gamma[\sum_{j=1}^{K} \mu_j \tau]}.$$

A posteriori $\boldsymbol{\theta}_i | \boldsymbol{\mu}, \tau, \boldsymbol{n}_i \overset{ind}{\sim} \text{Dirichlet}(\boldsymbol{n}_i + \boldsymbol{\mu}\tau), \qquad \boldsymbol{\theta}_i \in C_i, i = 1, \ldots, I,$
where

$$f_{\boldsymbol{\theta}_i|\boldsymbol{\mu},\tau,\boldsymbol{n}} = \frac{\frac{\Gamma[\sum_{j=1}^{K}(n_{ij}+\mu_j\tau)]}{\prod_{j=1}^{K}\Gamma(n_{ij}+\mu_j\tau)} \prod_{j=1}^{K} \theta_{ij}^{n_{ij}+\mu_j\tau-1}}{\int_{\boldsymbol{\theta}_i \in C_i} \frac{\Gamma[\sum_{j=1}^{K}(n_{ij}+\mu_j\tau)]}{\prod_{j=1}^{K}\Gamma(n_{ij}+\mu_j\tau)} \prod_{j=1}^{K} \theta_{ij}^{n_{ij}+\mu_j\tau-1} d\boldsymbol{\theta}_i}$$

$$= \frac{\frac{\Gamma[\sum_{j=1}^{K}(n_{ij}+\mu_j\tau)]}{\prod_{j=1}^{K}\Gamma(n_{ij}+\mu_j\tau)} \prod_{j=1}^{K} \theta_{ij}^{n_{ij}+\mu_j\tau-1}}{C(\boldsymbol{n}_i + \boldsymbol{\mu}\tau)}.$$

3 Computations

It is straightforward to generate samples from M1; see Nandram (1998). In fact, using the Griddy Gibbs sampler, it can be done easier than the method in Nandram (1998). We present a new method for the order restrictions of $\boldsymbol{\mu}$ and $\boldsymbol{\theta}$ into two parts for model M2.

3.1 Sampling $\boldsymbol{\theta}$ in M2

In the first part of new method, the posterior of $\boldsymbol{\theta}$ has a recognizable distribution, which is the Dirichlet distribution with the order restriction. Sedransk et al. (1985) provided an efficient algorithm to generate random vectors from the constrained density. However, instead of drawing samples directly from the Dirichlet distribution with the order restriction, we present a direct sampling from truncated gamma distributions as from truncated Dirichlet distribution. Nadarajah and Kotz (2006) offered a method for truncated gamma distributions.

Method1 : To draw $\boldsymbol{\theta} = (\theta_1, \ldots, \theta_K) \sim \text{Dirichlet}(\alpha_1, \ldots, \alpha_K), \boldsymbol{\theta} \in C,$

denote $\boldsymbol{\beta} = (\beta_1, \ldots, \beta_K)$,

If $0 \leqslant \theta_1 \leqslant \theta_2 \leqslant \ldots \leqslant \theta_m \geqslant \ldots \geqslant \theta_K$, the mode is θ_m,

$0 \leqslant \beta_1 \leqslant \beta_2 \leqslant \ldots \leqslant \beta_m \geqslant \ldots \geqslant \beta_K$, the mode is β_m,

1. Draw $\beta_m \sim \text{Gamma}(\alpha_m, 1)$, where $0 \leqslant \beta_m < \infty$,
2. Draw $\beta_{m-1} \sim \text{Truncated Gamma}(\alpha_{m-1}, 1)$, where $0 \leqslant \beta_{m-1} \leqslant \beta_m$,

$\ldots \beta_1 \sim \text{Truncated Gamma}(\alpha_1, 1)$, where $0 \leqslant \beta_1 \leqslant \beta_2$,

3. Draw $\beta_{m+1} \sim$ Truncated Gamma$(\alpha_{m+1}, 1)$, where $0 \leqslant \beta_{m+1} \leqslant \beta_m$,
$\ldots \beta_K \sim$ Truncated Gamma$(\alpha_K, 1)$, where $0 \leqslant \beta_K \leqslant \beta_{K-1}$.

Then,

$$\theta_1 = \frac{\beta_1}{\beta_1 + \beta_2 + \ldots + \beta_K}, \ldots, \theta_{K-1} = \frac{\beta_{K-1}}{\beta_1 + \beta_2 + \ldots + \beta_K}, \theta_K = 1 - \sum_{i=1}^{K-1} \theta_i.$$

3.2 Gibbs Sampling for μ and τ

In the second part of new method, we present Gibbs sampling, a Markov chain Monte Carlo (MCMC) algorithm, for $\boldsymbol{\mu}$ with an order restriction. We present the modified Gibbs sampler for $\boldsymbol{\mu} \in C_\mu$ and τ. The joint posterior density is

$$\pi(\boldsymbol{\theta}, \boldsymbol{\mu}, \tau|\boldsymbol{n}) \propto \prod_{i=1}^{I} \{ \frac{\prod_{j=1}^{K} \theta_{ij}^{n_{ij}+\mu_j\tau-1} I_C I_{C_\mu}}{D(\boldsymbol{\mu}\tau) C(\boldsymbol{\mu}\tau)} \} \frac{1}{(1+\tau)^2},$$

where

$$C(\boldsymbol{\mu}\tau) = \int_{\boldsymbol{\theta}_i \in C} \frac{\Gamma(\sum_{j=1}^{K} \mu_j \tau)}{\prod_{j=1}^{K} \Gamma(\mu_j \tau)} \prod_{j=1}^{K} \theta_{ij}^{\mu_j\tau-1} d\boldsymbol{\theta}_i.$$

There is no recognizable conditional distribution of $\boldsymbol{\mu}$ and τ to generate samples. We use a griddy Gibbs sampling (See Nandram 1998) to draw $\boldsymbol{\mu}$ and τ from $\pi(\boldsymbol{\mu}, \tau|\boldsymbol{n})$ after integrating with respect to $\boldsymbol{\theta}$, we get

$$\pi(\boldsymbol{\mu}, \tau|\boldsymbol{n}) \propto \prod_{i=1}^{I} \{ \frac{D(\boldsymbol{\mu}\tau + \boldsymbol{n}_i) C(\boldsymbol{\mu}\tau + \boldsymbol{n}_i)}{D(\boldsymbol{\mu}\tau) C(\boldsymbol{\mu}\tau)} \} \frac{I_{C_\mu}}{(1+\tau)^2}$$
$$\propto \prod_{i=1}^{I} \{ \frac{\int_{\boldsymbol{\theta}_i \in C} \prod_{j=1}^{K} \theta_{ij}^{\mu_j\tau+n_{ij}-1} d\boldsymbol{\theta}_i}{\int_{\boldsymbol{\theta}_i \in C} \prod_{j=1}^{K} \theta_{ij}^{\mu_j\tau-1} d\boldsymbol{\theta}_i} \} \frac{I_{C_\mu}}{(1+\tau)^2}.$$

Chen and Shao (1997) mentioned that the importance sampling could be used to estimate the ratio, $\frac{\int_{\boldsymbol{\theta}_i \in C} \prod_{j=1}^{K} \theta_{ij}^{\mu_j\tau+n_{ij}-1} d\boldsymbol{\theta}_i}{\int_{\boldsymbol{\theta}_i \in C} \prod_{j=1}^{K} \theta_{ij}^{\mu_j\tau-1} d\boldsymbol{\theta}_i}$. We consider Dirichlet $(r\bar{n_j})$, where r is an adjustable ratio and $\bar{n_j} = \frac{\sum_{i=1}^{I} n_{ij}}{I}$. More details can be found in appendix.

Method 2:
1. Draw τ from $\pi(\tau|\boldsymbol{\mu}, \boldsymbol{n})$.
2. For j from m-1 to 1, draw μ_j from $\pi(\mu_j|\boldsymbol{\mu}^{(-j)}, \tau, \boldsymbol{n})$,

$$\text{where } 0 < \mu_j < \min\{\mu_{j+1}, \frac{1-\sum_{t=1,t\neq m,t\neq j}^{K}\mu_t}{2}\}.$$

3. For j from m+1 to K, draw μ_j from $\pi(\mu_j|\boldsymbol{\mu}^{(-j)}, \tau, \boldsymbol{n})$,

$$\text{where } 0 < \mu_j < \min\{\mu_{j-1}, \frac{1-\sum_{t=1,t\neq m,t\neq j}^{K}\mu_t}{2}\}.$$

4. Get $\mu_m = 1 - \sum_{j=1,j\neq m}^{K}\mu_j$; repeat Step 1 to Step 4 to get converged MCMC samples,

$$\mu^{(-j)} = (\mu_1, \ldots, \mu_{j-1}, \mu_{j+1}, \ldots, \mu_K).$$

4 Application to BMI

4.1 Body Mass Index

In our application, we use a selected subset of the female BMI data from NHANES III, where we use only the female BMI data from the 35 largest counties with a population at least 500,000. Our goal is to estimate the proportions of the BMI levels. Table 1 gives an illustration of the female BMI data of a few counties, where it can be seen that the cell probability is largest for the normal range and other probabilities roughly tail off on both sides to form the unimodal order restriction. Indeed, there are violations in some counties in the earliest and latest cells.

For large population counties, we consider that people randomly fall into five BMI categorical levels, which are underweight, normal, overweight, obese1 and obese2. Thus, for each county, the BMI counts can be assumed to follow a multinomial distribution because each individual person can be assumed to exist independently. Figure 1 shows a histogram of all BMI values for females aggregated into a single large sample. It can be clearly seen that the unimodal order restriction holds. Because the data in the individual counties are generally sparse, it is difficult to tell whether the unimodal order restrictions holds, a way to improve posterior inference. However, it is sensible to assume that the same unimodal restriction holds within all the counties. Therefore, we can use multinomial distributions to model the female BMI counts.

4.2 MCMC Convergence

We run 20,000 MCMC iterations, take 10,000 as a 'burn-in' and use every 10th to obtain 1,000 converged posterior samples. Table 2 gives the effective sample size of the parameters $\boldsymbol{\mu}, \tau$ for the model with the order restriction and the general model. The effective sample sizes are almost 1,000. Table 3 gives the p-values of the Geweke test for the parameters (Cowles and Carlin, 1996). The p-values are all large, so

Table 1 US female BMI data

State ID	County ID	BMI_lvl1	BMI_lvl2	BMI_lvl3	BMI_lvl4	BMI_lvl5
4	13	3	40	37	13	4
6	1	1	36	38	15	1
6	19	3	20	49	13	5
6	37	2	145	174	77	14
6	59	1	29	31	16	3
…	…	…	…	…	…	…

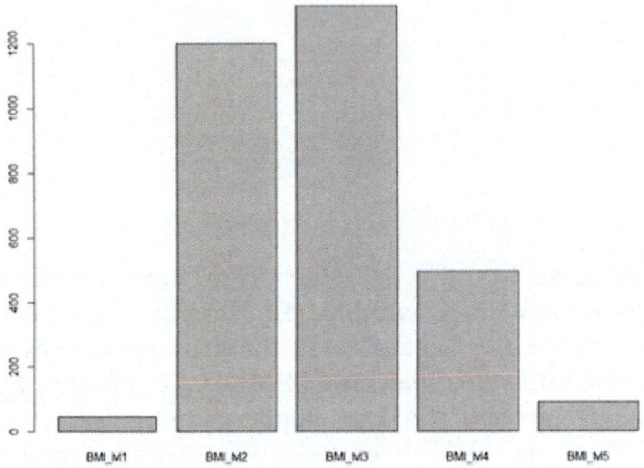

Fig. 1 Total counts for 35 counties

Table 2 Effective sizes

Models	μ_1	μ_2	μ_3	μ_4	μ_5	τ
W. order	974	1000	1000	1000	1000	1000
W/O order	859	1000	1000	971	1000	1032

Table 3 Geweke diagnostics

Models	μ_1	μ_2	μ_3	μ_4	μ_5	τ
W. order	0.4275	0.3221	0.2376	0.0895	0.3784	0.1393
W/O order	0.8352	0.785	0.6931	0.4425	0.3692	0.8983

we cannot reject that null hypothesis that the MCMC is stationary. Then, posterior samples can be used for the further inference (Table 4).

In Fig. 2, posterior densities of μ show a nice pattern and μ_3 is centered at the largest value. It means that our samples from μ posterior densities have an order restriction. It matches our model assumptions. But we notice that there is an overlap

Table 4 $\log(CPOi)$ for each county by model

County	Size	M1	M2	County	Size	M1	M2
1	97	−10.33	−9.73	18	61	−7.88	−7.78
2	91	−8.71	−7.87	19	52	−7.72	−7.72
3	90	−12.50	−13.40	20	64	−8.21	−8.17
4	412	−14.38	−13.36	21	49	−11.64	−13.40
5	80	−9.13	−8.26	22	77	−8.83	−8.62
6	66	−9.32	−9.64	23	50	−7.04	−6.42
7	62	−7.87	−7.37	24	70	−8.41	−7.91
8	53	−7.70	−7.16	25	64	−8.90	−9.98
9	73	−8.37	−7.69	26	60	−8.65	−7.91
10	81	−8.35	−8.54	27	48	−9.75	−9.78
11	98	−10.94	−10.42	28	52	−7.59	−7.11
12	84	−8.87	−9.13	29	75	−7.72	−7.21
13	217	−12.35	−12.56	30	82	−9.51	−9.29
14	72	−8.93	−8.81	31	75	−9.02	−8.31
15	87	−10.22	−9.54	32	102	−9.98	−10.20
16	101	−9.03	−8.12	33	129	−10.02	−8.84
17	99	−10.79	−10.48	34	84	−8.79	−7.85
				35	92	−9.31	−9.81

[1] *Note* Shaded Area: The model without order restrictions (M1)
Unshaded Area: The model with any order restrictions (M2)

Fig. 2 Posterior density of μ

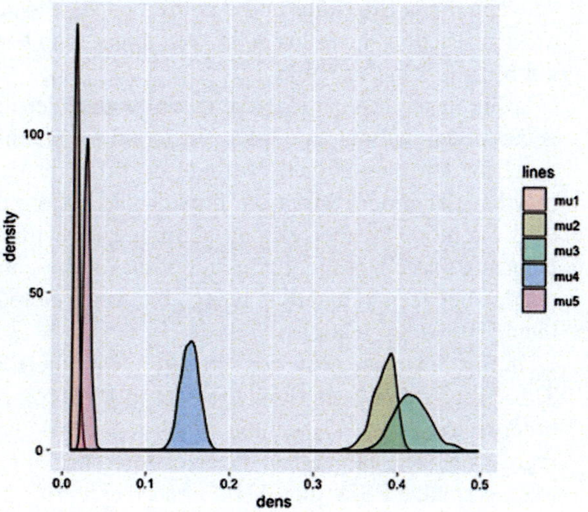

between μ_2 and μ_3. It is apparent that $\mu_2 \leqslant \mu_3$ may not be appropriate for BMI counts. The order restriction assumption may be too strong in this case.

4.3 Model Comparison

We compute the estimated cell probabilities for each county and their variances, which are the posterior sample means and posterior standard deviations of parameter $\boldsymbol{\theta}$. In Table 5 (appendix), we show their posterior means, standard deviations and coefficients of variation. We notice that the posterior means from the model with order restrictions (M2) have lower variances compared with the general model (M1). Generally, we have higher accuracy for the estimation of the cell probabilities $\boldsymbol{\theta}$. But for parameters θ_1 and θ_5 in some counties, such as the second county in Table 4, the model with order restriction does not gain precision on them. This is expected because the extreme cells are generally sparse. In general, many of the coefficients of variation are small enough to declare that the posterior means are reliable.

In Fig. 3, the top panel is the model with order restrictions and the bottom panel is the model without any order restriction for the same county, county 1. It can be seen from the plots of the posterior densities of the θ's that $\boldsymbol{\theta}$ in this county has an order restriction. Our unimodal assumption for this county holds. However in the first density (top panel) and the second density (bottom panel), there are overlaps between θ_2 and θ_3. It means that the order restriction may not hold for this county. The overlap between θ_1 and θ_5 is acceptable, since there is no direct comparison between them. Specially in the bottom panel, the densities from the model without order restriction show that θ_2 is even larger than θ_3. Our unimodal assumption may not be proper in this county.

n Fig. 4, the top panel is the model with order restrictions and the bottom panel is the model without any order restriction for another county, county 3. Plots of the posterior densities of the θ's without any order restriction show that $\boldsymbol{\theta}$s in each county may have an order restriction. It can be seen from the second density (bottom) that θ_3 is the mode for the cell probabilities even without order restriction assumption. It means that our unimodal assumption in this county is valid. Like in Fig. 3, the overlap between θ_1 and θ_5 is acceptable, since there is no direct comparison between them.

In Fig. 5, we use posterior standard deviation (PSD) to generate regression lines. Those regression lines show the overall PSD comparison between the model with order restrictions (M2) and the model without order restriction (M1). If the slope of regression line is larger than the black reference line whose slope is one, it means that M2 has smaller PSDs than M1. For each cell probability θ shown in different colors, the slope is larger than the reference lines. Therefore, we gain higher precision on estimation of cell probabilities among 35 counties.

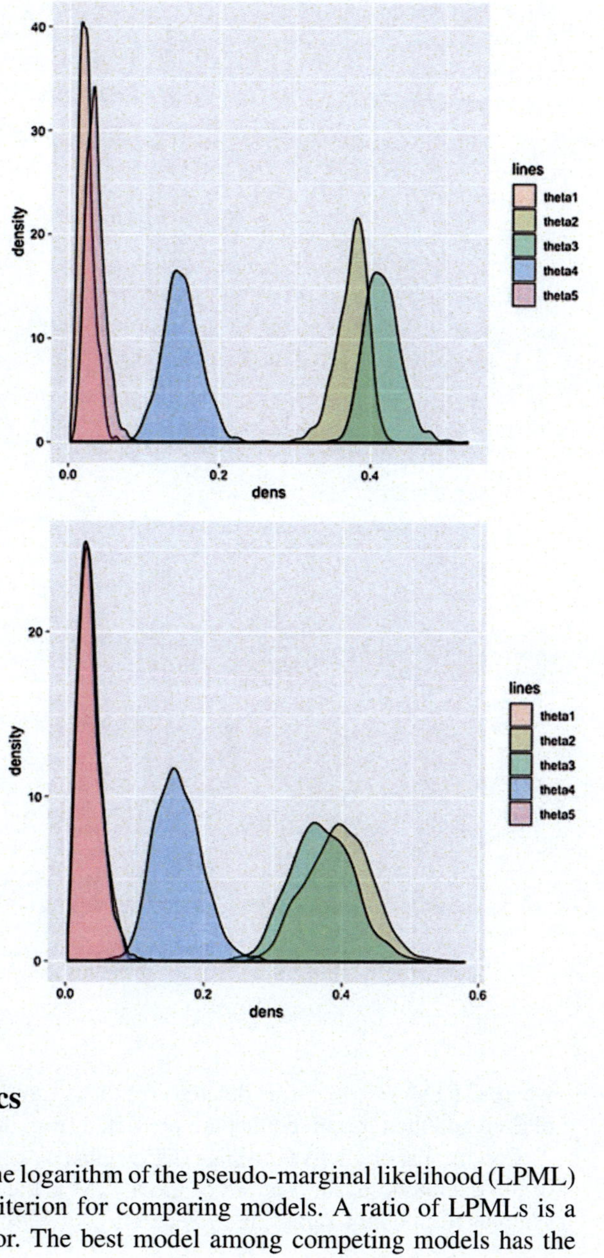

Fig. 3 Posterior densities of $\boldsymbol{\theta}$ for county 1

5 Bayesian Diagnostics

In the Bayesian framework, the logarithm of the pseudo-marginal likelihood (LPML) is a well-known Bayesian criterion for comparing models. A ratio of LPMLs is a surrogate for the Bayes factor. The best model among competing models has the largest LPML,

$$LPML = \sum_{i=1}^{I} \log(CPO_i),$$

Fig. 4 Posterior densities of $\boldsymbol{\theta}$ for county 3

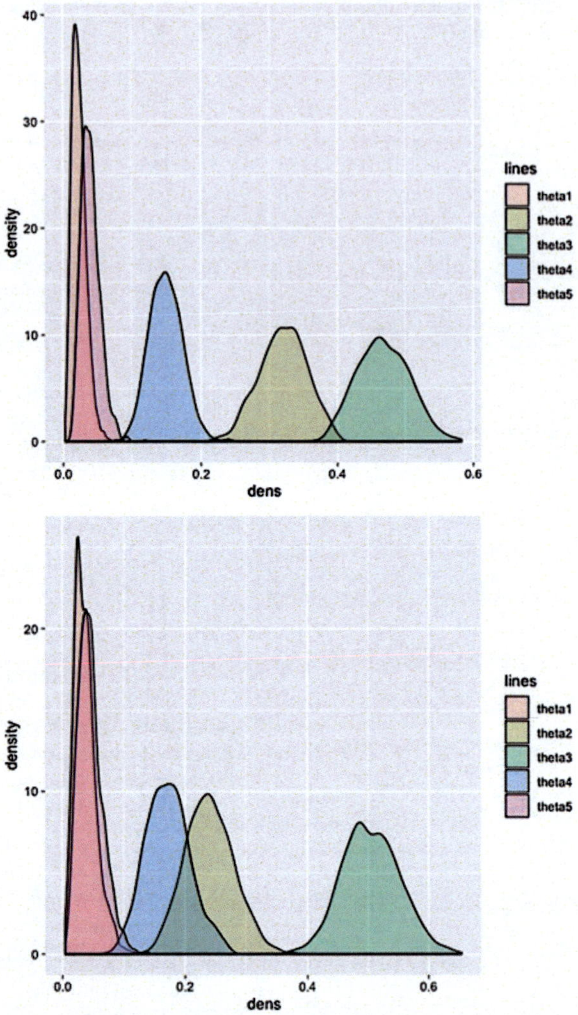

where $CPO_i = P(n_i \mid n_i \text{ is deleted})$ for the ith county. Essentially, the ith county is deleted and then its cell counts are predicted from the remaining counties.

Conditional predictive ordinate (CPO) can be estimated using the harmonic mean of the likelihood of the vectors of the n_{ij} (M is the number of converged posterior samples from Gibbs sampling in Sect. 3).

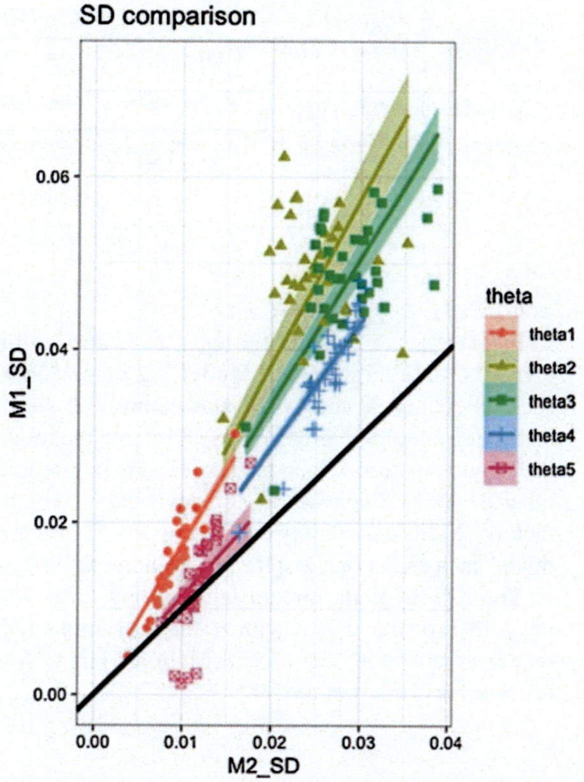

Fig. 5 Posterior standard deviation comparisons of $\boldsymbol{\theta}$s'

$$
\begin{aligned}
C\hat{P}O_i &= \left[\frac{1}{M}\sum_{h=1}^{M}\frac{1}{f(n_i|\boldsymbol{\mu}^{(h)},\tau^{(h)})}\right]^{-1} \\
&= \left[\frac{1}{M}\sum_{h=1}^{M}\frac{\prod_{j=1}^{K} n_{ij}!}{n_{i.}!}\,\frac{\int_{\boldsymbol{\theta}_i\in C}\prod_{j=1}^{K}\theta_{ij}^{\mu_j^{(h)}\tau^{(h)}-1}d\boldsymbol{\theta}_i}{\int_{\boldsymbol{\theta}_i\in C}\prod_{j=1}^{K}\theta_{ij}^{\mu_j^{(h)}\tau^{(h)}+n_{ij}-1}d\boldsymbol{\theta}_i}\right]^{-1}.
\end{aligned}
$$

As mentioned by Sedransk et al. (1985), a different importance sampling could be used to estimate the ratio,

$$
\frac{\int_{\boldsymbol{\theta}_i\in C}\prod_{j=1}^{K}\theta_{ij}^{\mu_j^{(h)}\tau^{(h)}-1}d\boldsymbol{\theta}_i}{\int_{\boldsymbol{\theta}_i\in C}\prod_{j=1}^{K}\theta_{ij}^{\mu_j^{(h)}\tau^{(h)}+n_{ij}-1}d\boldsymbol{\theta}_i}.
$$

CPO_i can be estimated as

$$\widehat{CPO}_{i\,(M2)} = \left[\frac{1}{M}\sum_{h=1}^{M}\frac{\prod_{j=1}^{K} n_{ij}!}{n_{i.}!}\frac{D(\boldsymbol{\mu}^{(h)}\tau^{(h)})C(\boldsymbol{\mu}^{(h)}\tau^{(h)})}{D(\boldsymbol{n}_i+\boldsymbol{\mu}^{(h)}\tau^{(h)})C(\boldsymbol{n}_i+\boldsymbol{\mu}^{(h)}\tau^{(h)})}\right]^{-1}$$

$$= \left[\frac{1}{M}\sum_{h=1}^{M}\frac{\prod_{j=1}^{K} n_{ij}!}{n_{i.}!}\frac{\int_{\boldsymbol{\theta}_i\in C}\prod_{j=1}^{K}\theta_{ij}^{\mu_{(h)}\tau_{(h)}-1}d\boldsymbol{\theta}_i}{\int_{\boldsymbol{\theta}_i\in C}\prod_{j=1}^{K}\theta_{ij}^{n_{ij}+\mu_{(h)}\tau_{(h)}-1}d\boldsymbol{\theta}_i}\right]^{-1}$$

$$= \left[\frac{1}{M}\sum_{h=1}^{M}\frac{\prod_{j=1}^{K} n_{ij}!}{n_{i.}!}\int_{\boldsymbol{\theta}_i\in C}\frac{\prod_{j=1}^{K}\theta_{ij}^{\mu_{(h)}\tau_{(h)}-1}}{\prod_{j=1}^{K}\theta_{ij}^{n_{ij}+\mu_{(h)}\tau_{(h)}-1}}\frac{\prod_{j=1}^{K}\theta_{ij}^{n_{ij}+\mu_{(h)}\tau_{(h)}-1}}{\int_{\boldsymbol{\theta}_i\in C}\prod_{j=1}^{K}\theta_{ij}^{n_{ij}+\mu_{(h)}\tau_{(h)}-1}}d\boldsymbol{\theta}_i\right]^{-1},$$

and $LPML \approx \sum_{i=1}^{I}\log(C\hat{P}O_i)$.

In Table 4, we present the $\log(CPO)$ for the individual counties. Most of the $\log(CPO)$ are larger under Model M2 than Model M1. However, a few counties, such as county 3, county 21 and county 32, show that the model without order restriction should be preferred. We assume the mode of cell probabilities should be at the third position, which may not hold for some counties. Perhaps, one can consider other Bayesian diagnostics (e.g., deviance information criterion—DIC—and Bayes factors). Mode uncertainty can be considered in future study to create a more flexible model such as the one inspired by Nandram (1997).

The LPML of the model without any order restriction (M1) is −326.76, and the LPML of the model with order restrictions (M2) is −319.11, which is larger than the model without order restrictions. The LPML suggests the model with order restrictions is the best model.

6 Conclusion

Hierarchical Bayesian multinomial Dirichlet models can be used to make inference for small areas. We have proposed the model with order restrictions to increase the accuracy of the estimation for the parameters. We have also shown how to generate samples from posterior distributions with order restrictions. We have significantly increased the precision of the estimation of cell probabilities for cells 2, 3 and 4 for the female BMI data. This is true for most of the counties. We have also shown difficulties in assessing model fit using Bayesian diagnostic measures ($\log(CPO)$) under order restriction; we believe that this is an open problem.

However, as shown in Fig. 3, the same unimodal assumption may be too strong. Some counties have more people in BMI level 2 than level 3, for instance, county 1; some counties have opposite situations. It seems that the mode of cell probabilities in each county is not fixed. Nandram and Sedransk (1995) and Nandram, Sedransk and Smith (1997) presented a good discussion about unimodal order restriction in a stratified population. They made inference about the proportion of firms belonging to each of several classes when there are unimodal order relations among the proportions. In this paper, the hyperparameters are specified and they did not have a small area estimation problem; our problem is much more difficult. They discussed an extension of their approach, which is the uncertain modal positions for their case. So for our model, one possible solution is considering uncertainty of modal position of the cell probabilities. Uncertain modal positions may be more robust for our BMI data. But again this is a much more difficult problem.

In the future, we might want to make the correlation structure of the Dirichlet distribution (all components are negatively correlated) more flexible. This can be done by using multivariate logit models with similar unimodal order restrictions as are studied in this paper. Also, we can use more flexible prior distributions such as Dirichlet process on the cell probabilities; this is a difficult problem with the unimodal order restrictions.

7 Appendix

7.1 *Details of Gibbs Sampling for* μ *and* τ

Since $\boldsymbol{\mu} \in C_\mu = \{\boldsymbol{\mu} = (\mu_1, \ldots, \mu_K) : \mu_1 \leqslant \ldots \leqslant \mu_m \geqslant \ldots \geqslant \mu_K, 0 \leqslant \mu_j \leqslant 1\}$,

for j from $m-1$ to 1, we have $\sum_{t=1}^{K} \mu_t = 1$,

$$\mu_m = 1 - \sum_{t=1, t\neq m}^{K} \mu_t \geqslant \mu_j,$$

$$1 - \Big(\sum_{t=1, t\neq m, t\neq j}^{K} \mu_t\Big) - \mu_j \geqslant \mu_j,$$

$$\mu_j \leqslant \frac{1 - \sum_{t=1, t\neq m, t\neq j}^{K} \mu_t}{2} \text{ and } 0 \leqslant \mu_j \leqslant \mu_{j+1}.$$

Therefore,

$$0 < \mu_j < \min\{\mu_{j+1}, \frac{1 - \sum_{t=1, t\neq m, t\neq j}^{K} \mu_t}{2}\}.$$

Similarly, for j from $m+1$ to K, $0 < \mu_j < \min\{\mu_{j-1}, \frac{1 - \sum_{t=1, t\neq m, t\neq j}^{K} \mu_t}{2}\}$.

Draw $\boldsymbol{\mu}$ and τ from $\pi(\boldsymbol{\mu}, \tau | \boldsymbol{n})$,

$$\pi(\boldsymbol{\mu}, \tau|\boldsymbol{n}) \propto \prod_{i=1}^{I}\{\frac{D(\boldsymbol{\mu}\tau + \boldsymbol{n}_i)C(\boldsymbol{\mu}\tau + \boldsymbol{n}_i)}{D(\boldsymbol{\mu}\tau)C(\boldsymbol{\mu}\tau)}\}\frac{I_{C_\mu}}{(1+\tau)^2}$$

$$\propto \prod_{i=1}^{I}\{\frac{\int_{\boldsymbol{\theta}_i \in C}\prod_{j=1}^{K}\theta_{ij}^{\mu_j\tau+n_{ij}-1}d\boldsymbol{\theta}_i}{\int_{\boldsymbol{\theta}_i \in C}\prod_{j=1}^{K}\theta_{ij}^{\mu_j\tau-1}d\boldsymbol{\theta}_i}\}\frac{I_{C_\mu}}{(1+\tau)^2}$$

$$\propto \prod_{i=1}^{I}\{\frac{\int_{\boldsymbol{\theta}_i \in C}\prod_{j=1}^{K}\theta_{ij}^{\mu_j\tau+r\bar{n}_j-r\bar{n}_j+n_{ij}-1}d\boldsymbol{\theta}_i}{\int_{\boldsymbol{\theta}_i \in C}\prod_{j=1}^{K}\theta_{ij}^{\mu_j\tau+r\bar{n}_j-r\bar{n}_j-1}d\boldsymbol{\theta}_i}\}\frac{I_{C_\mu}}{(1+\tau)^2}$$

$$\propto \prod_{i=1}^{I}\{\frac{\int_{\boldsymbol{\theta}_i \in C}\prod_{j=1}^{K}\theta_{ij}^{\mu_j\tau-r\bar{n}_j+n_{ij}}\frac{\prod_{j=1}^{K}\theta_{ij}^{r\bar{n}_j-1}}{D(r\bar{n}_j)C(r\bar{n}_j)}d\boldsymbol{\theta}_i}{\int_{\boldsymbol{\theta}_i \in C}\prod_{j=1}^{K}\theta_{ij}^{\mu_j\tau-r\bar{n}_j}\frac{\prod_{j=1}^{K}\theta_{ij}^{r\bar{n}_j-1}}{D(r\bar{n}_j)C(r\bar{n}_j)}d\boldsymbol{\theta}_i}\}\frac{I_{C_\mu}}{(1+\tau)^2}.$$

Draw samples $\boldsymbol{\theta}^{(q)} \sim \text{Dirichlet}(r\bar{n}_j)$ for Monte Carlo integration,

$$\pi(\boldsymbol{\mu}, \tau|\boldsymbol{n}) \propto \prod_{i=1}^{I}\{\frac{\sum_{q=1}^{M}\prod_{j=1}^{K}\theta_j^{(q)n_{ij}-r\bar{n}_j+\mu_j\tau}}{\sum_{q=1}^{M}\prod_{j=1}^{K}\theta_j^{(q)-r\bar{n}_j+\mu_j\tau}}\}\frac{I_{C_\mu}}{(1+\tau)^2}$$

$$\propto \prod_{i=1}^{I}\{\sum_{q=1}^{M}\prod_{j=1}^{K}\theta_j^{(q)n_{ij}}\frac{\prod_{j=1}^{K}\theta_j^{(q)-r\bar{n}_j+\mu_j\tau}}{\sum_{q=1}^{M}\prod_{j=1}^{K}\theta_j^{(q)-r\bar{n}_j+\mu_j\tau}}\}\frac{I_{C_\mu}}{(1+\tau)^2}$$

$$\propto \prod_{i=1}^{I}\{\sum_{q=1}^{M}w_q\prod_{j=1}^{K}\theta_j^{(q)n_{ij}}\}\frac{I_{C_\mu}}{(1+\tau)^2},$$

where

$$w_q = \frac{\prod_{j=1}^{K}\theta_j^{(q)-r\bar{n}_j+\mu_j\tau}}{\sum_{q=1}^{M}\prod_{j=1}^{K}\theta_j^{(q)-r\bar{n}_j+\mu_j\tau}}.$$

Note that we only need to generate samples $\boldsymbol{\theta}$ once from a Dirichlet distribution with the order restriction, which does not depend on $\boldsymbol{\mu}$ and τ. This is an efficient method.

7.2 Model Comparison

Table 5 Posterior mean, standard deviation and coefficient of variation

	θ_1			θ_2			θ_3			θ_4			θ_5		
Counties	PM	PSD	CV	PM	PSD	CV	PM	PSD	CV	PM	PSD	CV	PM	PSD	CV
1	0.0234	0.0103	0.4402	0.3751	0.0203	0.0541	0.4157	0.0243	0.0585	0.1505	0.0248	0.1648	0.0354	0.0128	0.3616
	0.0356	0.0169	0.4747	0.3959	0.0465	0.1175	0.3709	0.0449	0.1211	0.163	0.0341	0.2092	0.0345	0.0175	0.5072
2	0.0152	0.0085	0.5592	0.3726	0.0243	0.0652	0.4275	0.0279	0.0653	0.162	0.026	0.1605	0.0227	0.0106	0.4670
	0.0198	0.0128	0.6465	0.3807	0.0448	0.1177	0.4011	0.0431	0.1075	0.1892	0.0366	0.1934	0.0092	0.0089	0.9674
3	0.0233	0.0104	0.4464	0.3203	0.0347	0.1083	0.4678	0.0383	0.0819	0.1489	0.0241	0.1619	0.0397	0.0138	0.3476
	0.0374	0.0184	0.4920	0.24	0.0396	0.1650	0.5051	0.0476	0.0942	0.1719	0.0358	0.2083	0.0455	0.0198	0.4352
4	0.0079	0.0039	0.4937	0.3583	0.0189	0.0527	0.4224	0.0204	0.0483	0.1787	0.0165	0.0923	0.0327	0.0077	0.2355
	0.0078	0.0044	0.5641	0.35	0.0227	0.0649	0.4182	0.0237	0.0567	0.1916	0.0188	0.0981	0.0324	0.0084	0.2593
5	0.0158	0.0087	0.5506	0.3625	0.0231	0.0637	0.4151	0.0264	0.0636	0.174	0.027	0.1552	0.0326	0.0123	0.3773
	0.0213	0.0149	0.6995	0.3544	0.0473	0.1335	0.3778	0.0487	0.1289	0.216	0.0407	0.1884	0.0306	0.0167	0.5458
6	0.012	0.0085	0.7083	0.3662	0.0254	0.0694	0.4195	0.0303	0.0722	0.1515	0.0277	0.1828	0.0507	0.0177	0.3491
	0.0146	0.0128	0.8767	0.3649	0.051	0.1398	0.3798	0.0527	0.1388	0.1722	0.0404	0.2346	0.0685	0.027	0.3942
7	0.0118	0.0079	0.6695	0.3634	0.0259	0.0713	0.4192	0.0293	0.0699	0.1749	0.0298	0.1704	0.0307	0.0127	0.4137
	0.0149	0.0138	0.9262	0.3594	0.0534	0.1486	0.3737	0.0528	0.1413	0.2276	0.0451	0.1982	0.0245	0.0173	0.7061
8	0.0174	0.0094	0.5402	0.3743	0.0274	0.0732	0.4419	0.0323	0.0731	0.1392	0.0281	0.2019	0.0272	0.013	0.4779
	0.0304	0.0202	0.6645	0.3748	0.0542	0.1446	0.4314	0.0571	0.1324	0.1488	0.0405	0.2722	0.0145	0.0142	0.9793
9	0.0166	0.0092	0.5542	0.3694	0.0236	0.0639	0.4184	0.0279	0.0667	0.1704	0.0292	0.1714	0.0253	0.0114	0.4506
	0.0242	0.0158	0.6529	0.3806	0.0504	0.1324	0.3731	0.0508	0.1362	0.2115	0.0424	0.2005	0.0105	0.0102	0.9714
10	0.0106	0.0072	0.6792	0.3743	0.0217	0.0580	0.4202	0.0252	0.0600	0.1758	0.0272	0.1547	0.0189	0.0093	0.4921
	0.0116	0.0109	0.9397	0.3883	0.048	0.1236	0.3847	0.047	0.1222	0.215	0.0396	0.1842	0.0004	0.0022	5.5000
11	0.023	0.0104	0.4522	0.3575	0.0229	0.0641	0.4041	0.0268	0.0663	0.1751	0.0272	0.1553	0.0402	0.0138	0.3433
	0.0351	0.0161	0.4587	0.3545	0.0421	0.1188	0.3595	0.0437	0.1216	0.209	0.0363	0.1737	0.0419	0.018	0.4296

(continued)

Table 5 (continued)

	θ_1			θ_2			θ_3			θ_4			θ_5		
12	0.0108	0.0071	0.6574	0.3898	0.0237	0.0608	0.4419	0.0272	0.0616	0.1297	0.0252	0.1943	0.0278	0.0118	0.4245
	0.0119	0.011	0.9244	0.4146	0.0483	0.1165	0.4307	0.0483	0.1121	0.1238	0.0334	0.2698	0.0189	0.0133	0.7037
13	0.0127	0.006	0.4724	0.3894	0.0146	0.0375	0.4099	0.0171	0.0417	0.1524	0.0214	0.1404	0.0357	0.0103	0.2885
	0.0138	0.0078	0.5652	0.4367	0.0321	0.0735	0.364	0.0311	0.0854	0.1507	0.024	0.1593	0.0348	0.0119	0.3420
14	0.0215	0.0105	0.4884	0.3845	0.0209	0.0544	0.4243	0.0262	0.0617	0.1444	0.0268	0.1856	0.0254	0.0113	0.4449
	0.0357	0.0193	0.5406	0.4264	0.0521	0.1222	0.3747	0.0514	0.1372	0.1524	0.0378	0.2480	0.0108	0.0104	0.9630
15	0.0194	0.0098	0.5052	0.3604	0.0283	0.0785	0.4381	0.0309	0.0705	0.1405	0.0243	0.1730	0.0417	0.014	0.3357
	0.03	0.0157	0.5233	0.3411	0.0449	0.1316	0.4355	0.0467	0.1072	0.1475	0.0361	0.2447	0.046	0.0189	0.4109
16	0.0141	0.0078	0.5532	0.3686	0.0258	0.0700	0.4378	0.0301	0.0688	0.1581	0.0253	0.1600	0.0214	0.0094	0.4393
	0.0182	0.0124	0.6813	0.3573	0.0447	0.1251	0.4345	0.046	0.1059	0.1814	0.0352	0.1940	0.0086	0.009	1.0465
17	0.0309	0.0125	0.4045	0.3559	0.0258	0.0725	0.4281	0.0303	0.0708	0.1591	0.0247	0.1552	0.026	0.0109	0.4192
	0.0511	0.0193	0.3777	0.3313	0.0433	0.1307	0.4183	0.0444	0.1061	0.1827	0.0356	0.1949	0.0165	0.0117	0.7091
18	0.012	0.0079	0.6583	0.378	0.022	0.0582	0.4141	0.0248	0.0599	0.1638	0.0293	0.1789	0.0321	0.0128	0.3988
	0.0154	0.0134	0.8701	0.425	0.0556	0.1308	0.3388	0.0522	0.1541	0.1956	0.0426	0.2178	0.0252	0.0175	0.6944
19	0.0181	0.0103	0.5691	0.3769	0.0226	0.0600	0.4162	0.0257	0.0617	0.1664	0.0305	0.1833	0.0223	0.0117	0.5247
	0.0304	0.0193	0.6349	0.4237	0.0575	0.1357	0.3426	0.0554	0.1617	0.2027	0.0472	0.2329	0.0005	0.0025	5.0000
20	0.0114	0.0075	0.6579	0.3547	0.0319	0.0899	0.4644	0.0374	0.0805	0.1445	0.0247	0.1709	0.025	0.0115	0.4600
	0.0143	0.0134	0.9371	0.3042	0.0504	0.1657	0.5057	0.0553	0.1094	0.1644	0.0404	0.2457	0.0115	0.0116	1.0087
21	0.0355	0.0159	0.4479	0.3745	0.0213	0.0569	0.4071	0.0254	0.0624	0.1428	0.0289	0.2024	0.04	0.0154	0.3850
	0.0745	0.0303	0.4067	0.4398	0.0623	0.1417	0.2993	0.0539	0.1801	0.1428	0.0412	0.2885	0.0437	0.024	0.5492
22	0.0162	0.0092	0.5679	0.3853	0.0197	0.0511	0.4222	0.0251	0.0595	0.1467	0.0271	0.1847	0.0295	0.0126	0.4271
	0.0231	0.015	0.6494	0.4384	0.0513	0.1170	0.3673	0.0495	0.1348	0.1509	0.036	0.2386	0.0202	0.0141	0.6980
23	0.0124	0.0084	0.6774	0.3689	0.0275	0.0745	0.4336	0.0314	0.0724	0.1515	0.0294	0.1941	0.0336	0.0138	0.4107
	0.0173	0.0167	0.9653	0.3652	0.057	0.1561	0.4053	0.0583	0.1438	0.1839	0.0446	0.2425	0.0284	0.0203	0.7148

(continued)

Table 5 (continued)

	θ_1			θ_2			θ_3			θ_4			θ_5		
24	0.0117	0.0076	0.6496	0.3629	0.0247	0.0681	0.4171	0.0294	0.0705	0.173	0.0285	0.1647	0.0353	0.0138	0.3909
	0.0141	0.0133	0.9433	0.364	0.0498	0.1368	0.3731	0.0485	0.1300	0.2158	0.0432	0.2002	0.033	0.0185	0.5606
25	0.0175	0.0099	0.5657	0.3944	0.0205	0.0520	0.4304	0.0253	0.0588	0.1366	0.0276	0.2020	0.021	0.0107	0.5095
	0.0255	0.0168	0.6588	0.4673	0.0568	0.1215	0.3769	0.0542	0.1438	0.13	0.0361	0.2777	0.0004	0.002	5.0000
26	0.0226	0.0116	0.5133	0.3676	0.0267	0.0726	0.4336	0.0308	0.0710	0.1457	0.0263	0.1805	0.0306	0.0119	0.3889
	0.0401	0.0217	0.5411	0.3571	0.053	0.1484	0.4178	0.0538	0.1288	0.1608	0.0417	0.2593	0.0242	0.0169	0.6983
27	0.0233	0.0117	0.5021	0.3356	0.0352	0.1049	0.4516	0.0387	0.0857	0.162	0.0298	0.1840	0.0275	0.0125	0.4545
	0.0464	0.0259	0.5582	0.2428	0.0524	0.2158	0.4785	0.0587	0.1227	0.2177	0.048	0.2205	0.0146	0.0136	0.9315
28	0.018	0.0098	0.5444	0.3758	0.0227	0.0604	0.4179	0.0258	0.0617	0.1605	0.0297	0.1850	0.0277	0.0128	0.4621
	0.0311	0.0216	0.6945	0.4109	0.0577	0.1404	0.3531	0.0559	0.1583	0.1903	0.0455	0.2391	0.0147	0.0148	1.0068
29	0.0114	0.0072	0.6316	0.3756	0.0246	0.0655	0.4338	0.0301	0.0694	0.1542	0.0271	0.1757	0.025	0.0112	0.4480
	0.0129	0.0111	0.8605	0.3832	0.0488	0.1273	0.4146	0.0476	0.1148	0.1784	0.0372	0.2085	0.0109	0.0106	0.9725
30	0.0153	0.0085	0.5556	0.3484	0.0283	0.0812	0.4254	0.0318	0.0748	0.1877	0.0297	0.1582	0.0231	0.0107	0.4632
	0.0219	0.0144	0.6575	0.3101	0.0445	0.1435	0.4074	0.0479	0.1176	0.2504	0.0427	0.1705	0.0102	0.0104	1.0196
31	0.0208	0.0098	0.4712	0.373	0.0225	0.0603	0.4159	0.0277	0.0666	0.1608	0.0286	0.1779	0.0295	0.0122	0.4136
	0.0337	0.0188	0.5579	0.3953	0.0475	0.1202	0.3633	0.0483	0.1329	0.1858	0.0396	0.2131	0.0218	0.0142	0.6514
32	0.0097	0.0064	0.6598	0.3412	0.0292	0.0856	0.439	0.0326	0.0743	0.184	0.0265	0.1440	0.026	0.0111	0.4269
	0.0102	0.0086	0.8431	0.2919	0.0413	0.1415	0.4469	0.045	0.1007	0.2346	0.0384	0.1637	0.0163	0.0114	0.6994
33	0.0128	0.0073	0.5703	0.3653	0.0234	0.0641	0.4228	0.0254	0.0601	0.172	0.0249	0.1448	0.0271	0.0103	0.3801
	0.015	0.0105	0.7000	0.3586	0.0386	0.1076	0.4082	0.0394	0.0965	0.1982	0.0309	0.1559	0.02	0.0111	0.5550
34	0.0156	0.0085	0.5449	0.37	0.0261	0.0705	0.4451	0.0317	0.0712	0.1414	0.0244	0.1726	0.0278	0.0109	0.3921
	0.0217	0.0142	0.6544	0.3584	0.0479	0.1336	0.4466	0.0492	0.1102	0.1532	0.0362	0.2363	0.0201	0.0141	0.7015
35	0.015	0.0082	0.5467	0.3914	0.022	0.0562	0.4409	0.0258	0.0585	0.1342	0.0246	0.1833	0.0184	0.01	0.5435
	0.02	0.0129	0.6450	0.4227	0.0457	0.1081	0.4236	0.0448	0.1058	0.1334	0.0309	0.2316	0.0003	0.0013	4.3333

[2]*Note* Shaded Area: The model with order restrictions (M2)
Unshaded Area: The model without any order restriction (M1)

References

Chen, M.-H., & Shao, Q.-M. (1997). On monte carlo methods for estimating ratios of normalizing constants. *The Annals of Statistics, 25*(4), 1563–1594.

Cowles, M. K., & Carlin, B. P. (1996). Markov chain monte carlo convergence diagnostics: A comparative review. *Journal of the American Statistical Association, 91*(434), 883–904.

Dunson, D. B., & Neelon, B. (2003). Bayesian inference on order-constrained parameters in generalized linear models. *Biometrics, 59*(2), 286–295.

Gelfand, A. E., Smith, A. F. M., & Lee, T.-M. (1992). Bayesian analysis of constrained parameter and truncated data problems using gibbs sampling. *Journal of the American Statistical Association, 87*(418), 523–532.

Heck, D. W., & Davis-Stober, C. P. (2019). Multinomial models with linear inequality constraints: Overview and improvements of computational methods for bayesian inference. *Journal of Mathematical Psychology, 91*, 70–87.

Malinovsky, Y. & Rinott, Y. (2010). Prediction of ordered random effects in a simple small area model. *Statistica Sinica*, 697–714.

Nadarajah, S., & Kotz, S. (2006). R programs for truncated distributions. *Journal of Statistical Software, 16*(1), 1–8.

Nandram, B. (1997). Bayesian inference for the best ordinal multinomial population in a taste test. In: Gatsonis, C., Hodges, J. S., Kass, R. E., McCulloch, R., Rossi, P., and Singpurwalla, N. D. (eds.) *Case studies in bayesian statistics*, pp. 399–418, New York, NY. Springer.

Nandram, B. (1998). A bayesian analysis of the three-stage hierarchical multinomial model. *Journal of Statistical Computation and Simulation, 61*(1–2), 97–126.

Nandram, B., Kim, D., & Zhou, J. (2019). A pooled bayes test of independence for sparse contingency tables from small areas. *Journal of Statistical Computation and Simulation, 89*(5), 899–926.

Nandram, B. & Sedransk, J. (1995). Bayesian inference for the mean of a stratified population when there are order restrictions. In *Case studies in bayesian statistics, Volume II*, pp. 309–322. Springer.

Nandram, B., Sedransk, J., & Smith, S. J. (1997). Order-restricted bayesian estimation of the age composition of a population of atlantic cod. *Journal of the American Statistical Association, 92*(437), 33–40.

Rao, J. & Molina, I. (2015). *Small Area Estimation*. Wiley Series in Survey Methodology. Wiley.

Sedransk, J., Monahan, J., & Chiu, H. Y. (1985). Bayesian estimation of finite population parameters in categorical data models incorporating order restrictions. *Journal of the Royal Statistical Society. Series B (Methodological), 47*(3), 519–527.

Wu, J., Meyer, M. C., & Opsomer, J. D. (2016). Survey estimation of domain means that respect natural orderings. *The Canadian Journal of Statistics / La Revue Canadienne de Statistique, 44*(4), 431–444.

A Hierarchical Bayesian Beta-Binomial Model for Sub-areas

Lu Chen and Balgobin Nandram

Abstract Many population-based surveys have binary responses from a large number of individuals in each household within small areas. An example is the second Nepal Living Standards Survey (NLSS II), in which binary data on health status (good versus poor) for each individual respondent from sampled households (sub-areas) are available in sampled wards (small areas). In order to estimate the finite population proportion of healthy individuals in each household, we propose a hierarchical Bayesian sub-area beta-binomial model. With many sub-areas and many small areas, as in our case, the exact method of Bayesian computation will be time-consuming. Hence, improved performance is needed to reduce the additional computational burden. Accordingly, two Bayesian computational methods, namely an exact method by Metropolis–Hastings sampler and an approximation method with a random sampler, are discussed and compared in the paper. The comparison illustrates that the approximation method fits the model efficiently as well. We apply our model to NLSS II data to show that the approximation method can provide reliable estimates just like the exact method.

Keywords Hierarchical Bayesian model · Markov chain Monte Carlo method · Weakly identifiable parameters

1 Introduction

The second Nepal Living Standard Survey (NLSS) II is based on a two-stage stratified sampling scheme. A random sample of wards (areas) was selected from six strata, and twelve households (sub-areas) were selected from each sampled ward. All individuals

L. Chen
National Institute of Statistical Sciences, Washington, DC 20006, USA

B. Nandram (✉)
Department of Mathematical Sciences, Worcester Polytechnic Institute, Worcester, MA 01609, USA
e-mail: balnan@wpi.edu

D. D. Hanagal et al. (eds.), *Applied Statistical Methods*, Springer Proceedings in Mathematics & Statistics 380, https://doi.org/10.1007/978-981-16-7932-2_2

in sampled household were interviewed. The object of interest is health status, a binary variable. We want to estimate the finite population proportion of healthy individuals in each household. To do so, we use the hierarchical Bayesian (HB) model with sub-area or household-specific random effect model and obtain indirect estimates for many sub-areas (i.e., small areas). Most of the sample surveys are designed to provide reliable "direct" estimates of interests for large areas or domains (e.g., state level and national level). However, direct estimates are not reliable for areas or domains for which only small samples or no samples are available.

Due to the hierarchical structure of the NLSS II data, we are particularly interested in small area models that can capture the property of a grouping structure of small sub-areas within areas (households within wards). Although the one-fold basic models are very popular and in common use in producing reliable estimates for small areas, they may not preserve the hierarchical structure of the data and make it inconsistent to estimate first (area) and second (sub-area) levels. In particular, the sampling designs of many population-based survey are two-stage stratified sampling as NLSS II. But if we use one-fold unit level model to fit the data, sub-area-level effects would be ignored. Yan and Sedransk (2007) studied the case that the data follow a normal model with a two-stage hierarchical structure while the fitted model has a one-stage hierarchical structure by using posterior predictive p-values. Yan and Sedransk (2010) discussed the ability to detect a three-stage model when a two-stage model is actually fitted.

Two-fold models are an important extension of basic small area models. Many authors have proposed such models. But most of them are for continuous data. Fuller and Goyeneche (1998) proposed a sub-area-level model which provides model-based estimates that account for the hierarchical structure of the data. Two-fold sub-area-level models were studied by Torabi and Rao (2014), Rao and Molina (2015), Chen and Nandram (2018), Erciulescu et al. (2019). This is an area-level model which extends the Fay and Herriot (1979) model to sub-area level. Two-fold nested error regression models were considered by Stukel and Rao (1997,1999).

The beta-binomial model is a commonly used model for the categorical data (Nandram, 1998; Rao & Molina, 2015). In this paper, we apply the two-fold beta-binomial sub-area model to estimate the finite population proportion of healthy individuals in each household covered by the NLSS II. Nandram (1998) gave a full Bayesian analysis of the multinomial model. He and Sun (1998) applied one-fold beta-binomial model on Turkey hunting problems in order to provide reliable estimates on hunter success rates. Nandram and Sedransk (1993) described a HB model to make inference about the finite population proportion under two-stage clustering sampling. You and Reiss (1999) extended the beta-binomial model to two-fold model and used Gibbs sampling to obtain the posterior estimates. Nandram and Chen (2016) showed that it is important to consider the sample design within each area and proposed a two-fold small area model so that it can capture the intracluster correlation at the first stage and the intercluster correlation at the second stage. It showed that the two-fold model is preferable over the one-fold one if the data have hierarchical structure. Lee et al. (2017) extended Nandram and Chen (2016) to accommodate heterogeneous correlations. They used a HB model to make posterior inference about the finite population proportion of each area accounting for intracluster correlations.

On the other hand, two-fold models can capture the heterogeneity between samples not only within areas but also within sub-areas. Many model-based estimation techniques for sampling variances have been considered in the literature, but most of them are area-level models (Wang & Fuller, 2003; You & Chapman, 2006; Erciulescu & Berg, 2014). Nandram and Chen (2016) studied a Bayesian model under heterogeneous sampling variance, which is more preferable than a homogeneous model.

The other side of our application is that there are numerous small areas (households and individuals) and MCMC methods cannot handle them efficiently which involve complicated integrals. Our model is similar to the model in Nandram and Chen (2016), but we apply the approximation method mentioned by Nandram (1998) and block Gibbs sampler to obtain posterior estimates rather than the usual Gibbs sampler since it is more efficient. We also provide a numerical example of NLSS II to show the accuracy of the approximation method.

In Sect. 2, we give a full description of the one-fold and two-fold beta-binomial models. We discuss and show that some parameters are weakly identified. Instead of using the standard Gibbs sampler to fit the model, we use the blocked Gibbs sampler. Then, we describe how to fit the model using approximation method with blocked Gibbs sampler. In Sect. 3, we give a detailed exposition of the exact and approximate computational schemes. In Sect. 4, we illustrate and compare the aforementioned computational schemes using the NLSS II data. Finally, in Sect. 5, we make concluding marks and provide pointers for future work.

2 Hierarchical Bayesian Small Area Models

In this section, we describe the one-fold and two-fold beta-binomial sub-area models in Bayesian framework.

We have a finite population of L small areas (wards), and within the ith area, there are N_i sub-areas (households). Within the jth sub-area, there are M_{ij} individuals. We assume that $\ell(< L)$ areas are included in the sample and a simple random sample of $n_i(< N_i)$ households are taken from the ith area (i.e., self-weighting). All individuals in sampled households are sampled. Here, we assume the survey weights are the same within all households in each area. Let y_{ijk}, $k = 1, \ldots, m_{ij}$, $j = 1, \ldots, n_i$, $i = 1, \ldots\ldots, \ell$ denote the binary responses, namely the health status corresponding to the kth individual belonging to the jth household within the ith area. Let $\underset{\sim}{y} = (y_{ijk},\ k = 1, \ldots, m_{ij}, j = 1, \ldots, n_i, i = 1, \ldots\ldots, \ell)'$. Let $y_{ij} = \sum_{k=1}^{m_{ij}} y_{ijk}$ be the total number of healthy individuals sampled from the jth household in the ith area, and m_{ij} is the total number of people who responded from the jth household in the ith area.

The primary interests discussed in this paper are the finite population proportions of healthy individuals within both sampled and non-sampled households among sampled wards, which are

$$P_{ij} = \frac{1}{M_{ij}} \sum_{k=1}^{M_{ij}} y_{ijk},\ j = 1, \ldots, N_i,\ i = 1, \ldots\ldots, \ell$$

and the finite population proportions of sampled wards which are

$$P_i = \frac{1}{N_i} \sum_{j=1}^{N_i} \sum_{k=1}^{M_{ij}} y_{ijk},\ i = 1, \ldots\ldots, \ell.$$

To make inference about the P_i and P_{ij}, we fit HB model to the data. The beta-binomial model can capture the two-stage stratified design. In Sect. 2.1, we describe the one-fold beta-binomial model in Bayesian paradigm. In Sect. 2.2, we give a full description of the two-fold beta-binomial model.

2.1 A One-Fold Beta-Binomial Model

First, we write down a brief review of the small area one-fold Bayesian beta-binomial model,

$$y_{ij}|p_i \overset{ind}{\sim} \text{Binomial}(m_{ij}, p_i)\ j = 1, ..., n_i,$$

$$p_i|\theta, \phi \overset{ind}{\sim} \text{Beta}(\theta\phi, (1-\theta)\phi)\ i = 1, ..., \ell,$$

$$\pi(\phi, \theta) = \frac{1}{(1+\phi)^2},$$

where $0 < \theta < 1$ are the mean of the beta random variable and $\phi > 0$ are the sum of the parameters of the standard beta distribution. θ and ϕ are independent. Note that there is no heterogeneity between households since the p_i do not have subscript j. Here, θ has a uniform prior and the prior of ϕ is proper. The joint posterior density of this model is proper, and with certain improper prior of θ, the joint posterior density is also proper (see appendix).

Applying Bayes' theorem, the joint posterior density is

$$\pi(\underset{\sim}{p}, \theta, \phi|\underset{\sim}{y}) \propto \prod_{i=1}^{l} \prod_{j=1}^{n_i} \left\{ p_i^{y_{ij}} (1-p_i)^{m_{ij}-y_{ij}} \right\} \prod_{i=1}^{l} \left\{ \frac{p_i^{\phi\theta-1} (1-p_i)^{(1-\theta)\phi-1}}{B(\theta\phi, (1-\theta)\phi)} \right\} \frac{1}{(1+\phi)^2}. \tag{1}$$

It is easy to see that the conditional posterior distribution of $\underset{\sim}{p}$ is beta distribution,

$$p_i|\theta, \phi, y_{ij} \sim \text{Beta}(\sum_{j=1}^{n_i} y_{ij} + \theta\phi, \sum_{j=1}^{n_i} (m_{ij} - y_{ij}) + (1-\theta)\phi), \ j = 1, \ldots, n_i, \ i = 1, \ldots, \ell.$$

After we integrate out p_i, we can get the joint posterior distribution

$$\pi(\theta, \phi|\underset{\sim}{y}) \propto \prod_{i=1}^{l} \left\{ \frac{B(\sum_{j=1}^{n_i} y_{ij} + \theta\phi, \ \sum_{j=1}^{n_i} (m_{ij} - y_{ij}) + (1-\theta)\phi)}{B(\theta\phi, (1-\theta)\phi)} \right\} \frac{1}{(1+\phi)^2}. \tag{2}$$

The one-fold model can be fitted by making random draws from the joint posterior distribution of θ and ϕ. Samples of p_i can be obtained using the multiplication rule once we get the samples of θ and ϕ.

2.2 A Two-Fold Beta-Binomial Model

The two-fold beta-binomial model is

$$y_{ij}|p_{ij} \overset{ind}{\sim} \text{Binomial}(m_{ij}, p_{ij}), \ j = 1, ..., n_i,$$

$$p_{ij}|\mu_i, \tau \overset{ind}{\sim} \text{Beta}(\mu_i\tau, (1-\mu_i)\tau), \ i = 1, ..., \ell,$$

$$\mu_i \overset{iid}{\sim} \text{Beta}(\theta\phi, (1-\theta)\phi),$$

$$\pi(\tau, \phi, \theta) \propto \frac{1}{(1+\tau)^2} \frac{1}{(1+\phi)^2},$$

where $0 < \mu_i < 1, \ 0 < \theta < 1, \ \tau > 0, \ \phi > 0, \ i = 1, \ldots\ldots, \ell$.

Applying Bayes' theorem and letting $\underset{\sim}{p} = (p_{ij}, \ j = 1, \ldots, n_i, \ i = 1, \ldots\ldots, \ell)'$ and $\underset{\sim}{\mu} = (\mu_i, \ i = 1, \ldots\ldots, \ell)'$, the joint posterior density is

$$\pi(\underset{\sim}{p}, \underset{\sim}{\mu}, \tau, \theta, \phi|\underset{\sim}{y}) \propto \prod_{i=1}^{l}\prod_{j=1}^{n_i} \left\{ p_{ij}^{y_{ij}} (1-p_{ij})^{m_{ij}-y_{ij}} \right\} \left\{ \frac{p_{ij}^{\mu_i\tau-1}(1-p_{ij})^{(1-\mu_i)\tau-1}}{B(\mu_i\tau, (1-\mu_i)\tau)} \right\}$$
$$\times \prod_{i=1}^{l} \left\{ \frac{\mu_i^{\phi\theta-1}(1-\mu_i)^{(1-\theta)\phi-1}}{B(\theta\phi, (1-\theta)\phi)} \right\} \frac{1}{(1+\tau)^2} \frac{1}{(1+\phi)^2}.$$

It is easy to see that the conditional posterior distribution of $\underset{\sim}{p}$ is beta distribution,

$$p_{ij}|\mu, \tau, y_{ij} \sim \text{Beta}(y_{ij} + \mu_i\tau, m_{ij} - y_{ij} + (1-\mu_i)\tau), \ j = 1, \ldots, n_i, \ i = 1, \ldots, \ell.$$

Accordingly, once samples are obtained from the joint posterior density of $\underset{\sim}{\mu}, \tau, \theta, \phi | \underset{\sim}{y}$, a sample of each p_{ij} is obtained. Then after integrating out $\underset{\sim}{p}$, we obtain

$$\pi(\underset{\sim}{\mu}, \tau, \theta, \phi | \underset{\sim}{y}) \propto \prod_{i=1}^{l} \prod_{j=1}^{n_i} \left\{ \frac{B(y_{ij} + \mu_i \tau, m_{ij} - y_{ij} + (1 - \mu_i)\tau)}{B(\mu_i \tau, (1 - \mu_i)\tau)} \right\}$$
$$\times \prod_{i=1}^{l} \left\{ \frac{\mu_i^{\phi\theta - 1}(1 - \mu_i)^{(1-\theta)\phi - 1}}{B(\theta\phi, (1 - \theta)\phi)} \right\} \frac{1}{(1 + \tau)^2} \frac{1}{(1 + \phi)^2} \quad (3)$$

3 Computation

Notice that the posterior distribution of parameters of the two-fold beta-binomial model does not exist in closed form in (3). Besides, if n_i is very large, the exact MCMC method would be very slow. Therefore, we apply the approximation method proposed by Nandram (1998) in our model to get fast but reliable computations.

3.1 Approximation Method

In this section, we propose the approximation method for beta-binomial sub-area model so that all parameters are easy to draw. Notice that the product of ratios of beta functions in the joint density (3) makes the computation difficult. That is,

$$g(\underset{\sim}{\mu}, \tau | \underset{\sim}{y}) \propto \prod_{i=1}^{l} \prod_{j=1}^{n_i} \left\{ \frac{B(y_{ij} + \mu_i \tau, m_{ij} - y_{ij} + (1 - \mu_i)\tau)}{B(\mu_i \tau, (1 - \mu_i)\tau)} \right\}$$

We apply the approximation method in Nandram (1998) to get the conditional posterior density of

$$\mu_i | \tau, \underset{\sim}{y} \sim \text{Beta}(\mu_i^{(a)} \tau^{(a)}, (1 - \mu_i^{(a)})\tau^{(a)}),\ i = 1, \ldots, \ell,$$

where $\mu_i^{(a)}$, $\tau^{(a)}$ are functions of τ. See the details in Appendix A.

By applying Bayes's theorem, we can get the approximate joint posterior density

$$\pi_a(\underset{\sim}{\mu}, \tau, \theta, \phi | \underset{\sim}{y}) \propto \pi_a(\underset{\sim}{\mu} | \tau, \underset{\sim}{y}) \pi_a(\tau | \underset{\sim}{y}) \pi(\underset{\sim}{\mu} | \theta, \phi) \pi(\tau, \theta, \phi)$$
$$= \prod_{i=1}^{l} \prod_{j=1}^{n_i} \left\{ \frac{B(y_{ij} + \hat{\alpha}_i \tau, m_{ij} - y_{ij} + (1 - \hat{\alpha}_i)\tau)}{B(\hat{\alpha}_i \tau, (1 - \hat{\alpha}_i)\tau)} \right\}$$

$$\times \prod_{i=1}^{l} \left\{ \frac{\mu_i^{\mu_i^{(a)}\tau^{(a)}-1}(1-\mu_i)^{(1-\mu_i^{(a)})\tau^{(a)}-1}}{B(\mu_i^{(a)}\tau^{(a)}, (1-\mu_i^{(a)})\tau^{(a)})} \right\}$$
$$\times \prod_{i=1}^{l} \left\{ \frac{\mu_i^{\phi\theta-1}(1-\mu_i)^{(1-\theta)\phi-1}}{B(\theta\phi, (1-\theta)\phi)} \right\} \frac{1}{(1+\tau)^2} \frac{1}{(1+\phi)^2},$$

where $\hat{\alpha}_i = \frac{\sum_{j=1}^{n_i} y_{ij}+1}{\sum_{j=1}^{n_i} m_{ij}+2}$.

It is easy to see that the approximate conditional posterior distribution of $\mu_i|\tau, \theta, \phi, \underset{\sim}{y}$ follows beta distribution, that is,

$$\mu_i|\tau, \theta, \phi, \underset{\sim}{y} \sim \text{Beta}(\mu_i^{(a)}\tau^{(a)} + \theta\phi, (1-\mu_i^{(a)})\tau^{(a)} + (1-\theta)\phi), \ j = 1, \ldots, n_i, \ i = 1, \ldots, \ell.$$

Then, we can integrate out $\underset{\sim}{\mu}$ easily and the joint posterior density of $\tau, \theta, \phi|\underset{\sim}{y}$ is

$$\pi(\tau, \theta, \phi|\underset{\sim}{y}) \propto \prod_{i=1}^{l}\prod_{j=1}^{n_i} \left\{ \frac{B(y_{ij} + \hat{\alpha}_i\tau, m_{ij} - y_{ij} + (1-\hat{\alpha}_i)\tau)}{B(\hat{\alpha}_i\tau, (1-\hat{\alpha}_i)\tau)} \right\}$$
$$\times \prod_{i=1}^{l} \left\{ \frac{B(\mu_i^{(a)}\tau^{(a)} + \theta\phi, (1-\mu_i^{(a)})\tau^{(a)} + (1-\theta)\phi)}{B(\theta\phi, (1-\theta)\phi)} \right\} \frac{1}{(1+\tau)^2} \frac{1}{(1+\phi)^2}.$$

Now, we use Gibbs sampler to draw samples for τ, θ, ϕ from the joint posterior density of $\tau, \theta, \phi|\underset{\sim}{y}$. Notice that $0 < \tau, \phi < \infty$, we make a transformation to $\eta_1 = \frac{1}{1+\tau}$ and $\eta_2 = \frac{1}{1+\phi}$ so that $0 < \eta_1, \eta_2 < 1$. The posterior density, $\pi(\eta_1, \theta, \eta_2|\underset{\sim}{y})$, is

$$\pi(\eta_1, \theta, \eta_2|\underset{\sim}{y}) \propto \prod_{i=1}^{l}\prod_{j=1}^{n_i} \left\{ \frac{B(y_{ij} + \hat{\alpha}_i\tau, m_{ij} - y_{ij} + (1-\hat{\alpha}_i)\tau)}{B(\hat{\alpha}_i\tau, (1-\hat{\alpha}_i)\tau)} \right\}$$
$$\times \prod_{i=1}^{l} \left\{ \frac{B(\mu_i^{(a)}\tau^{(a)} + \theta\phi, (1-\mu_i^{(a)})\tau^{(a)} + (1-\theta)\phi)}{B(\theta\phi, (1-\theta)\phi)} \right\}_{\tau=\frac{1-\eta_1}{\eta_1},\ \phi=\frac{1-\eta_2}{\eta_2}}.$$

Then, conditional on θ, η_2 and data, the posterior density of η_1 is

$$\pi(\eta_1|\theta, \eta_2, \underset{\sim}{y}) \propto \prod_{i=1}^{l}\prod_{j=1}^{n_i} \left\{ \frac{B(y_{ij} + \hat{\alpha}_i\tau, m_{ij} - y_{ij} + (1-\hat{\alpha}_i)\tau)}{B(\hat{\alpha}_i\tau, (1-\hat{\alpha}_i)\tau)} \right\}$$
$$\times \prod_{i=1}^{l} \left\{ \frac{B(\mu_i^{(a)}\tau^{(a)} + \theta\phi, (1-\mu_i^{(a)})\tau^{(a)} + (1-\theta)\phi)}{B(\theta\phi, (1-\theta)\phi)} \right\}_{\tau=\frac{1-\eta_1}{\eta_1},\ \phi=\frac{1-\eta_2}{\eta_2}}.$$

Using a univariate grid sampler, samples are drawn from the posterior density of η_1. Next, conditional on η_1, η_2 and data, the posterior density of θ is

$$\pi(\theta|\eta_1,\eta_2,\underset{\sim}{y}) \propto \prod_{i=1}^{l}\left\{\frac{B(\mu_i^{(a)}\tau^{(a)}+\theta\phi,(1-\mu_i^{(a)})\tau^{(a)}+(1-\theta)\phi)}{B(\theta\phi,(1-\theta)\phi)}\right\}_{\tau=\frac{1-\eta_1}{\eta_1},\ \phi=\frac{1-\eta_2}{\eta_2}}.$$

Again using the univariate grids, samples of the posterior density of θ are obtained. Next, conditional on η_1, θ, the posterior density of η_2 is

$$\pi(\eta_2|\eta_1,\theta,\underset{\sim}{y}) \propto \prod_{i=1}^{l}\left\{\frac{B(\mu_i^{(a)}\tau^{(a)}+\theta\phi,(1-\mu_i^{(a)})\tau^{(a)}+(1-\theta)\phi)}{B(\theta\phi,(1-\theta)\phi)}\right\}_{\tau=\frac{1-\eta_1}{\eta_1},\ \phi=\frac{1-\eta_2}{\eta_2}}.$$

Then, we use univariate grid sampler to draw samples for η_2. Once we get all samples for (η_1, θ, η_2), we transform η_1, η_2 back to τ, ϕ, respectively. We have always used 100 grids for η_1, θ, η_2.

Finally, conditional on (τ, θ, ϕ, y), the μ_i are independent and samples can be simply obtained from $\pi_a(\mu_i|\tau, \theta, \phi, \underset{\sim}{y})$, the beta distribution, which is very fast.

3.2 Exact Method

There are weakly identified parameters θ, ϕ in the model, which will cause poor mixing in the Gibbs sampler. Therefore, an alternative method to fit the model is desired. We use multiplication rule and Metropolis sampler to draw samples from the posterior density $\pi(\underset{\sim}{\mu}, \tau, \theta, \phi|\underset{\sim}{y})$. Here, we use Riemann midpoint rule to approximate the integration of μ_i with high efficiency and accuracy, which is much more close than the approximation we mentioned above.

First, we integrate out $\mu_i, i = 1, \ldots, \ell$, and the joint posterior density is

$$\pi(\tau,\theta,\phi|\underset{\sim}{y}) \propto \frac{1}{(1+\tau)^2}\frac{1}{(1+\phi)^2}\prod_{i=1}^{l}\left\{\int_0^1 g_i(\mu_i)f_i(\mu_i)d\mu_i\right\}$$

where $g_i(\mu_i) = \prod_{j=1}^{n_i}\left\{\frac{B(y_{ij}+\mu_i\tau, n_{ij}-y_{ij}+(1-\mu_i)\tau)}{B(\mu_i\tau,(1-\mu_i)\tau)}\right\}$ and $f_i(\mu_i) = \left\{\frac{\mu_i^{\phi\theta-1}(1-\mu_i)^{(1-\theta)\phi-1}}{B(\theta\phi,(1-\theta)\phi)}\right\}$.

Notice that $f_i(\mu_i)$ is a density function of a beta random variable and $g_i(\mu_i)$ is the ratio of two beta functions. We can integrate the μ_i one at a time since they are independent. So, we only need to consider one $\int_0^1 g_i(\mu_i)f_i(\mu_i)d\mu_i$, and others are in the same manner. Their product would be the complete integral.

Divide the interval (0,1) into R equal subintervals $[a_0, a_1], \ldots, [a_{R-1}, a_R]$ where $a_0 = 0, a_i = \frac{i}{R}$ and $a_R = 1$. Let $F(\cdot)$ denote the cdf corresponding to $f(\cdot)$, the beta density function. Then using Riemann midpoint sum and integration definition, it is easy to show that

$$\lim_{R\to\infty}\sum_{r=1}^{R} g_i\Big(\frac{a_{r-1}+a_r}{2}\Big)\{F_i(a_r)-F_i(a_{r-1})\} = \int_0^1 g_i(\mu_i)f_i(\mu_i)d\mu_i,\ i=1,\ldots,\ell.$$

Therefore, with fairly large R, $\sum_{r=1}^{R} g_i(\frac{a_{r-1}+a_r}{2})\{F_i(a_r)-F_i(a_{r-1})\} \approx \int_0^1 g_i(\mu_i) f_i(\mu_i)d\mu_i,\ i=1,\ldots,\ell$ with high accuracy.

We use Metropolis–Hastings algorithm to draw samples for τ, θ, ϕ simultaneously. We use a Metropolis step with $\pi_a(\tau, \theta, \phi|\underset{\sim}{y})$, obtained from the approximate method. We fit a multivariate Student's t distribution with ν degrees of freedom to the iterates, $(\log(\tau), \log(\frac{\theta}{1-\theta}), \log(\phi))$ from the approximate method explained later as the proposal density in the Metropolis step. It is standard to tune the Metropolis step by varying ν in order to increase the sampling efficiency of a Metropolis chain. We can adjust ν to change the shape of the proposal Student's t distribution so that it can closely resemble the actual posterior distribution of the parameters.

Then, conditional on τ, θ, ϕ and data, we draw samples for μ_i from

$$\pi(\mu_i|\tau,\theta,\phi,\underset{\sim}{y}) \propto \frac{1}{(1+\tau)^2}\frac{1}{(1+\phi)^2} g_i(\mu_i)f_i(\mu_i)\prod_{j\neq i}^{l}\left\{\int_0^1 g_j(\mu_j)f_j(\mu_j)d\mu_j\right\}$$

using the univariate grid sampler.

4 Numerical Example

In Sect. 4.1, we describe the NILL II and auxiliary data. In Sect. 4.2, we present model-based estimation and prediction results. The comparisons between the approximation methods and the exact method are discussed based on the NILL II example.

4.1 Nepal Living Standards Survey II

The performance of our method is studied using the second Nepal Living Standard Survey (NLSS II), conducted in the years 2003–2004. The main objective of NLSS II is to track changes and progress about national living standards and social indicators of the Nepalese population. It is an integrated survey which covers samples from the whole country and runs throughout the year.

The NLSS II gathers information on a variety of aspects. It has collected data on demographics, housing, education, health, fertility, employment, income, agricultural activity, consumption and various other areas. The sampling design of NLSS II is two-stage stratified sampling. Nepal is stratified into PSUs, and within each PSU, a number of households (sub-area) are selected. All household members in the sample were interviewed.

Table 1 Nepal distribution of wards and households in the sample

Strata	Mountains	Kathmandu	Urban hills	Rural hills	Urban Tarai	Rural Tarai	Total
PSU	32	34	28	96	34	102	326
Households	384	408	336	1,152	408	1,224	3,912
Individuals	1,949	1,954	1,467	5,755	2,104	7,034	20,263

In detail, NLSS II has records for 20,263 individuals from 3,912 households (subareas) from 326 PSUs (areas) from a population of 60,262 households and about two million Nepalese. A sample of PSUs was selected from strata using the probability proportional to size (PPS) sampling, and 12 households were systematically selected from each PSU. Therefore, the survey is self-weighed, but some adjustments were made after conducting the survey for non-response or missing data. For simplicity, in this paper, we assume all sampled individuals have the same weight. Table 1 shows the distribution of all sampled persons by stratum.

According to the 2001 census data, only about 0.091% of households and only 0.904% of PSU were sampled. NLSS II was designed for providing reliable estimates only at stratum level or even larger areas than stratum. It cannot give estimates in small area (PSU or household level) since the sample sizes are too small. Therefore, we need to use statistical models to fit the available data and find reliable estimates in small areas. In our study, we choose the binary variable, health status, from the health section of the questionnaire.

4.2 Numerical Comparison

We use data from NLSS II to illustrate our two-fold beta-binomial model. We predict the household proportions of members in good health for 18,924 households (sampled and non-sampled). This analysis is based on 1,224 sample households from 102 wards (PSUs) in strata 6. Our primary purposes are to show that our two-fold beta-binomial model can provide good estimates and to compare the approximate method with the exact method when there are random effects at the household level.

In Figs. 1, 2 and 3, we compare, respectively, the posterior means (PMs), posterior standard deviations (PSDs) and posterior coefficient of variations (CVs) in the household level as our primary purpose. We can see that the PMs and PSDs are very close. That is, all points lie close on the 45 degree line through the origin, respectively. The CVs are a little bit spread out, but all points still lie on the line. Overall, these approximations are quite acceptable still. In Figs. 4, 5 and 6, we compare, respectively, the PMs, PSDs and posterior CVs in the ward level as our primary purpose. Notice that all plots are thinner than those in the household level due to the smaller size of wards. But the PSDs in wards are much smaller than those in household level. This is reasonable since larger sample sizes to estimate the finite proportion of members with good health in the wards. All plots are still very good. Notice that

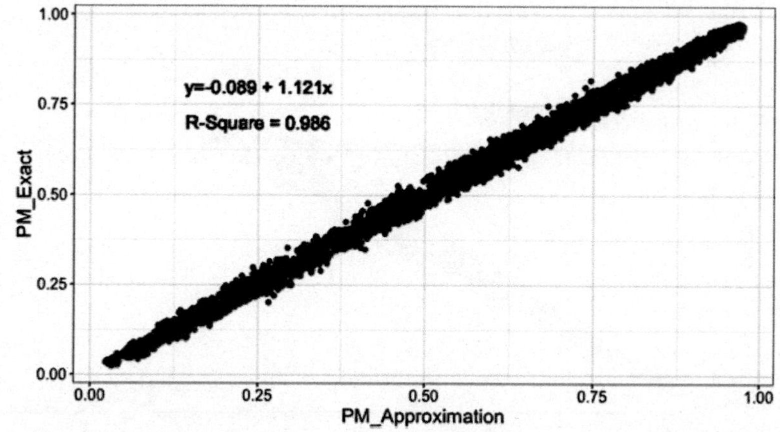

Fig. 1 Comparison of the posterior means (PMs) of the household proportions by the approximate method and the exact method

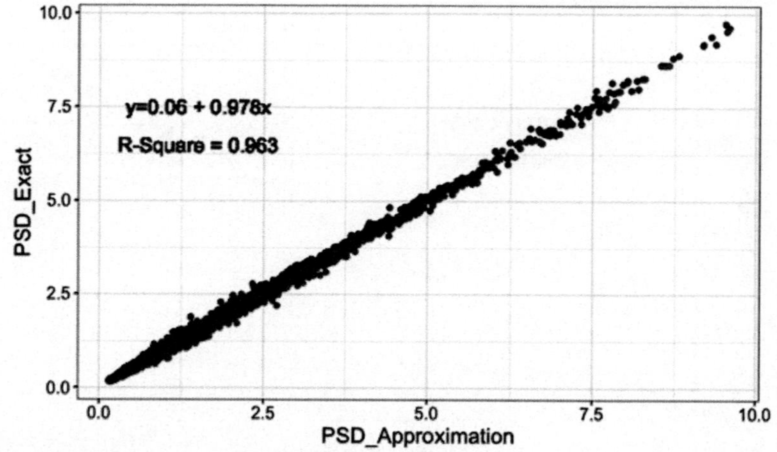

Fig. 2 Comparison of the posterior standard deviations (PSDs) of the household proportions by the approximate method and the exact method

other two plots of PSDs and CVs are more spread out. But again the approximate method and the exact method are reasonably close.

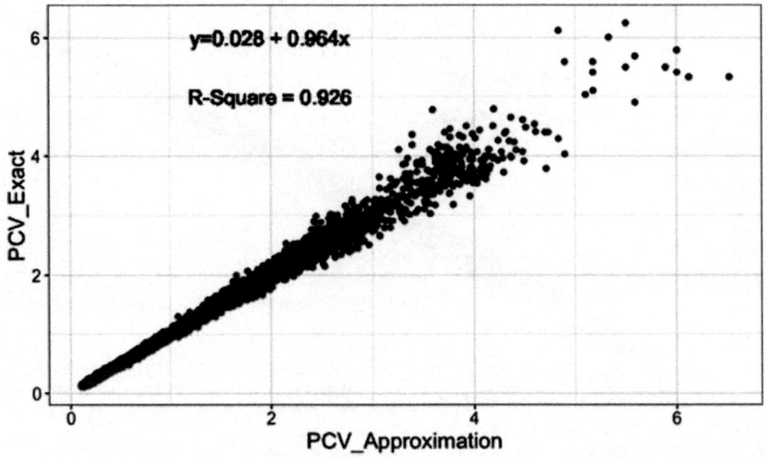

Fig. 3 Comparison of the posterior coefficient of variations (CVs) of the household proportions by the approximate method and the exact method

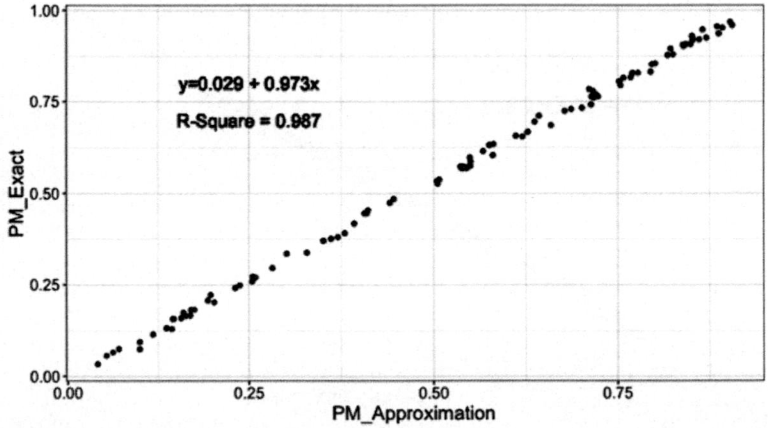

Fig. 4 Comparison of the posterior means (PMs) of the ward proportions by the approximate method and the exact method

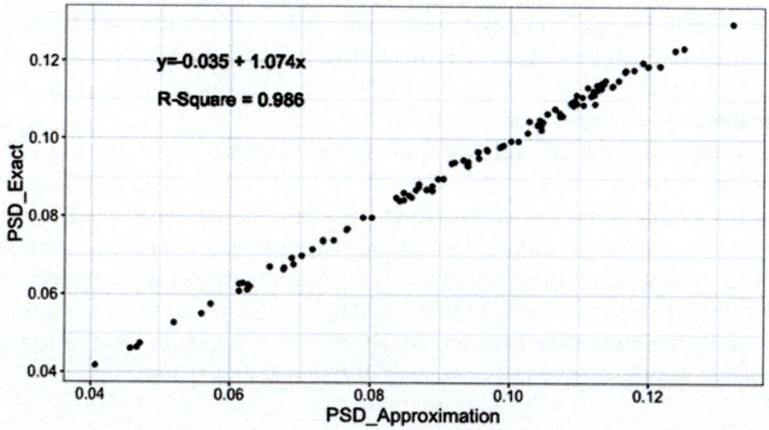

Fig. 5 Comparison of the posterior standard deviations (PSDs) of the ward proportions by the approximate method and the exact method

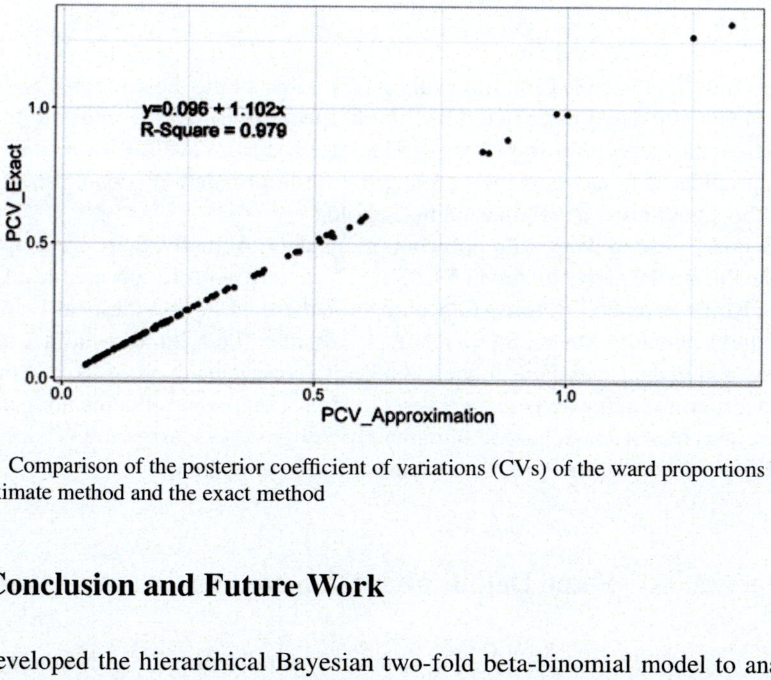

Fig. 6 Comparison of the posterior coefficient of variations (CVs) of the ward proportions by the approximate method and the exact method

5 Conclusion and Future Work

We developed the hierarchical Bayesian two-fold beta-binomial model to analyze binary data based on a two-stratified sampling survey. However, a one-fold model ignores the hierarchical structure of such data and assumed that they are homogeneous among households. Therefore, we applied the two-fold model into the data and it can capture the heterogeneous relationships among households. In order to fit the two-fold model, we applied the method proposed by Nandram (1998) to approximate the conditional posterior density of μ, which is easy to draw. In the meanwhile, we discuss the exact method by using M̃etropolis sampler and numerical integration in detail as well. We perform a Bayesian predictive inference for the finite population proportion of each household. An illustrative example of NLSS II is presented in order to compare the approximation method and the exact method. It shows that when there are a large number of areas and sub-areas, the approximation method can be accurate and it can also provide reasonable estimates.

There are many future works on two-fold small area model. First, in this paper, we assume equal survey weights. However, after data were collected, the selection probabilities are usually adjusted for various characteristics or based on non-response as well. Incorporating those survey weights into the model is also very important. Generally, we need to consider these weights into the model. NLSS II is a national-wide and population-based survey. We should rescale sample weights to sum to an equivalent sample size. That is, we consider adjusted weight as

$$w_{ijk}^{*} = \hat{n}\Big(\frac{w_{ijk}}{\sum_{i=1}^{\ell}\sum_{j=1}^{n_i}\sum_{k=1}^{m_{ij}} w_{ijk}}\Big), \; \hat{n} = \frac{(\sum_{i=1}^{\ell}\sum_{j=1}^{n_i}\sum_{k=1}^{m_{ij}} w_{ijk})^2}{\sum_{i=1}^{\ell}\sum_{j=1}^{n_i}\sum_{k=1}^{m_{ij}} w_{ijk}^2}$$

as an equivalent sample (Potthoff et al., 1992). Introducing the sampling weights, we can obtain a updated normalized likelihood function. Based the updated likelihood function and same prior in the two-fold model, we can have full Bayesian analysis on the updated model and then project the finite population proportion of family members with good health in each household.

Second, we can work with polychotomous data. Actually, there are 4 options in the health status questionnaire in NLSS II. Third, the two-fold sub-area-level models can also be extended to three-fold models if the data have additional hierarchical structure; actually, the NLSS II has this structure (households within wards and wards within districts). Fourth, there are health covariates in NLSS II, and these can be incorporated using logistic regression models. Finally, introducing nonparametric prior, Dirichlet process, instead of parametric prior we discussed in the paper, might be able to make our method more robust to its specifications.

Appendix A Some Details about Approximation of $\pi(\mu_i|\tau)$

In the appendix A, the details of $\mu_i^{(a)}$ and $\tau^{(a)}$ are discussed. See Nandram (1998) as well. A beta approximation for the conditional posterior distribution of $\mu_i|\tau, \underset{\sim}{y}$ is utilized, that is,

$$\mu_i|\tau, \underset{\sim}{y} \overset{ind}{\sim} \text{Beta}(\mu_i^{(a)}\tau^{(a)}, (1-\mu_i^{(a)})\tau^{(a)}),$$

where $\mu_i^{(a)}$ and $\tau^{(a)}$ are in the following forms.

First, $\mu^{(a)} = \hat{\mu}_i$, where $\hat{\mu}_i = \tilde{\mu}_{ik}/(\tilde{\mu}_{i1} + \tilde{\mu}_{i2})$, $k = 1, 2$ and

$$\tilde{\mu}_{ik} = \begin{cases} \hat{\hat{\mu}}_{ik}, & 0 < \hat{\hat{\mu}}_{ik} < 1 \\ \hat{\alpha}_{ik}, & \hat{\hat{\mu}}_{ik} \leq 0 \text{ or } \geq 1 \end{cases}, \quad k = 1, 2,$$

where $\hat{\hat{\mu}}_{i1} = \hat{\alpha}_i + B_{i1}^{-1}(A_{i1} - \bar{A}_i)$, $\hat{\hat{\mu}}_{i2} = 1 - \hat{\alpha}_i + B_{i2}^{-1}(A_{i2} - \bar{A}_i)$.

We have

$$B_{i1} = \tau \sum_{j=1}^{n_i} \left\{ \hat{\alpha}_i^{-1} - \left(\hat{\alpha}_i^{-1} + \frac{y_{ij}}{\tau} \right)^{-1} \right\},$$

$$B_{i2} = \tau \sum_{j=1}^{n_i} \left\{ (1 - \hat{\alpha}_i)^{-1} - \left((1 - \hat{\alpha}_i)^{-1} + \frac{m_{ij} - y_{ij}}{\tau} \right)^{-1} \right\}$$

and

$$A_{i1} = \tau \sum_{j=1}^{n_i} \ln\left(1 + \frac{y_{ij}}{\tau \hat{\alpha}_i}\right), \quad A_{i2} = \tau \sum_{j=1}^{n_i} \ln\left(1 + \frac{m_{ij} - y_{ij}}{\tau(1 - \hat{\alpha}_i)}\right), \quad \bar{A}_i = \frac{B_{i1}^{-1} A_{i1} + B_{i2}^{-1} A_{i2}}{B_{i1}^{-1} + B_{i2}^{-1}},$$

$$\hat{\hat{\mu}}_{i1} = \hat{\alpha}_i + B_{i1}^{-1}(A_{i1} - \bar{A}_i), \; \hat{\hat{\mu}}_{i2} = 1 - \hat{\alpha}_i + B_{i2}^{-1}(A_{i2} - \bar{A}_i).$$

Second, let

$$d_{ikk'} = \begin{cases} \mu^{(a)2}_{ik}(B_{ik}^{-1} - \nu_{ik}^2), & k = k' \\ \mu_{ik}^{(a)} \mu_{ik'}^{(a)} / \nu_{ik} \nu_{ik'} - 1, & k \neq k' \end{cases},$$

and

$$I_{ikk'} = \begin{cases} 1, & d_{ikk'} > 0 \\ 0, & d_{ikk'} \leq 0 \end{cases},$$

$k = 1, 2$, we take

$$\tau^{(a)} = \frac{1}{l} \sum_{i=1}^{l} \frac{d_{i11} I_{i11} + d_{i12} I_{i12} + d_{i21} I_{i21} + d_{i22} I_{i22}}{I_{i11} + I_{i12} + I_{i21} + I_{i22}},$$

where $\nu_{i1} = B_{i1}^{-1} / \sqrt{B_{i1}^{-1} + B_{i2}^{-1}}$ and $\nu_{i2} = B_{i2}^{-1} / \sqrt{B_{i1}^{-1} + B_{i2}^{-1}}$.

Appendix B Propriety of the One-Fold Model

In the proof of the propriety of (2.1), we incorporate a more general small area one-fold Bayesian beta-binomial model,

$$y_{ij}|p_i \overset{ind}{\sim} \text{Binomial}(m_{ij}, p_i) \; j = 1, ..., n_i,$$

$$p_i|\theta, \phi \overset{ind}{\sim} \text{Beta}(\theta\phi, (1-\theta)\phi) \; i = 1, ..., \ell,$$

$$\pi(\theta) \propto \frac{1}{\theta^s(1-\theta)^s}, \; , s \geq 0,$$

$$\pi(\phi) = \frac{1}{(1+\phi)^2},$$

where $0 < \theta < 1$ and $\tau > 0$. Note that when $s = 0$, the prior of θ is uniform distribution, which is same with the model in our paper. When $s \geq 1$, the prior of θ is improper.

Similarly, we can get the joint posterior density,

$$\pi(\underset{\sim}{p}, \theta, \phi|\underset{\sim}{y}) \propto \prod_{i=1}^{l}\prod_{j=1}^{n_i}\left\{p_i^{y_{ij}}(1-p_i)^{m_{ij}-y_{ij}}\right\}\prod_{i=1}^{l}\left\{\frac{p_i^{\phi\theta-1}(1-p_i)^{(1-\theta)\phi-1}}{B(\theta\phi, (1-\theta)\phi)}\right\}\frac{1}{\theta^s(1-\theta)^s}\frac{1}{(1+\phi)^2}. \tag{4}$$

We show that under the condition that $s < \ell + 1$, the joint posterior (B.1) is proper. It is easy to see that the conditional posterior distribution of $\underset{\sim}{p}$ is beta distribution,

$$p_i|\theta, \phi, y_{ij} \sim \text{Beta}(\sum_{j=1}^{n_i} y_{ij} + \theta\phi, \sum_{j=1}^{n_i}(m_{ij} - y_{ij}) + (1-\theta)\phi), \; j = 1, \ldots, n_i, \; i = 1, \ldots, \ell.$$

After we integrate out p_i, we can get the joint posterior distribution

$$\pi(\theta, \phi|\underset{\sim}{y}) \propto \prod_{i=1}^{l}\left\{\frac{B(\sum_{j=1}^{n_i} y_{ij} + \theta\phi, \; \sum_{j=1}^{n_i}(m_{ij} - y_{ij}) + (1-\theta)\phi)}{B(\theta\phi, (1-\theta)\phi)}\right\}\frac{1}{\theta^s(1-\theta)^s}\frac{1}{(1+\phi)^2}. \tag{5}$$

Based on the relationship between beta function and gamma function and the property of gamma function, we can obtain the joint posterior density (B.2) to

$$\pi(\theta, \phi|\underset{\sim}{y}) \propto \prod_{i=1}^{l}\left\{\left(\frac{\Gamma(\sum_{j=1}^{n_i} y_{ij} + \theta\phi)}{\Gamma(\theta\phi)}\right)\left(\frac{\Gamma(\sum_{j=1}^{n_i}(m_{ij} - y_{ij}) + (1-\theta)\phi)}{\Gamma((1-\theta)\phi)}\right)\Bigg/\left(\frac{\Gamma(\sum_{j=1}^{n_i} m_{ij} + \phi)}{\Gamma(\phi)}\right)\right\}$$
$$\times \frac{1}{\theta^s(1-\theta)^s}\frac{1}{(1+\phi)^2}$$

$$\propto \prod_{i=1}^{l} \left\{ \frac{\prod_{t=0}^{\sum_{j=1}^{n_i} y_{ij}-1} (t+\theta\phi) \prod_{t=0}^{\sum_{j=1}^{n_i} (m_{ij}-y_{ij})-1} (t+(1-\theta)\phi)}{\prod_{t=0}^{\sum_{j=1}^{n_i} m_{ij}-1} (t+\phi)} \right\} \frac{1}{\theta^s(1-\theta)^s} \frac{1}{(1+\phi)^2}. \quad (6)$$

Since $0 < \theta < 1$, we can bound (B.3) to obtain

$$\begin{aligned} \pi(\theta, \phi|\underset{\sim}{y}) &\leq M\theta^{\ell}(1-\theta)^{\ell} \times \frac{1}{\theta^s(1-\theta)^s} \frac{1}{(1+\phi)^2} \\ &= M\theta^{\ell-s}(1-\theta)^{\ell-s} \frac{1}{(1+\phi)^2}. \end{aligned}$$

So,

$$\int_0^1 \int_0^{\infty} \pi(\theta, \phi|\underset{\sim}{y}) d\phi d\theta \leq M \int_0^1 \theta^{\ell-s}(1-\theta)^{\ell-s} d\theta \int_0^{\infty} \frac{1}{(1+\phi)^2} d\phi.$$

When $s < \ell + 1$, the beta function $B(\ell - s + 1, \ell - s + 1)$ is finite, that is,

$$\begin{aligned} \int_0^1 \int_0^{\infty} \pi(\theta, \phi|\underset{\sim}{y}) d\phi d\theta &\leq M \int_0^1 \theta^{\ell-s}(1-\theta)^{\ell-s} d\theta \int_0^{\infty} \frac{1}{(1+\phi)^2} d\phi \\ &\leq MB(\ell - s + 1, \ell - s + 1) < \infty. \end{aligned}$$

References

Central Bureau of Statistics Thapathali. Nepal living Standards Survey 2003/04. Statistical Report Volume 1, Kathmandu, Nepal.

Chen, L., & Nandram, B. (2018). Bayesian Logistic Regression Model for Sub-Areas. In V. A. Alexandria (Ed.), *JSM Proceedings, Bayesian Statistical Science Section* (pp. 966–994). American Statistical Association.

Erciulescu, A. L., & Berg, E. (2014). Small Area Estimates for the Conservation Effects Assessment Project. In: *Frontiers of Hierarchical Modeling in Observational Studies, Complex Surveys and Big Data: A Conference Honoring Professor Malay Ghosh, College Park, MD; Women in Statistics Conference, Cary, NC.*

Erciulescu, A. L., Cruze, N. B., & Nandram, B. (2019). Model-based county level crop estimates incorporating auxiliary sources of information. *Journal of the Royal Statistical Society: Series A (Statistics in Society), 182*(1), 283–303.

Fay, R. E., & Herriot, R. A. (1979). Estimates of income for small places: An application of James-stein procedures to census data. *Journal of the American Statistical Association, 74,* 269–277.

Fuller, W. A., & Goyeneche, J. J. (1998). Estimation of the state variance component (unpublished).

He, Z., & Sun, D. (1998). Hierarchical Bayes estimation of hunting success rates. *Environmental and Ecological Statistics, 5*(3), 223–236.

Lee, D., Nandram, B., & Kim, D. (2017). Bayesian predictive inference of a proportion under a two-fold small area model with heterogeneous correlations. *Survey Methodology, 17*(17), 69–92.

Nandram, B. (1998). A Bayesian analysis of the three-stage hierarchical multinomial model. *Journal of Statistical Computation and Simulation, 61,* 97–126.

Nandram, B. (2000). Bayesian Generalized Linear Models for Inference about Small Areas. In: D. K. Dey, S.K. Ghosh, & B.K. Mallick (Eds), *In Generalized linear models: A Bayesian Perspective*, Chapter 6, Marcel Dekker, pp. 91–114.

Nandram, B. (2015). Bayesian predictive inference of a proportion under a two-fold small area model. *Journal of Official Statistics, 32*(1), 187–207.

Nandram, B. (1989). Discrimination between complementary log-log and logistic model for ordinal data. *Communications in Statistics, Theory and Methods, 18,* 2155–2164.

Nandram, B., & Chen, L. (2016). A Bayesian model for small areas under heterogeneous sampling variances. *Survey Methodology*, submitted.

Nandram, B., & Chen, M.-H. (1996). Reparameterizing the generalized linear model to accelerate gibbs sampler convergence. *Journal of Statistical Computation and simulation, 54,* 129–144.

Nandram, B., & Choi, J. W. (2010). A Bayesian analysis of body mass index data from small domains under nonignorable nonresponse and selection. *Journal of the American Statistical Association, 105,* 120–135.

Nandram, B., & Erhardt, E. (2005). Fitting Bayesian two-stage generalized linear models using random samples via the sir algorithm. *Sankhya, 66,* 733–755.

Nandram, B., & Sedransk, J. (1993). Bayesian predictive inference for a finite population proportion: two-stage cluster sampling. *Journal of the Royal Statistical Society. Series B (Methodological)*, 399–408.

Potthoff, R. F., Woodbury, M. A., & Manton, K. G. (1992). Equivalent sample size and equivalent degrees of freedom refinements for inference using survey weights under superpopulation models. *Journal of the American Statistical Association, 87,* 383–396.

Rao, J. N. K., & Molina, I. (2015). *Small Area Estimation* (2nd ed.). New York: Wiley.

Stukel, D. M., & Rao, J. N. K. (1997). Estimation of regression models with nested error structure and unequal error variances under two and three stage cluster sampling. *Statistics & Probability Letters, 35*(4), 401–407.

Stukel, D. M., & Rao, J. N. K. (1997). On small-area estimation under two-fold nested error regression models. *Journal of Statistical Planning and Inference, 78*(1–2), 131–147.

Torabi, M., & Rao, J. N. K. (2014). On small area estimation under a sub-area level model. *Journal of Multivariate Analysis, 127,* 36–55.

Wang, J., & Fuller, W. A. (2003). The mean squared error of small area predictors constructed with estimated area variances. *Journal of the American Statistical Association, 98*(463), 716–723.

Yan, G., & Sedransk, J. (2007). Bayesian diagnostic techniques for detecting hierarchical structure. *Bayesian Analysis, 2*(4), 735–760.

Yan, G., & Sedransk, J. (2010). A note on Bayesian residuals as a hierarchical model diagnostic technique. *Statistical Papers, 51*(1), 1.

You, Y., & Chapman, B. (2006). Small area estimation using area level models and estimated sampling variances. *Survey Methodology, 32,* 97–103.

You, Y., & Reiss, P. (1999). Hierarchical Bayes small area estimation of response rates for an expenditure survey. In: *Proceedings of the Survey Methods Section*, Statistical Society of Canada, pp. 123–128.

Hierarchical Bayes Inference from Survey-Weighted Small Domain Proportions

Priyanka Anjoy and Hukum Chandra

Abstract The demand for acceptable disaggregate-level statistics from sample surveys has grown substantially over the past few decades due to their extensive and varied use in public and private sectors. It is the main endeavour of 'small area estimation (SAE)' approach to produce sound prediction of a target statistic for small domains to answer the problem of small sample sizes. The traditional survey estimation approaches are not suitable enough for generating disaggregate or small domain-level estimates because of sample size problem; hence, SAE becomes the suitable choice. This chapter describes a hierarchical Bayes (HB) approach of small area estimation for survey-weighted proportions. The HB approach has advantage over the frequentist framework due to its relative flexibility in handling complex models as well as provides quick and easier mean squared error estimation. The HB technique is applied for estimating the district-level employment proportions in the state of Odisha using the 2011–12 employment and unemployment survey of National Sample Survey Office of India and the 2011 Population Census.

Keywords Generalized mixed model · Hierarchical Bayes · Small area estimation · Survey-weight

1 Introduction

Sample surveys are designed to provide reliable estimates for domain parameters of interest based on only part of the population (i.e. sample). Direct estimators (i.e. estimators that use only the sample data from the domain of interest) based on any sampling design produce adequate estimates for parameters of interest for larger domains. On many occasions, however, the interest is in estimating parameters for domains that contain only a small number of sample observations or sometimes no sample observations. Nationally designed large-scale sample surveys generally leave certain domains with very few even zero sample observations. The term 'small area' or 'small domain' is used to describe such domains whose sample sizes are not large

P. Anjoy (✉) · H. Chandra
ICAR-Indian Agricultural Statistics Research Institute, New Delhi, India

D. D. Hanagal et al. (eds.), *Applied Statistical Methods*, Springer Proceedings in Mathematics & Statistics 380, https://doi.org/10.1007/978-981-16-7932-2_3

enough to allow sufficiently precise direct estimation. When direct estimation is not possible, one has to search for an efficient alternative for producing small area estimates. Model-assisted approaches are design-based while assisted by models that result in more accurate design-unbiased estimates but still suffer from instability in case of small sample sizes. Therefore, model-dependent or model-based approaches are widely preferred and extensively used in producing acceptable small domain statistics (Rao, 2003). Eventually, model-based estimators suffer from design bias problem but their overall accuracy measures (mean squared errors) remain small. Hence, model-based technique has received attention in constructing indirect estimators for small domains. An indirect estimator for a domain 'borrow strength' from related domains by using the values of study variable *y* from related domain and/or time periods. This is the process by which 'effective sample size' of a particular domain increases and thus reduces the standard error of indirect estimator. Generally, estimator with less standard error or per cent 'coefficient of variation' (%CV) is more reliable and useful than others. Hence, in the absence of sufficient domain-specific sample sizes model-based indirect estimators are preferred over direct estimators. Indirect estimation process brings the values of survey variable into estimation process through either implicit or explicit linking models which link to related auxiliary variables drawn from administrative register or population census. Traditional indirect estimators based on implicit linking models include synthetic, composite and James–Stein estimators (Rao & Molina, 2015). Traditional indirect estimators are generally design biased and inconsistent too. However, if the linking model is approximately true, design bias will be smaller as well as will lead to smaller mean squared error (MSE) as compared to a direct estimator. Indirect estimation method based on explicit linking models is termed as 'small area models' which works under mixed modelling framework incorporating area-specific random effect that accounts for between area variations. The explicit linking models or 'mixed effect' models are generally highlighted and utilized in most of the small area literatures and real-life applications concerning small domain estimation problems. Such models are more efficient than traditional indirect approach. Indirect estimation procedure is substitute term for small area estimation (SAE) technique, whereas mixed effect models or small area models are in the centre.

The demand for efficient and reliable small area statistics for various official parameters and indicators has increased potentially since past few years from model-based SAE techniques. Various government and international agencies are stressing on the need and availability of representative small area statistics for policy-making purposes. United Nations Sustainable Development Agenda committed with 17 global goals to 'Leave no one behind' has also marked the developmental strategy through availing and utilizing disaggregate-level statistics in its several programmes and plans aiming at inclusive development of all regions globally. Nationwide large-scale surveys are designed to generate reliable estimates at higher geographical or macro-level, and such estimates cannot reflect the variations which are available at local or micro-levels. This restricts targeting of heterogeneity at disaggregated or regional level and also limits the scope for monitoring and evaluation of parameters locally within and across administrative units. In this context, model-based SAE techniques are attempted to meet the indispensable need of reliable disaggregate-level official statistics from the existing survey data without incurring extra budgetary expenditure. The industry standard for SAE is to use either unit- or area-level models (Battese et al., 1988; Fay & Herriot, 1979). In the first case, these models are for the individual survey observations and include area effects, while in the latter case these models are used to smooth out the variability in the unstable area-level direct estimates. Area-level modelling is usually employed when unit-level data is not available, or, as is often the case, where model covariates (e.g. census variables) are available only in aggregate form. Fay–Herriot (FH) model is the pioneering example of area-level small area model in small area literatures. For continuous survey variable, the FH model is the popular choice and widely used in practice, but when survey data is binary or in counts and particularly the target of inference is small area proportions rather than means or totals, then implementation of FH model which is based on linear mixed modelling structure may often lead to unrealistic estimates (Chandra et al., 2018). Hence, the potential alternative is generalized linear mixed model (GLMM), basically logistic linear mixed model or logistic normal mixed model is used for binary data, and log linear mixed model is used for count data.

Two basic approaches for drawing inferences about the small area parameters of interest are known to be popular: the empirical best prediction method is based on frequentist idea to estimate unknown model parameters, and the hierarchical Bayes (HB) approach assumes particular prior distributions for the hyper-parameters to obtain posterior quantities of the parameter of interest. The HB approach has the flexibility to deal with complex SAE model as it overcomes the difficulties of analytical MSE estimation in frequentist set-up and provides quick and easier posterior variance computation based on Markov chain Monte Carlo (MCMC) simulation. See for example, Jiang and Lahiri (2001), You and Zhou (2011), Torabi (2012), Liu et al. (2014), Rao and Molina (2015), Pratesi and Salvati (2016), Chandra et al. (2017, 2019) for frequentist and Bayesian related studies and various real life applications. Among the broad range of MCMC simulation methods, one algorithm, Gibbs sampling, has been increasingly used in applied Bayesian analyses (You & Zhou, 2011). The appeal of Gibbs sampling is that it can be used to estimate posterior distributions by drawing sample values randomly from the full conditional distributions for each of the individual parameters (i.e. the conditional distribution of a parameter given the other parameters and the observed data). On many occasions, the full conditional distributions do not have closed form; in such cases, some rejection sampling algorithm, such as the Metropolis–Hastings (M–H) algorithm within the Gibbs sampler, can be used. In addition to MCMC, HB estimation can also be implemented using alternative approximate methods including Laplace's method and Gauss–Hermite quadrature (see for details Rao & Molina, 2015).

Standard model-based approaches (in either frequentist or Bayesian framework) to the analysis often ignore the sampling mechanism. The GLMM model implicitly considers equal probability sampling (SRSWR) within each small area and thus ignores the survey weight (Chandra et al., 2019). But, this may result in potentially large biases in the final estimates. In FH model, for estimation of small area population mean, direct design-based estimators are modeled directly and the survey variance of the associated direct estimator is introduced into the model via the design-based errors. The Horvitz–Thompson estimator and weighted Hájek estimator are the structures here to incorporate survey design information (Hidiroglou & You, 2016). However, this method for continuous data requires extension for binary or count data for estimating more representative small area proportions. Consequently, the strategic idea is to modelling survey-weighted proportions. For example, Liu et al. (2014) have carried forward this idea in a HB framework; Chandra et al. (2019) elaborated the idea of using survey weight under a spatial version of GLMM in a frequentist framework. Certainly, the survey-weighted estimation has gained lot of attention from survey practitioners due its capability and flexibility in incorporating complex survey design information.

2 Hierarchical Bayesian Framework

Let us consider a finite population U of size N which is partitioned into D distinct small areas or simply areas. The set of population units in area i is denoted as U_i with known size N_i, such that $U = \cup_{i=1}^{D} U_i$ and $N = \sum_{i=1}^{D} N_i$. A sample s of size n is drawn from population U using a probabilistic mechanism. This resulted in sample $s_i$in area i with size n_i, so that $s = \cup_{i=1}^{D} s_i$ and $n = \sum_{i=1}^{D} n_i$. Assume that y_{ij} be the value of target variable y for unit j ($j = 1, \ldots, n_i$) in small area i. The target variable with values y_{ij} has binary response, taking value either 1 or 0. Our aim is to estimate the small domain proportions $P_i = N_i^{-1} \sum_{j=1}^{N_i} y_{ij}$. When the sample s is drawn following a complex survey design, then with each unit y_{ij} in small area i certain design weight w_{ij} is also attached, which is alternatively known as survey weights or sampling weights.

The basic area-level FH model combines direct aggregate (e.g. district)-level survey estimates with the available auxiliary variables obtained from various secondary sources, e.g. census or administrative records. Thus, the model has two components, sampling model for the direct survey estimates and linking model to incorporate area-specific auxiliary variables through linear regression framework. Consider p_i be the direct survey estimator for the parameter of interest P_i. In FH model, it is customary to assume that

$$p_i = P_i + e_i; \ i = 1, \ldots, D,$$

where e_i's are independent sampling error associated with direct estimator p_i. Sampling error e_i is assumed to have zero mean and known sampling variance σ_e^2. The linking model of P_i attempts to relate area-specific auxiliary variables and random effect component

$$P_i = \mathbf{x}_i' \boldsymbol{\beta} + v_i; i = 1, \ldots, D,$$

where $\mathbf{x}_i$ represent matrix of area-specific auxiliary variables, $\boldsymbol{\beta}$ is the regression coefficient or fixed effect parameter vector and v_i is the area-specific random effect, independent and identically distributed as $E(v_i) = 0$ and $\text{var}(v_i) = \sigma_v^2$. Random area-specific effects are included in the linking model to account for between areas dissimilarities. A possible limitation of the FH model is that when the target of inference is proportion, then assuming linear linking model with normal random effects may lead to unreliable and unrealistic estimates. Since P_i takes value in the range 0 to 1, to overcome the problem of unusual estimates (value <0 or >1), logistic or logit link function is preferred. Thus, the resultant model is logistic linear mixed model. Another class is log linear mixed models applicable for count data. Thus, the expression of linking model for GLMM is

$$g(P_i) = \mathbf{x}_i'\boldsymbol{\beta} + v_i; i = 1, \ldots, D,$$

where the linking function $g(.)$ is *logit* for binary data and *log* for count data. Random effect v_i has the same assumptions as in FH model. Working under HB set-up, for estimation of small area proportions certain prior distributions are assumed for the hyper-parameters. For estimating small area proportions P_i, the sampling and linking models of hierarchical Bayes Fay–Herriot (HB FH) and hierarchical Bayes generalized linear mixed model (HB GLMM) are represented as

HB Fay–Herriot Model

$$p_i | P_i \sim N(P_i, \sigma_{ei}^2),\ i = 1, \ldots, D \text{ and}$$
$$P_i | \boldsymbol{\beta}, \sigma_v^2 \sim N(\mathbf{x}_i'\boldsymbol{\beta}, \sigma_v^2),\ i = 1, , D$$

HB Generalized Linear Mixed Model

$$p_i | P_i \sim N(P_i, \sigma_{ei}^2),\ i = 1, \ldots, D\ and$$
$$logit(P_i) | \boldsymbol{\beta}, \sigma_v^2 \sim N(\mathbf{x}_i'\boldsymbol{\beta}, \sigma_v^2),\ i = 1, \ldots, D$$

Following standard literature, prior choice for $\boldsymbol{\beta}$ is usually taken to be $N(0, \sigma_0^2)$ and for σ_v^2 prior choice is $IG(a_0, b_0)$, (IG stands for inverse gamma) where σ_0^2 is set to be very large (say, 10^6) and very small value for a_0 and b_0 (usually $a_0 = b_0 \to 0$) to reflect lack of prior knowledge about variance parameters (Rao & Molina, 2015; You & Zhou, 2011; Anjoy et al., 2019). Then, inferences about the small area parameter of interest are drawn from posterior distribution. Posterior mean is taken as the point estimate of the parameter and posterior variance as a measure of the uncertainty associated with the estimate. For HB FH model, the required full conditional distributions for the Gibbs sampler are given as

$$P_i | \boldsymbol{\beta}, \sigma_v^2, p_i \sim N\left(\frac{\sigma_v^2}{\sigma_v^2 + \sigma_{ei}^2} p_i + \frac{\sigma_{ei}^2}{\sigma_v^2 + \sigma_{ei}^2}\mathbf{x}_i^T\boldsymbol{\beta}, \frac{\sigma_{ei}^2\sigma_v^2}{\sigma_v^2 + \sigma_{ei}^2}\right),$$

$$\beta \mid P_i, \sigma_v^2 \sim N\left(\left(\sum_{i=1}^{D} \mathbf{x}_i \mathbf{x}_i'\right)^{-1}\left(\sum_{i=1}^{D} \mathbf{x}_i P_i\right), \sigma_v^2\left(\sum_{i=1}^{D} \mathbf{x}_i \mathbf{x}_i'\right)^{-1}\right),$$

$$\sigma_v^2 \mid \boldsymbol{\beta}, P_i, \sim IG\left(a+\frac{D}{2}, b+\frac{\sum_{i=1}^{D} (P_i - \mathbf{x}_i'\boldsymbol{\beta})^2}{2}\right).$$

For HB GLMM, the full conditional distributions for the Gibbs sampler are given as

$$P_i \mid \boldsymbol{\beta}, \sigma_v^2, p_i \propto \frac{1}{P_i (1 - P_i)\sqrt{\sigma_{ei}^2 \sigma_v^2}} \exp\left(-\frac{(p_i - P_i)^2}{2\sigma_{ei}^2} - \frac{\left(logit(P_i) - \mathbf{x}_i'\boldsymbol{\beta}\right)^2}{2\sigma_v^2}\right),$$

$$\boldsymbol{\beta} \mid P_i, \sigma_v^2 \sim N\left(\left(\sum_{i=1}^{D} \mathbf{x}_i \mathbf{x}_i'\right)^{-1}\left(\sum_{i=1}^{D} \mathbf{x}_i logit(P_i)\right), \sigma_v^2\left(\sum_{i=1}^{D} \mathbf{x}_i \mathbf{x}_i'\right)^{-1}\right),$$

$$\sigma_v^2 \mid \boldsymbol{\beta}, P_i, \sim IG\left(a+\frac{D}{2}, b+\frac{\sum_{i=1}^{D} logit\left((P_i) - \mathbf{x}_i'\boldsymbol{\beta}\right)^2}{2}\right).$$

It is believed that survey-weighted direct estimates used for HB modelling purpose have the potentiality to reduce the bias or design error of the final estimates. Consider sample s of size n is drawn from population U using a complex design or at least unequal probability scheme. Let p_{ij} be the selection probability attached to jth sampling unit y_{ij} in the area i. The basic design weight can be given by $w_{ij} = (n_i p_{ij})^{-1}$. These weights can be adjusted to account for non-response and/or auxiliary information (Hidiroglou & You, 2016). Normalized survey weights d_{ij} may also be constructed, $d_{ij} = w_{ij}\left(\sum_j w_{ij}\right)^{-1}$. Liu et al. (2014) and Anjoy et al. (2019) have considered HB modelling of survey-weighted small area proportions, where the expression of design effect (deff) was computed based on the ratio of underlying sampling design used and assumption of simple random sampling design, such as deff term devoid of population parameters easy to calculate based on sample data mainly, design weight, sample sizes, sample size at particular stage in case of multi-stage design, etc.

Let $p_{i.uw}$ be the direct survey unweighted estimator for small area proportion P_i,

$p_{i.uw} = (n_i)^{-1} \sum_{j=1}^{n_i} y_{ij}$ and the variance of $p_{i.uw}$ is given as $\sigma_{ei.uw}^2 = n_i^{-1} P_i (1 - P_i)$.

The survey-weighted estimator denoted as $p_{i.sw}$ and its variance are expressed as

$$p_{i.sw} = \left(\sum_{j=1}^{n_i} w_{ij}\right)^{-1} \sum_{j=1}^{n_i} w_{ij} y_{ij} \text{ and the variance}$$

$$\sigma^2_{ei.sw} = \left(\sum_{j=1}^{N_i} w_{ij}\right)^{-2} \left\{\sum_{j=1}^{N_i} w_{ij}(w_{ij}-1)(y_{ij}-P_i)^2\right\} = \sigma^2_{ei.uw} deff_i.$$

Design effect is obtained as $deff_i = n_i\left(\sum_{j=1}^{N_i} w_{ij}\right)^{-2} \sum_{j=1}^{N_i} w_{ij}(w_{ij}-1)$. HB model for survey-weighted direct small area proportions is represented as follows

HB GLMM for Survey-weighted Proportions

$$p_{i.sw} \mid P_i \sim N(P_i, \sigma^2_{ei.sw}),\ i = 1, \ldots, D \text{ and}$$
$$logit(P_i) \mid \boldsymbol{\beta},\ \sigma_v^2 \sim N(\mathbf{x}_i'\boldsymbol{\beta},\ \sigma_v^2),\ i = 1, \ldots, D$$

The assumption of unknown sampling variance in this HB structure has been explored in Liu et al. (2014) and Anjoy et al. (2019). The smoothened sampling variance is also often a choice. Additionally, these papers have reported the case of assuming Beta (Beta-I) distribution for sampling model which may be quite logical as for small samples normality assumption might be questionable. For the above model, the required full conditional distributions for the Gibbs sampler are given as

$$P_i \mid \boldsymbol{\beta}, \sigma_v^2, p_{i.sw} \propto \frac{1}{P_i(1-P_i)\sqrt{\sigma^2_{ei.sw}\sigma_v^2}} \exp\left(-\frac{(p_{i.sw}-P_i)^2}{2\sigma^2_{ei.sw}} - \frac{\left(logit(P_i)-\mathbf{x}_i'\boldsymbol{\beta}\right)^2}{2\sigma_v^2}\right),$$

$$\beta \mid P_i, \sigma_v^2 \sim N\left(\left(\sum_{i=1}^{D}\mathbf{x}_i\mathbf{x}_i'\right)^{-1}\left(\sum_{i=1}^{D}\mathbf{x}_i logit(P_i)\right), \sigma_v^2\left(\sum_{i=1}^{D}\mathbf{x}_i\mathbf{x}_i'\right)^{-1}\right),$$

$$\sigma_v^2 \mid \boldsymbol{\beta}, P_i, \sim IG\left(a+\frac{D}{2},\ b+\frac{\sum_{i=1}^{D}\left(logit(P_i)-\mathbf{x}_i'\boldsymbol{\beta}\right)^2}{2}\right).$$

In the next section, we show a motivating application of HB GLMM for survey-weighted proportions to labour force indicator data in India.

3 Application

National Sample Survey Office (NSSO) under the Ministry of Statistics and Programme Implementation, Government of India, conducts the employment and unemployment surveys at periodic intervals to obtain estimates of level parameters for various labour force characteristics at the national as well as state level (employment and unemployment situation in India). These statistical indicators on labour market are critical inputs to the government think tanks and other policy-makers for framing development-oriented policies and decision-making at various levels, within the government or outside. 68th round of NSSO survey on employment status was conducted during July 2011 to June 2012. Based upon reference period and economic activities, four different estimates of the labour force indicators have been obtained in 68th round survey report of NSSO. These are termed as labour force indicators in usual status (us)/principal status (ps) taking principal activity only in previous 1 year, us/ps + subsidiary status (ss) taking principal and subsidiary activities together in previous 1 year, current weekly status (one week reference period in persons) and current daily status (each of the 7 days preceding the date of survey in person days). Accordingly, status of a person in the labour force would be (i) working/employed, (ii) not working but available for work/unemployed and (iii) neither working nor available/inactive. Labour force participation rate (LFPR) is obtained as (no. of employed + unemployed persons) * 100/total population. Worker population ratio (WPR) is obtained as (no. of employed persons) * 100/total population. LFPR determines the ratio of population engaged in economic activity or labour force, whether WPR is the determinant of employment rate. WPR by age, sex in rural and urban areas gives a picture of detailed employment portfolio across India and in various states.

The sampling design used in the 2011–12 employment and unemployment survey (2011–12 EUS) of NSSO is stratified multi-stage random sampling with districts as strata, the census villages in the rural sector as first stage units and households as the ultimate stage units. This survey is designed and conducted to produce reliable estimates at macro or higher geographical (e.g. nation and state) level. Due to small sample sizes, this survey data cannot be used directly to generate reliable micro-level or local (e.g. district or further disaggregation)-level estimates using traditional direct survey estimation methods. The 2011–12 EUS data for rural areas of Odisha state in India comprises 371 sample villages spread in total 30 districts covering 2965 sample households. Different districts of the state are considered as small areas. Sample sizes in each small area were varying from a minimum of 220 to maximum of 726 with an average of 434. Table 1 presents distribution of sample persons based

Table 1 Distribution of sample from the 2011–12 EUS data in Odisha

Sample range	All age			More than 15 years age		
	Total sample	Employed	Not employed	Total sample	Employed	Not employed
Minimum	220	80	98	153	80	28
Average	434	180	254	316	179	137
Maximum	726	322	462	510	318	277
Sum	13,031	5401	7630	9473	5376	4097

on NSSO survey for usual status (ps + ss) in labour force indicator. Note that usual status determines that the reference period of the survey is previous 1 year.

The SAE technique is employed to obtain sufficiently precise estimate of employment rate at district level which otherwise may not be representable for areas with small sample sizes using direct estimation approach. Survey variable y_{ij} is binary whether a sample person (unit) is working/employed or not. The survey-weights w_{ij} associated with the variable of interest y_{ij} were available from survey data. The auxiliary variable used for HB SAE has been obtained from the 2011 Population Census. After initial scrutiny of the pools of auxiliary variables in census data, four auxiliary variables have been retained for final analysis based on stepwise regression taking survey-weighted direct estimates of employment proportions as dependent variable. These variables are proportion of Scheduled Caste (SC) population, proportion of Scheduled Tribe (ST) population, worker population ratio and literacy rate. The selected covariates can very well be connected with observed employment ratio, as for example, SC and ST people are socially marginalized in many regions; lower literacy rates reduce the chances of salaried, self or other type of employment. For computing HB estimates for employment rate (P_i), we have considered prior for σ_v^2 as $IG(0.01, 0.01)$ and distribution of $\boldsymbol{\beta}$ has been taken to be $N(0, 10^6)$. Survey-weighted direct estimates ($p_{i.sw}$) and its variance ($\sigma_{ei.sw}^2$) are used as input for the sampling model. Gibbs sampling method is implemented with three independent chains each of length 10,000; the first 5000 iterations are deleted as 'burn-in' periods. The results presenting the district-wise estimates of employment proportions and percentage CV along with 95% confidence (credible) interval generated by HB GLMM method for survey-weighted proportions are furnished in Table 2.

It is worth noting that choice of prior distributions plays a crucial role in Bayesian analysis, because inferences drawn from posterior densities depend on wide range of prior distributions. Improper prior densities such as usual choice of $\sigma_v^2 \sim IG\,(0.001,\ 0.001)$ may not necessarily lead to proper limiting posterior distributions. As a result, posterior inferences are sensitive to setting a small value like 0.001, indicated from the studies of Gelman (2006). Various non-informative prior distributions of σ_v^2 have been suggested in Bayesian literature including a uniform density on variance parameter; see, for example, Gelman (2006), Souza et al. (2009) and references therein. Non-informative prior distributions are intended to allow Bayesian inference for parameters about which not much is known beyond the data included in the analysis at hand. Therefore, use of informative prior for the hyper-parameters needs to be examined in detail. Choice of improper or non-informative prior may also be problematic due to small amount of data under various parameterization processes, and therefore, selection of suitable distributions for the hyper-parameters with detailed check on posterior inferences can be a potential researchable issue.

Small area estimates of employment proportions in Table 2 reveal that the average employment proportions in rural Odisha is 0.43. Employment rate was varying from minimum 30% to maximum 59%. For above 15 age, employment proportions were

Table 2 District-wise estimates of employment proportions, percentage CV (%CV) along with 95% credible interval based on survey-weighted HB GLMM

Districts	All age				More than 15 years age			
	Estimates	%CV	Lower	Upper	Estimates	%CV	Lower	Upper
Bargarh	0.50	5.67	0.45	0.56	0.65	5.26	0.58	0.71
Jharsuguda	0.52	6.81	0.45	0.59	0.74	4.89	0.67	0.81
Sambalpur	0.53	7.13	0.45	0.60	0.74	5.90	0.65	0.82
Debagarh	0.49	7.73	0.42	0.57	0.71	6.47	0.62	0.80
Sundargarh	0.48	6.38	0.42	0.54	0.71	4.95	0.64	0.78
Kendujhar	0.43	6.41	0.37	0.48	0.58	6.00	0.51	0.65
Mayurbhanj	0.52	5.11	0.47	0.57	0.73	3.97	0.68	0.79
Baleshwar	0.39	6.91	0.33	0.44	0.53	6.98	0.46	0.60
Bhadrak	0.30	7.32	0.25	0.34	0.41	7.35	0.35	0.47
Kendrapara	0.32	7.58	0.28	0.37	0.45	7.49	0.39	0.52
Jagatsinghapur	0.35	7.46	0.30	0.40	0.49	7.31	0.42	0.56
Cuttack	0.38	6.51	0.33	0.42	0.53	5.79	0.47	0.59
Jajapur	0.34	6.95	0.30	0.39	0.50	5.92	0.44	0.56
Dhenkanal	0.38	6.95	0.33	0.43	0.50	7.08	0.43	0.57
Anugul	0.42	6.38	0.37	0.47	0.56	6.20	0.49	0.63
Nayagarh	0.34	7.74	0.29	0.39	0.45	7.84	0.38	0.52
Khordha	0.30	8.12	0.25	0.35	0.40	8.35	0.33	0.46
Puri	0.34	7.49	0.29	0.39	0.44	7.69	0.38	0.51
Ganjam	0.42	5.76	0.37	0.47	0.62	4.99	0.56	0.68
Gajapati	0.52	6.25	0.46	0.58	0.79	3.82	0.73	0.84
Kandhamal	0.52	6.71	0.45	0.58	0.77	4.37	0.71	0.84
Baudh	0.59	5.80	0.52	0.65	0.83	3.15	0.78	0.89
Subarnapur	0.51	8.17	0.43	0.60	0.69	6.62	0.60	0.78
Balangir	0.46	7.24	0.40	0.53	0.67	5.62	0.60	0.74
Nuapada	0.43	8.34	0.36	0.50	0.61	7.81	0.52	0.71
Kalahandi	0.38	6.90	0.32	0.43	0.57	6.39	0.50	0.64
Rayagada	0.41	8.04	0.34	0.47	0.59	7.90	0.50	0.68
Nabarangapur	0.42	7.18	0.36	0.48	0.59	6.78	0.51	0.67
Koraput	0.44	7.59	0.38	0.51	0.69	5.84	0.61	0.77
Malkangiri	0.46	8.74	0.39	0.54	0.67	7.47	0.57	0.77

varying from 0.40 to 0.83, with an average of 0.61. Regional disparity pattern in unemployment rate can be seen across the districts; specifically, there are 14 districts having employment proportions below the state average. These aspirant districts need special emphasis through state and central various developmental plans.

4 Concluding Remarks

There are number of issues that warrant further investigation. In this paper, we assume that district-specific random effects are independent. Spatial dependence among neighbouring districts may also be taken into account within the same Bayes modeling framework to improve the model-based estimates. Concept of spatial correlation and spatial nonstationarity under the HB paradigm needs further investigation to improve the GLMM-based estimates. Baldermann et al. (2018) have recently described spatial nonstationarity concept for explaining spatial variability between areas, but their model is for unit-level data. Hence, scope remains to develop area-level version of spatial nonstationary model under Bayesian approach. In frequentist framework, Chandra et al. (2017) have devised the concept of spatial nonstationarity in area-level version of GLMM (NSGLMM) for small area estimation of proportions. It needs to be looked upon using HB structure and requires investigation oriented towards choice of Gibbs sampling procedures and along with various prior choices. Again, a few literatures are available on small area procedures considering informativeness and exploring its impact. Current approach of HB GLMM for survey-weighted proportions may be extended for survey variable with three or more categories yielding survey-weighted HB multinomial linear mixed model.

It is the first instance of using employment–unemployment survey data of NSSO for producing adequate employment statistics at district level in Odisha state. Generated labour force indicators will aid the policy-makers to find the loophole in existing policies as well as framing new strategies to uplift the socio-economic conditions of people. However, more detailed analysis definitely will assist the government organizations, particularly focusing aspirant districts. As soon the availability of new data set on employment situation, it will be easy to compare the disaggregated level figures and unemployment trend. Employment data may very well be coupled with poverty, food insecurity proportion data at disaggregate level, to come up with combined developmental plans. Generating gender-specific WPR and combining them at district level is another good option to carry on, which may simultaneously reveal the gender gap in employment–unemployment situation also.

Acknowledgements This chapter is a tribute to **Dr. Hukum Chandra** who was a passionate researcher in the field of 'sample surveys' and specifically 'small area estimation'. He has left for heavenly abode before the final acceptance of the chapter.

References

Anjoy, P., Chandra, H., & Basak, P. (2019). Estimation of disaggregate-level poverty incidence in Odisha under area-level hierarchical Bayes small area model. *Social Indicators Research, 144*, 251–273.

Baldermann, C., Salvati, N., & Schmid, T. (2018). Robust small area estimation under spatial non-stationarity. *International Statistical Review, 86*, 136–159.

Battese, G. E., Harter, R. M., & Fuller, W. A. (1988). An error-component model for prediction of country crop areas using survey and satellite data. *Journal of the American Statistical Association, 83*, 28–36.

Chandra, H., Salvati, N., & Chambers, R. (2017). Small area prediction of counts under a nonstationary spatial model. *Spatial Statistics, 20*, 30–56.

Chandra, H., Salvati, N., & Chambers, R. (2018). Small area estimation under a spatially non-linear model. *Computational Statistics and Data Analysis, 126*, 19–38.

Chandra, H., Chambers, R., & Salvati, N. (2019). Small area estimation of survey weighted counts under aggregated level spatial model. *Survey Methodology, 45*, 31–59.

Employment and Unemployment Situation in India. NSS Report No. 554(68/10/1), NSSO, Ministry of Statistics and Programme Implementation, Govt. of India.

Fay, R. E., & Herriot, R. (1979). Estimates of income for small places: An application of James-Stein procedures to census data. *Journal of the American Statistical Association, 74*, 269–277.

Gelman, A. (2006). Prior distributions for variance parameters in hierarchical models. *Bayesian Analysis, 1*, 515–533.

Hidiroglou, M. A., & You, Y. (2016). Comparison of unit level and area level small area estimators. *Survey Methodology, 42*, 41–61.

Jiang, J., & Lahiri, P. (2001). Empirical best prediction for small area inference with binary data. *Annals of the Institute of Statistical Mathematics, 53*, 217–243.

Liu, B., Lahiri, P., & Kalton, G. (2014). Hierarchical Bayes modeling of survey-weighted small area proportions. *Survey Methodology, 40*, 1–13.

Rao, J. N. K. (2003). *Small area estimation.* Wiley.

Rao, J. N. K., & Molina, I. (2015). *Small area estimation* (2nd ed.). Wiley.

Pratesi, M., & Salvati, N. (2016). *Introduction on measuring poverty at local level using small area estimation methods, Pratesi/analysis of poverty data by small area estimation.* Wiley.

Souza, D. F., Moura, F. A. S., & Migon, H. S. (2009). Small area population prediction via hierarchical models. *Survey Methodology, 35*, 203–214.

Torabi, M. (2012). Hierarchical Bayes estimation of spatial statistics for rates. *Journal of Statistical Planning and Inference, 142*, 358–365.

You, Y., & Zhou, M. Q. (2011). Hierarchical Bayes small area estimation under a spatial model with application to health survey data. *Survey Methodology, 37*, 25–37.

Efficiency of Ranked Set Sampling Design in Goodness of Fit Tests for Cauchy Distribution

Ehsan Zamanzade and M. Mahdizadeh

Abstract In this work, we describe some goodness of fit tests for Cauchy distribution using ranked set sampling (RSS) design. The powers of the developed tests in RSS are compared with their counterparts in simple random sampling (SRS) for both perfect and imperfect ranking cases. It is found that the test based on Kullback–Leibler distance is much more powerful than its competitors in most of the cases considered. It also controls type **I** error well in the case of imperfect ranking.

Keywords Imperfect ranking · Power · Monte Carlo simulation · Ranked set sampling

1 Introduction

Ranked set sampling (RSS), proposed by McIntyre (1952) in agricultural context, is an efficient alternative to simple random sampling (SRS), which is applicable in many practical situations where obtaining exact measurements of the sample units is far harder or expensive than ranking them in a set of small size. These situations typically happen in forestry (Halls & Dell, 1966), environmental monitoring (Kwam, 2003), clinical trials and genetic quantitative trait loci mappings (Chen, 2007), educational studies (Wang et al., 2017), and medicine (Zamanzade & Wang, 2017; Mahdizadeh & Zamanzade, 2019).

To draw a ranked set sample of size n, the researcher first determines set size k and cycle size r such that $n = k \times r$. He then draws a simple random sample of size $r \times k^2$ from the population of interest and randomly divides them into $r \times k$ samples

E. Zamanzade (✉)
Faculty of Mathematics and Statistics, Department of Statistics, University of Isfahan, Isfahan 81746-73441, Iran
e-mail: e.zamanzade@sci.ui.ac.ir

M. Mahdizadeh
Department of Statistics, Hakim Sabzevari University, P.O. Box 397, Sabzevar, Iran
e-mail: mahdizadeh.m@hsu.ac.ir

D. D. Hanagal et al. (eds.), *Applied Statistical Methods*, Springer Proceedings in Mathematics & Statistics 380, https://doi.org/10.1007/978-981-16-7932-2_4

(sets) each of size k. Each set of size k is then ranked in an increasing magnitude. The ranking process in this step is done without obtaining the exact values of the units in different sets and using an inexpensive method such as eye inspection, personal judgment, or a freely available covariate. Finally, the researcher selects r units with rank 1 from first r sets and r units with rank 2 from the second r sets. The process continues until r units with rank k ... and measures all of them. The resulting ranked set sample is given by $\{X_{[i]j} : i = 1, \ldots, k; j = 1, \ldots, r\}$, where $X_{[i]j}$ is the actual measurement of the unit with judgment rank i in a set of size k in the jth cycle. Here, the term "*judgment rank*"is used to accentuate that the ranking process is done without actual measurement of the units in the set, and therefore, it is likely to observe errors in ranking. This situation is called imperfect ranking.

Because of importance and applicability of RSS in many practical situations, several researchers have worked on this research field and virtually most standard statistical problems have been addressed including but not limited to estimation of the cumulative distribution function (CDF) (Stokes & Sager, 1988; Kvam & Samaniego, 1994; Duembgen & Zamanzade, 2020), estimation of the population proportion (Zamanzade & Mahdizadeh, 2017, 2018), estimation of the population mean (Takahasi & Wakimoto, 1968; Frey, 2011), estimation of the population variance (Stokes, 1980; MacEachern et al., 2002), odds ratio (Samawi & Al-Saleh, 2013), logistic regression (Samawi et al., 2017, 2018, 2020), and statistical control quality charts (Al-Omari & Haq, 2011; Haq & Al-Omari, 2014; Haq et al., 2014).

A random variable X follows Cauchy distribution with location parameter μ and scale parameter σ if its probability density function is given by

$$f_0(x; \mu, \sigma) = \frac{1}{\sigma\pi\left(1 + \left(\frac{x-\mu}{\sigma}\right)^2\right)}, \quad x \in \mathbb{R}. \tag{1}$$

The corresponding CDF is also given by

$$F_0(x; \mu, \sigma) = \frac{1}{\pi} \arctan\left(\frac{x-\mu}{\sigma}\right) + \frac{1}{2}, \quad x \in \mathbb{R}. \tag{2}$$

The mean and variance of a random variable with Cauchy distribution do not exist. Due to its fatter tails than standard normal distribution, Cauchy distribution is more successful than normal distribution in modeling data with extreme values. In this work, we describe several goodness of fit tests for Cauchy distribution in RSS and compare their powers with their counterparts in SRS. We also discuss the problem of type **I** error inflation in goodness of fit tests based on RSS in the presence of errors in rankings.

2 Goodness of Fit Test for Cauchy Distribution

In this section, we describe different goodness of fit tests for Cauchy distribution based on SRS and RSS sampling designs. Suppose that one is interested in testing the null hypothesis

$$H_0 : F(x) = F_0(x; \mu, \sigma), \quad x \in \mathbb{R},$$

for some $\mu \in \mathbb{R}$ and $\sigma \in \mathbb{R}^+$, against the alternative hypothesis

$$H_1 : F(x) \neq F_0(x; \mu, \sigma), \quad x \in \mathbb{R},$$

for any $\mu \in \mathbb{R}$ and $\sigma \in \mathbb{R}^+$, where $F_0(x; \mu, \sigma)$ is given in Eq. (2).

Let $X_1, \ldots, X_n$ be a simple random sample of size n from the population of interest with CDF F and $X_{(1)}, \ldots, X_{(n)}$ be the corresponding order statistics. Let $\hat{F}_{srs}(t) = \sum_{i=1}^{n} \mathbb{I}(X_i \leqslant t)/n$ be the empirical distribution function (EDF) based on a simple random sample. Mahdizadeh and Zamanzade (2017, 2020) utilized the estimators

$$\hat{\mu}_{srs} = \begin{cases} \frac{1}{2}\left(X_{(n/2)} + X_{(n/2+1)}\right) & \text{if } n \text{ is even} \\ X_{((n+1)/2)} & \text{if } n \text{ is odd} \end{cases},$$

and

$$\hat{\sigma}_{srs} = \frac{1}{2}\left(\hat{Q}_{srs,0.75} - \hat{Q}_{srs,0.25}\right),$$

for μ and σ in developing goodness of fit tests for Cauchy distribution, where $\hat{Q}_{srs,p}$ is the pth sample quantile based on a simple random sample. We use these estimators, because it was shown in the literature that the resulting tests based on the estimators have good powers. Below, we describe some well-known and powerful tests for Cauchy distribution based on a simple random samples, one by one.

- Kolmogorov–Smirnov test statistic

$$\begin{aligned} KS &= \sup_t \left| \hat{F}_{srs}(t) - F_0\left(t, \hat{\mu}_{srs}, \hat{\sigma}_{srs}\right) \right| \\ &= \max_{1 \le i \le n} \left(\max\left\{ \frac{i}{n} - F_0\left(X_{(i)}, \hat{\mu}_{srs}, \hat{\sigma}_{srs}\right), F_0\left(X_{(i)}, \hat{\mu}_{srs}, \hat{\sigma}_{srs}\right) - \frac{i-1}{n} \right\} \right) \end{aligned}$$

- Anderson–Darling test statistic

$$A^2 = \int_{-\infty}^{+\infty} \left\{\hat{F}_{srs}(t) - F_0\left(t, \hat{\mu}_{srs}, \hat{\sigma}_{srs}\right)\right\} F_0\left(t, \hat{\mu}_{srs}, \hat{\sigma}_{srs}\right)^{-1}$$
$$\left\{1 - F_0\left(t, \hat{\mu}_{srs}, \hat{\sigma}_{srs}\right)\right\}^{-1} dF_0\left(t, \hat{\mu}_{srs}, \hat{\sigma}_{srs}\right)$$
$$= -\frac{2}{n}\sum_{i=1}^{n}\left[\left(i - \frac{1}{2}\right)\log\left\{F_0\left(X_{(i)}, \hat{\mu}_{srs}, \hat{\sigma}_{srs}\right)\right\}\right.$$
$$\left. + \left(n - i + \frac{1}{2}\right)\log\left\{1 - F_0\left(X_{(i)}, \hat{\mu}_{srs}, \hat{\sigma}_{srs}\right)\right\}\right] - n.$$

- Cramer-von Mises test statistic

$$W^2 = n\int_{-\infty}^{+\infty} \left\{\hat{F}_{srs}(t) - F_0\left(t, \hat{\mu}_{srs}, \hat{\sigma}_{srs}\right)\right\}^2 dF_0\left(t, \hat{\mu}_{srs}, \hat{\sigma}_{srs}\right)$$
$$= \sum_{i=1}^{n}\left\{F_0\left(X_{(i)}, \hat{\mu}_{srs}, \hat{\sigma}_{srs}\right) - \frac{i - \frac{1}{2}}{n}\right\}^2 + \frac{1}{12n},$$

- Zhang (2002)'s test statistics

$$Z_k = \max_{1\le i\le n}\left(\left(i - \frac{1}{2}\right)\log\left\{\frac{i - \frac{1}{2}}{nF_0\left(X_{(i)}, \hat{\mu}_{srs}, \hat{\sigma}_{srs}\right)}\right\}\right.$$
$$\left. + \left(n - i + \frac{1}{2}\right)\log\left\{\frac{n - i + \frac{1}{2}}{n\left\{1 - F_0\left(X_{(i)}, \hat{\mu}_{srs}, \hat{\sigma}_{srs}\right)\right\}}\right\}\right),$$
$$Z_A = -\sum_{i=1}^{n}\left(\frac{\log\left\{F_0\left(X_{(i)}, \hat{\mu}_{srs}, \hat{\sigma}_{srs}\right)\right\}}{n - i + \frac{1}{2}} + \frac{\log\left\{1 - F_0\left(X_{(i)}, \hat{\mu}_{srs}, \hat{\sigma}_{srs}\right)\right\}}{i - \frac{1}{2}}\right),$$
$$Z_C = \sum_{i=1}^{n}\left(\log\left\{\frac{F_0\left(X_{(i)}, \hat{\mu}_{srs}, \hat{\sigma}_{srs}\right)^{-1} - 1}{(n - \frac{1}{2})/(i - \frac{3}{4}) - 1}\right\}\right)^2$$

- Test statistic based on Kullback–Leibler (KL) distance

$$D = -H_{n,m} - \frac{1}{n}\sum_{i=1}^{n}\log\left(f_0\left(X_{(i)}, \hat{\mu}_{srs}, \hat{\sigma}_{srs}\right)\right),$$

where $H_{n,m}$ is Vasicek's entropy estimator which has the form $H_{n,m} = \frac{1}{n}\sum_{i=1}^{n} \log\left\{\frac{n}{2m}\left(X_{(i+m)} - X_{(i-m)}\right)\right\}$, $X_{(i)} = X_{(1)}$ for $i < 1$, $X_{(i)} = X_{(n)}$ for $i > n$ and the integer $m \le n/2$ is the window size.

Remark 1 It is important to note that each of the above goodness of fit tests rejects the null hypothesis that the parent distribution follows a Cauchy distribution at the significance level α, when the observed test statistic is larger than the corresponding $100(1 - \alpha)$ percentile of the null distribution of the test statistic. It is worth mentioning that the distributions of the above test statistics under assumption that the parent distribution is Cauchy do not depend on the unknown parameters μ and σ, and hence, the true type **I** error rates of the above tests are exactly the same as their nominal level α.

Let $\{X_{[i]j} : i = 1, \ldots, k; j = 1, \ldots, r\}$ be a ranked set sample of size $n = r \times k$ with set size k from a population with CDF F. Let $\hat{F}_{rss}(t) = \sum_{i=1}^{k} \sum_{j=1}^{r} \mathbb{I}\left(X_{[i]j} \leqslant t\right)/n$ be the EDF based on the ranked set sample, and let $Z_1, \ldots, Z_n$ be an ordered ranked set sample which is obtained by putting RSS sample units in ascending order. In order to develop goodness of fit tests for Cauchy distribution, we propose to estimate the parameter μ and σ by

$$\hat{\mu}_{rss} = \begin{cases} \frac{1}{2}\left(Z_{n/2} + Z_{n/2+1}\right) & \text{if } n \text{ is even} \\ Z_{(n+1)/2} & \text{if } n \text{ is odd} \end{cases},$$

and

$$\hat{\sigma}_{srs} = \frac{1}{2}\left(\hat{Q}_{rss,0.75} - \hat{Q}_{rss,0.25}\right),$$

where $\widehat{Q}_{rss,p}$ is the pth sample quantile based on the ordered ranked set sample.

Therefore, the RSS counterpart of each of the above goodness of fit tests can be proposed as

- Kolmogorov–Smirnov test statistic

$$\begin{aligned} KS &= \sup_t \left|\hat{F}_{rss}(t) - F_0\left(t, \hat{\mu}_{rss}, \hat{\sigma}_{rss}\right)\right| \\ &= \max_{1 \le i \le n} \left(\max\left\{\frac{i}{n} - F_0\left(Z_i, \hat{\mu}_{rss}, \hat{\sigma}_{rss}\right), F_0\left(Z_i, \hat{\mu}_{rss}, \hat{\sigma}_{rss}\right) - \frac{i-1}{n}\right\}\right) \end{aligned}$$

- Anderson–Darling test statistic

$$\begin{aligned} A^2 &= \int_{-\infty}^{+\infty} \left\{\hat{F}_{rss}(t) - F_0\left(t, \hat{\mu}_{rss}, \hat{\sigma}_{rss}\right)\right\} \\ &\quad F_0\left(t, \hat{\mu}_{rss}, \hat{\sigma}_{rss}\right)^{-1} \left\{1 - F_0\left(t, \hat{\mu}_{rss}, \hat{\sigma}_{rss}\right)\right\}^{-1} dF_0\left(t, \hat{\mu}_{rss}, \hat{\sigma}_{rss}\right) \\ &= -\frac{2}{n} \sum_{i=1}^{n} \left[\left(i - \frac{1}{2}\right) \log\left\{F_0\left(Z_i, \hat{\mu}_{rss}, \hat{\sigma}_{rss}\right)\right\}\right. \\ &\quad \left. + \left(n - i + \frac{1}{2}\right) \log\left\{1 - F_0\left(Z_i, \hat{\mu}_{rss}, \hat{\sigma}_{rss}\right)\right\}\right] - n. \end{aligned}$$

- Cramer-von Mises test statistic

$$W^2 = n \int_{-\infty}^{+\infty} \left\{ \hat{F}_{rss}(t) - F_0\left(t, \hat{\mu}_{rss}, \hat{\sigma}_{rss}\right) \right\}^2 dF_0\left(t, \hat{\mu}_{rss}, \hat{\sigma}_{rss}\right)$$
$$= \sum_{i=1}^{n} \left\{ F_0\left(Z_i, \hat{\mu}_{rss}, \hat{\sigma}_{rss}\right) - \frac{i - \frac{1}{2}}{n} \right\}^2 + \frac{1}{12n},$$

- Zhang (2002)s test statistics

$$Z_k = \max_{1 \le i \le n} \left(\left(i - \frac{1}{2}\right) \log \left\{ \frac{i - \frac{1}{2}}{nF_0\left(Z_i, \hat{\mu}_{rss}, \hat{\sigma}_{rss}\right)} \right\} + \left(n - i + \frac{1}{2}\right) \log \left\{ \frac{n - i + \frac{1}{2}}{n\left\{1 - F_0\left(Z_i, \hat{\mu}_{rss}, \hat{\sigma}_{rss}\right)\right\}} \right\} \right),$$

$$Z_A = -\sum_{i=1}^{n} \left(\frac{\log\left\{F_0\left(Z_i, \hat{\mu}_{rss}, \hat{\sigma}_{rss}\right)\right\}}{n - i + \frac{1}{2}} + \frac{\log\left\{1 - F_0\left(Z_i, \hat{\mu}_{rss}, \hat{\sigma}_{rss}\right)\right\}}{i - \frac{1}{2}} \right),$$

$$Z_C = \sum_{i=1}^{n} \left(\log \left\{ \frac{F_0\left(Z_i, \hat{\mu}_{rss}, \hat{\sigma}_{rss}\right)^{-1} - 1}{(n - \frac{1}{2})/(i - \frac{3}{4}) - 1} \right\} \right)^2$$

- Test statistic based on Kullback–Leibler (KL) distance

$$D = -H_{n,m} - \frac{1}{n} \sum_{i=1}^{n} \log\left(f_0\left(Z_i, \hat{\mu}_{rss}, \hat{\sigma}_{rss}\right)\right),$$

where $H_{n,m}$ is Vasicek's entropy estimator which has the form $H_{n,m} = \frac{1}{n} \sum_{i=1}^{n} \log\left\{\frac{n}{2m}\left(Z_{i+m} - Z_{i-m}\right)\right\}$, $Z_i = Z_1$ for $i < 1$, $Z_i = Z_n$ for $i > n$ and the integer $m \le n/2$ is the window size.

Note that large values of each of the test statistics described in above can be regarded as an indication that the parent distribution does not follow a Cauchy distribution and therefore the null hypothesis is rejected at significance level α if its observed value is larger than $100\,(1 - \alpha)$ percentile of the null distribution.

Remark 2 It is worth mentioning that the distribution of each of the goodness of fit tests of Cauchy distribution in RSS does not depend on unknown parameters μ and σ, but it still depends on both ranking quality and imperfect ranking model. As the ranking quality and imperfect ranking model are not known in practice, $100\,(1 - \alpha)$ percentile of the test statistics in RSS under the null hypothesis is obtained when the perfect ranking is assumed. It is of course, in this case, the type **I** error will be inflated if the ranking is not perfect.

3 Power Comparison

In this section, we compare the power of different goodness of fit tests in SRS and RSS for Cauchy distribution. To do so, we set $n \in \{10, 20, 30, 50\}$, $k \in \{2, 5\}$, and for each configuration of (n, k), we have generated 10,000 random samples from SRS and RSS designs. Therefore, we can assess performance of different goodness of fit tests when k is increased and n is fixed and vice versa. To generate a ranked set sample, we have used fraction of random ranking model proposed by Frey et al. (2007). In this model, it is assumed that the sample unit with judgment rank i is identified correctly with probability ρ, and it is identified randomly with probability $1 - \rho$. Therefore, the CDF of the ith judgment order statistic is given by

$$F_{[i]}(t) = \rho F_{(i)}(t) + (1 - \rho) F(t),$$

where $F_{[i]}(t)$ and $F_{(i)}(t)$ are CDFs of the ith judgment and ith true order statistics and the parameter ρ controls quality of ranking. In this simulation, we select ρ from the set $\rho \in \{1, 0.75, 0.5\}$. So, perfect ranking case ($\rho = 1$), good ranking case ($\rho = 0.75$), and moderate ranking case ($\rho = 0.5$) are all included in our study.

The powers of the goodness of fit tests for Cauchy distribution in SRS and RSS designs are reported against the 15 different alternative distributions. The alternative distributions are as: Student's t distribution with 3 and 5 degrees of freedom (t_3, t_5), standard normal ($N(0, 1)$), standard logistic ($Lo(0, 1)$), standard Laplace ($La(0, 1)$), standard Gumbel ($Gu(0, 1)$), standard exponential ($E(1)$), standard uniform ($U(0, 1)$) distributions, beta distribution with parameters 0.5 and 0.5 ($B(0.5, 0.5)$), beta distribution with parameters 2 and 1 ($B(2, 1)$), gamma distribution with scale parameter 1 and shape parameter 0.5 ($G(0.5)$), gamma distribution with scale parameter 1 and shape parameter 2 ($G(2)$), Weibull distribution with scale parameter 1 and shape parameter 0.8 ($W(0.8)$), and Weibull distribution with scale parameter 1 and shape parameter 1.4 ($W(1.4)$). We have also included standard Cauchy distribution ($C(0, 1)$) in our comparison set. Thus, it is possible to observe amount of inflation in type **I** error in RSS when the ranking is not perfect.

Remark 3 To calculate the test statistic based on KL distance (D), one has to determine window size m subject to sample size n. In this simulation, we follow Mahdizadeh and Zamanzade (2017)'s guideline, which suggests selecting the window size resulting in smallest critical value. Therefore, for sample sizes 10, 20, 30, and 50, the window sizes producing minimum critical values are 2, 4, 8, and 20, respectively.

Here, for brevity, we only present results in RSS for $n = 10$, $k \in \{2, 5\}$, and $\rho \in \{0.75, 1\}$ in Tables 1 and 2 and in SRS for $n \in \{10, 20\}$ in Table 3. The complete simulation results are available upon request from the first author.

Table 1 presents simulation results for perfect ranking case ($\rho = 1$), $n = 10$, and $k \in \{2, 5\}$. We observe from this table that the test based on KL distance is in the most powerful test. It is also the only unbiased test in the sense that its power is

Table 1 Powers of different goodness of fit tests for Cauchy distribution in RSS with $n = 10$, $k \in \{2, 5\}$, and $\rho = 1$. For each alternative distribution, the test with the highest power is denoted by asterisk

Test	k	$C(0,1)$	t_3	t_5	$N(0,1)$	$Lo(0,1)$	$La(0,1)$	$Gu(0,1)$	$E(1)$	$U(0,1)$	$B(0.5,0.5)$	$B(2,1)$	$G(0.5)$	$C(2)$	$W(0.8)$	$W(1.4)$
D	2	0.052	0.131*	0.163*	0.234*	0.185*	0.109 *	0.249*	0.442*	0.538*	0.808*	0.482*	0.671*	0.320*	0.539*	0.346*
KS		0.050	0.035	0.033	0.041	0.034	0.032	0.063	0.225	0.091	0.253	0.111	0.496	0.104	0.361	0.103
A^2		0.049	0.017	0.015	0.020	0.017	0.017	0.029	0.119	0.052	0.156	0.055	0.312	0.052	0.211	0.055
W^2		0.051	0.032	0.033	0.038	0.032	0.032	0.059	0.177	0.088	0.224	0.094	0.389	0.092	0.278	0.091
Z_K		0.054	0.016	0.014	0.017	0.013	0.015	0.031	0.152	0.047	0.168	0.063	0.409	0.060	0.272	0.058
Z_A		0.051	0.021	0.019	0.028	0.020	0.018	0.042	0.165	0.096	0.303	0.096	0.403	0.075	0.276	0.077
Z_C		0.052	0.015	0.012	0.022	0.013	0.012	0.027	0.102	0.083	0.271	0.075	0.263	0.049	0.174	0.053
D	5	0.047	0.146*	0.185*	0.262*	0.205*	0.121*	0.265*	0.502*	0.611*	0.862*	0.521*	0.730*	0.360*	0.585*	0.382*
KS		0.050	0.042	0.047	0.051	0.047	0.041	0.073	0.282	0.101	0.268	0.132	0.590	0.131	0.424	0.119
A^2		0.049	0.017	0.016	0.021	0.020	0.016	0.030	0.131	0.051	0.146	0.056	0.352	0.059	0.228	0.053
W^2		0.051	0.047	0.049	0.053	0.049	0.040	0.075	0.230	0.115	0.267	0.125	0.480	0.123	0.342	0.114
K_Z		0.046	0.012	0.010	0.012	0.011	0.011	0.022	0.142	0.031	0.115	0.046	0.425	0.049	0.265	0.045
Z_A		0.050	0.021	0.021	0.029	0.023	0.017	0.043	0.186	0.100	0.329	0.099	0.451	0.081	0.302	0.080
Z_C		0.048	0.008	0.009	0.014	0.012	0.008	0.018	0.089	0.059	0.223	0.048	0.249	0.038	0.153	0.035

Table 2 Powers of different goodness of fit tests for Cauchy distribution in RSS with $n = 10$, $k \in \{2, 5\}$, and $\rho = 0.75$. For each alternative distribution, the test with the highest power is denoted by asterisk

Test	k	$C(0,1)$	t_3	t_5	$N(0,1)$	$Lo(0,1)$	$La(0,1)$	$Gu(0,1)$	$E(1)$	$U(0,1)$	$B(0.5,0.5)$	$B(2,1)$	$G(0.5)$	$C(2)$	$W(0.8)$	$W(1.4)$
D	2	0.058	0.134*	0.172*	0.231*	0.189*	0.118*	0.244*	0.445*	0.552*	0.808*	0.463*	0.659*	0.328*	0.530*	0.346*
KS		0.054	0.034	0.035	0.039	0.036	0.035	0.063	0.220	0.094	0.272	0.117	0.485	0.105	0.355	0.103
A^2		0.050	0.014	0.014	0.017	0.015	0.014	0.027	0.116	0.046	0.151	0.052	0.296	0.047	0.195	0.048
W^2		0.053	0.033	0.034	0.040	0.036	0.035	0.059	0.184	0.093	0.240	0.106	0.387	0.093	0.283	0.088
K_Z		0.050	0.013	0.012	0.015	0.015	0.012	0.025	0.143	0.044	0.168	0.063	0.380	0.051	0.254	0.051
Z_A		0.048	0.019	0.016	0.023	0.022	0.015	0.036	0.156	0.085	0.287	0.088	0.376	0.066	0.259	0.065
Z_C		0.050	0.010	0.011	0.015	0.014	0.009	0.023	0.094	0.063	0.232	0.060	0.234	0.039	0.150	0.040
D	5	0.055	0.163*	0.207*	0.276*	0.219*	0.135*	0.278*	0.506*	0.600*	0.856*	0.539*	0.721*	0.373*	0.596*	0.391*
KS		0.060	0.045	0.048	0.056	0.046	0.045	0.076	0.273	0.107	0.315	0.148	0.563	0.129	0.427	0.132
A^2		0.058	0.019	0.020	0.026	0.019	0.017	0.032	0.135	0.054	0.182	0.073	0.354	0.061	0.245	0.059
W^2		0.062	0.046	0.050	0.060	0.052	0.043	0.076	0.223	0.118	0.297	0.136	0.460	0.117	0.351	0.118
K_Z		0.053	0.012	0.016	0.020	0.013	0.014	0.029	0.158	0.041	0.179	0.068	0.432	0.055	0.292	0.056
Z_A		0.062	0.023	0.031	0.038	0.027	0.023	0.052	0.200	0.121	0.387	0.127	0.469	0.092	0.334	0.098
Z_C		0.055	0.012	0.016	0.023	0.015	0.011	0.027	0.112	0.083	0.307	0.081	0.287	0.051	0.190	0.054

Table 3 Powers of different goodness of fit tests for Cauchy distribution in SRS with $n \in \{10, 20\}$. For each alternative distribution, the test with the highest power is denoted by asterisk

Test	n	$C(0,1)$	t_3	t_5	$N(0,1)$	$Lo(0,1)$	$La(0,1)$	$Gu(0,1)$	$E(1)$	$U(0,1)$	$B(0.5,0.5)$	$B(2,1)$	$G(0.5)$	$C(2)$	$W(0.8)$	$W(1.4)$
D	10	0.051	0.114*	0.145*	0.210*	0.165*	0.104*	0.211*	0.403*	0.494*	0.774*	0.427*	0.641*	0.292*	0.495*	0.307*
KS		0.049	0.025	0.029	0.032	0.028	0.027	0.049	0.187	0.075	0.262	0.100	0.456	0.085	0.317	0.082
A^2		0.048	0.012	0.014	0.014	0.013	0.014	0.024	0.098	0.039	0.155	0.049	0.288	0.043	0.180	0.041
W^2		0.045	0.025	0.026	0.029	0.028	0.026	0.044	0.142	0.069	0.216	0.079	0.359	0.075	0.243	0.069
K_Z		0.047	0.009	0.010	0.011	0.009	0.010	0.020	0.112	0.036	0.169	0.051	0.355	0.039	0.221	0.042
Z_A		0.048	0.014	0.017	0.022	0.017	0.015	0.032	0.132	0.075	0.283	0.078	0.372	0.061	0.246	0.060
Z_C		0.048	0.008	0.009	0.013	0.009	0.008	0.016	0.075	0.048	0.208	0.047	0.219	0.030	0.131	0.032
D	20	0.053	0.342*	0.498*	0.740*	0.573*	0.334*	0.729*	0.924*	0.993*	1.000*	0.975*	0.978*	0.853*	0.941*	0.887*
KS		0.050	0.045	0.055	0.067	0.057	0.036	0.124	0.601	0.298	0.743	0.354	0.910	0.290	0.790	0.293
A^2		0.049	0.030	0.039	0.062	0.043	0.024	0.091	0.375	0.281	0.667	0.259	0.715	0.187	0.555	0.193
W^2		0.054	0.047	0.054	0.078	0.061	0.040	0.115	0.406	0.262	0.622	0.266	0.736	0.213	0.580	0.212
K_Z		0.047	0.029	0.034	0.056	0.040	0.022	0.121	0.651	0.344	0.796	0.403	0.931	0.319	0.830	0.326
Z_A		0.053	0.097	0.150	0.271	0.174	0.074	0.318	0.737	0.790	0.970	0.705	0.940	0.509	0.858	0.539
Z_C		0.050	0.058	0.094	0.190	0.114	0.046	0.202	0.509	0.684	0.933	0.554	0.779	0.336	0.636	0.366

always higher than its significance level $\alpha = 0.05$. All tests control type **I** error well, which is not surprising because of perfect ranking assumption. We also see from this table that although the power of D, KS, A^2, and W^2 increases with set size k while the total sample size n is fixed, the other three tests have irregular patterns.

Comparing simulation results in Table 1 with other results in the perfect ranking case, we found that although the pattern of the performance of different tests remains almost the same when the sample size n increases from 10 to 50, their powers increase.

Simulation results for imperfect ranking cases are presented in Table 2. We observe from this table that the test based on KL distance is the most powerful one in almost all considered cases. It is also interesting to note that amount of inflation in type **I** error in all considered cases is negligible, and therefore, one can use the critical values of test statistics under perfect ranking assumption in the imperfect ranking case with a high confidence.

Finally, it is evident from Tables 3 that the best test in RSS is generally more powerful than its counterpart in SRS.

4 Conclusion

In this chapter, we propose several goodness of fit tests for Cauchy distribution in RSS. We then compared the powers of different tests using a comprehensive simulation study for both perfect and imperfect ranking cases. It is found that while the test based on KL distance controls type **I** error well, it is significantly more powerful than its competitors in most considered cases.

References

Al-Omari, A. I., & Haq, A. (2011). Improved quality control charts for monitoring the process mean, using double-ranked set sampling methods. *Journal of Applied Statistics, 39*(4), 745–763.

Chen, Z. (2007). Ranked set sampling: Its essence and some new applications. *Environmental and Ecological Statistics, 14,* 355–363.

Duembgen, L., & Zamanzade, E. (2020). Inference on a distribution function from ranked set samples. *Annals Institute of Statistical Mathematics, 72,* 157–185.

Frey, J. (2011). A note on ranked-set sampling using a covariate. *Journal of Statistical Planning and Inference, 141,* 809–816.

Frey, J., Ozturk, O., & Deshpande, J. V. (2007). Nonparametric tests for perfect judgment rankings. *Journal of the American Statistical Association, 102*(478), 708–717.

Halls, L. K., & Dell, T. R. (1966). Trial of ranked-set sampling for forage yields. *Forest Science, 12,* 22–26.

Haq, A., & Al-Omari, A. I. (2014). A new Shewhart control chart for monitoring process mean based on partially ordered judgment subset sampling. *Quality and Quantity, 49*(3), 1185–1202.

Haq, A., Brown, J., Moltchanova, E., & Al-Omari, A. I. (2014). Effect of measurement error on exponentially weighted moving average control charts under ranked set sampling schemes. *Journal of Statistical Computation and Simulation, 85*(6), 1224–1246.

Kvam, P. H. (2003). Ranked set sampling based on binary water quality data with covariates. *Journal of Agricultural Biological and Environmental Statistics, 8,* 271–279.

Kvam, P. H., & Samaniego, F. J. (1994). Nonparametric maximum likelihood estimation based on ranked set samples. *Journal of the American Statistical Association, 89,* 526–537.

MacEachern, S. N., Ozturk, O., Wolfe, D. A., & Stark, G. V. (2002). A new ranked set sample estimator of variance. *Journal of the Royal Statistical Society: Series B, 64,* 177–188.

McIntyre, G. A. (1952). A method for unbiased selective sampling using ranked set sampling. *Australian Journal of Agricultural Research, 3,* 385–390.

Mahdizadeh, M., & Zamanzade, E. (2017). New goodness of fit tests for the Cauchy distribution. *Journal of Applied Statistics, 44*(6), 1106–1121.

Mahdizadeh, M., & Zamanzade, E. (2020). Goodness-of-fit tests for the Cauchy distribution with application to financial modelling. *Journal of King Saud University - Science, 31,* 1167–1174.

Mahdizadeh, M., & Zamanzade, E. (2019). Efficient body fat estimation using multistage pair ranked set sampling. *Statistical Methods in Medical Research, 28,* 223–234.

Samawi, H. M., Rochani, H., Linder, D., & Chatterjee, A. (2017). More efficient logistic analysis using moving extreme ranked set sampling. *Journal of Applied Statistics, 44*(4), 753–766.

Samawi, H. M., Helu, A., Rochani, H., Yin, J., Yu, L., & Vogel, R. (2018). Reducing sample size needed for accelerated failure time model using more efficient sampling methods. *Statistical Theory and Practice, 12*(3), 530–541.

Samawi, H. M., Yu, L., Rochani, H., & Vogel, R. (2020). Reducing sample size needed for Cox-proportional hazards model analysis using more efficient sampling method. *Communication of Statistics: Theory and Methods, 49,* 1281–1298.

Samawi, H. M., & Al-Saleh, M. F. (2013). Valid estimation of odds ratio using two types of moving extreme ranked set sampling. *Journal of the Korean Statistical Society, 42,* 17–24.

Stokes, S. L. (1980). Estimation of variance using judgment ordered ranked set samples. *Biometrics, 36,* 35–42.

Stokes, S. L., & Sager, T. W. (1988). Characterization of a Ranked-set sample with application to estimating distribution functions. *Journal of the American Statistical Association, 83,* 374–381.

Takahasi, K., & Wakimoto, K. (1968). On unbiased estimates of the population mean based on the sample stratified by means of ordering. *Annals Institute Statistical Mathematics., 20,* 1–31.

Wang, X., Lim, J., & Stokes, L. (2017). Using ranked set sampling with cluster randomized designs for improved inference on treatment effects. *Journal of the American Statistical Association, 111,* 1576–1590.

Zamanzade, E., & Mahdizadeh, M. (2017). A more efficient proportion estimator in ranked set sampling. *Statistics and Probability Letters, 129,* 28–33.

Zamanzade, E., & Wang, X. (2017). Estimation of population proportion for judgment post-stratification. *Computational Statistics and Data Analysis, 112,* 257–269.

Zamanzade, E., & Mahdizadeh, M. (2018). Estimating the population proportion in ranked set sampling with application to air quality monitoring. *Journal of Applied Statistics, 45*(3), 426–437.

Fuzzy Supply Chain Newsboy Problem Under Lognormal Distributed Demand for Bakery Products

M. R. Bhosale, Raosaheb Latpate, and Santosh Gitte

Abstract The daily demand of perishable items in the market is very volatile. The demand of perishable items is assumed lognormal distribution. The parameters of lognormal distribution are uncertain. Hence, fuzzy triangular numbers are used to overcome such problems. Single-period (newsboy) inventory model is used to obtain optimal order quantity, retailer profit, manufacturer profit and total supply chain profit under decentralized supply chain. At the end of the day, most of the remaining items are rotted. The data is collected from Koregaon Park, Pune Market, India, to study the methodology.

Keywords Newsboy problem · Fuzzy random variable · Supply chain model · Bakery products · Single cycle

1 Introduction

A Supply Chain is a system in which, manufacturer sends finished items to the customer. In this system, there are various components like manufacturer, distributor, supplier, retailer, etc. The main objective is to satisfy the customer demand and to generate profit itself. In practice, it is difficult to satisfy the customer demand. If the number of orders is less than the customer demand, then retailer met with loss. If the number of orders is greater than the customer demand, then retailer will meet with surplus inventory and have to sell items at minimum rate. Now a days, most of the manufacturers satisfied the customer demand with high level. In some system, all the entities are working independently and trying to optimize the main objective of

M. R. Bhosale
Department of Statistics, Shri Shahumandir Mahavidyalaya, Pune 411009, India

R. Latpate (✉)
Department of Statistics, Savitribai Phule Pune University, Pune 411007, India
e-mail: rvl@unipune.ac.in

S. Gitte
Department of Statistics, Mumbai University, Mumbai, India

D. D. Hanagal et al. (eds.), *Applied Statistical Methods*, Springer Proceedings in Mathematics & Statistics 380, https://doi.org/10.1007/978-981-16-7932-2_5

the supply chain. If all the entities are independent, then that system is nothing but decentralized system. If all the entities are working properly but decision regarding the system takes a authority person and all other remaining components follow the decision taken by main authority, then that system is nothing but centralized system. In reality, manufacturer receives material (raw material/semi-finished items) from several suppliers and after processing supplies to several warehouses and retailers. Therefore, supply chain is a real network. A supply chain is a network between a company and its suppliers to produce and distribute a specific product or service. The functions in a supply chain include product development, marketing, operations, distribution, finance and customer service. Many of the researchers studied the newsboy problem under fuzzy environment. Petrovic and Petrovic (1999) developed a model for the newsboy vendor problem under discrete fuzzy environment. Single inventory problem is known as newsboy problem.

Every day early in the morning, newsboy purchases newspapers from the wholesalers and he sells for whole day. If he purchased maximum number of papers and not sold out and he carries some unsold newspapers at the end of day, then he will have loss. And if he purchased less number of papers early in the morning and at the end of day customer having the demand, then the newsboy also has loss and is not able to satisfy the customer demand. Here, newsboy has loss on both ways. How much newspaper should he purchase early in the morning such that he will optimize his profit and also satisfy the customer demand. The demand of customer is uncertain. If the demand is uncertain, so it is a very difficult task in front of researcher to predict the demand, to obtain the statistical distribution of newly developed product and to satisfy the customers' wish. Here, the period of newspaper is only one day, and on the next day that newspaper do not have any value. So, the lifetime of item is very short. Therefore, to solve this type of problem, fuzzy theory has been developed (Zadeh, 1965). Fuzzy set theory is very useful in inventory to take the decision. This type of model is also called as newsboy vendor model or single supply chain model or one echelon supply chain model. Newsboy vendor model includes one retailer and one customer. This type of model is useful to optimize optimal order quantity, optimal inventory level, etc. Newsboy vendor model is a mathematical model which is very useful in operation management and applied economics for optimizing optimal inventory level, optimal order quantity, etc. This newsboy vendor model looks from 1888 which is used in central limit theorem to obtain the optimal cash reserves from the depositors.

In newsboy problem, the following two situations may arise:
(i) Demand X is more than order size Q. Then, he makes a profit of $Q(SP - CP)$, assuming there is no penalty for lost sales.
(ii) Demand X is less than order size Q. Then, $(X - Q)$ newspapers are left unsold at the end of the day, and he will get the salvage value, SV for each of these newspapers. The net profit is

$$X(SP) + (X - Q)SV - Q(CP)$$

If he buys Q newspapers at the beginning of the day, then the expected profit $E(P)$ (assuming SV=0) is:

$$E(P) = \sum_{X=0}^{Q}(X * SP - Q * CP)f(X) + \sum_{X=Q+1}^{\infty} Q(SP - CP)f(X)$$

$$E(P) = SP\left[\sum_{X=0}^{Q} X.f(X) + Q.\sum_{X=Q+1}^{\infty} f(X)\right] - Q.CP$$

where $f(X)$= probability density function (pdf) of demand, CP—purchase cost and SP—selling price.

Zadeh (1965) introduced the concept of fuzzy set theory. Kwakernaak (1978) defined fuzzy random variable, expectation and conditional expectation of fuzzy random variable and properties of fuzzy random variable. He also introduced fuzzy probability and fuzzy events. Puri and Ralescu (1986) explained the concept of fuzzy random variable and expectation of fuzzy random variable. He also discussed properties of integral set valued function. Feng (2001) developed the concept of variance and covariance of fuzzy random variable and discussed the application of variance and covariance. He also discussed the further applications in the correlation function and the criteria of mean squares calculus for fuzzy stochastic process. He extended results on fuzzy sets and systems and discussed independence of fuzzy random variable by means of convergence criterion in uniform Hausdorff metric. Latpate and Kurade (2020b) developed new optimization algorithm using fuzzy set theory and evolutionary algorithm, viz., fuzzy NSGA II for crude oil supply chain of India. Latpate and Kurade (2017) formulated new solution methodology, viz., fuzzy MOGA for solving multiobjective supply chain models. The uncertainty in various parameters was represented by triangular fuzzy number.

Hu et al. (2010) developed a model for two echelon supply chain systems with one retailer and one supplier for perishable items for centralized and decentralized system by using normal demand distribution. He also obtained optimal order quantity, retailer profit, manufacturer profit and total supply chain profit and has shown that optimal order quantity and total supply chain profit in centralized system are greater than decentralized system. Kurade and Latpate (2020) developed new economic order quantity probabilistic inventory models. Lognormal and generalized exponential distributed demand was assumed in the formulated model. For solving, they developed new solution methodology based on genetic algorithm. Zhang et al. (2014) developed a model for two echelon supply chain systems: ith one retailer and one supplier for perishable items for centralized and decentralized system by using uniform demand distribution. He also obtained optimal order quantity, retailer profit, manufacturer profit and total supply chain profit and has shown that optimal order quantity and total supply chain profit in centralized system are greater than decentralized system (Bhosale & Latpate, 2019; Latpate and Bajaj 2011a, b; Latpate & Bhosale, 2020a). Dutta et al. (2005) proposed methods for construction of membership function without using alpha-cut. He discussed the arithmetic operations of fuzzy members using alpha-cut. They also made the comparison between fuzzy arithmetic with and without using alpha-cut.

In the proposed model, we consider the demand of bakery products under decentralized decision-making system, which follows lognormal distribution because the support of lognormal distribution and demand should be same. Exact distribution is unknown. Hence, we assume the fuzzy random variable. The rest of the chapter includes preliminaries, and mathematical model is presented in Sect. 2. Numerical case study is presented in Sect. 3. The result and discussion are incorporated in Sect. 4. Section 5 gives the concluding remarks.

2 Preliminary Definitions and Mathematical Model

The following preliminary definitions are used for the newsboy vendor supply chain model.

Definition 1 *(Fuzzy Random Variable)* (Kwakernaak, 1978)
Let ($\Re$,μ,P) be the probability space and X be a random variable on ($\Re$,μ,P) with a probability density function $f(x)$. Fuzzy random variable $\widetilde{X}$ is a mapping from R to a family of fuzzy numbers, i.e., $\widetilde{X}$: x→ $\widetilde{X}_{(x)} \in F$, where F denotes fuzzy set. Fuzzy random variables are the random variables that are valued as fuzzy numbers.

For given $\alpha \in (0, 1]$, suppose that the α-cut $\widetilde{X}(x)_\alpha$ of a number $\widetilde{X}(x)$ is $\widetilde{X}(x)_\alpha = \left[\widetilde{X}(x)_\alpha^-, \widetilde{X}(x)_\alpha^+\right]$. Let $\widetilde{X}(x)_\alpha^-, \widetilde{X}(x)_\alpha^+$ denote the left end point and right end point of the α-cut $\widetilde{X}(x)_\alpha$ of $\widetilde{X}(x)$, where $\widetilde{X}(x)_\alpha^-, \widetilde{X}(x)_\alpha^+$ are real valued random variables.

Definition 2 *(Fuzzy Expectation)* (Kwakernaak, 1978)
The fuzzy expectation of the fuzzy random variable $\widetilde{X}$ is defined as:

$$E(\widetilde{X}) = \bigcup_{\alpha\in[0,1]} \alpha\left[\int_R \widetilde{X}_\alpha^- dp, \int_R \widetilde{X}_\alpha^+ dp\right]$$

By using the definition of alpha-cut, then definition of fuzzy expectation becomes:

$$[E(\widetilde{X})] = \bigcup_{\alpha\in[0,1]} \alpha[E(\widetilde{X})] = \bigcup_{\alpha\in[0,1]} \alpha\left[[E(\widetilde{X})]_\alpha^-, [E(\widetilde{X})]_\alpha^+\right]$$

Because the $\widetilde{X}_\alpha^-$ and $\widetilde{X}_\alpha^+$ are real valued random variable, their respective expectations are:

$$E(\widetilde{X}) = \bigcup_{\alpha\in[0,1]} \alpha\left[[E(\widetilde{X})]_\alpha^-, [E(\widetilde{X})]_\alpha^+\right] \tag{1}$$

Definition 3 *(Signed Distance of Fuzzy Expectation)* (Chang, 2004)
Let $\widetilde{A}$ be a fuzzy number with α-cut $\widetilde{A}_\alpha = \left[\widetilde{A}_\alpha^-, \widetilde{A}_\alpha^+\right]$, then signed distance of fuzzy

number $\widetilde{A}$ is $d(\widetilde{A}, 0)=\frac{1}{2}\int_0^1(\widetilde{A}_\alpha^- + \widetilde{A}_\alpha^+)d\alpha$. Then, the signed distance of fuzzy expectation $E(\widetilde{X})$ is

$$d(E(\widetilde{X}), 0) = \frac{1}{2}\int_0^1 \left[E(\widetilde{X}_\alpha^-) + E(\widetilde{X}_\alpha^+)\right] d\alpha \tag{2}$$

2.1 Notations and Assumptions

The following notations and assumptions are used for model formulation.

Notations:

s—manufacturers' wholesale price per unit.
p—manufacturers' production cost per unit.
n—retailers' holding cost per unit.
m—retailers' shortage cost per unit.
q—retail price per unit.
r—defective rate.
f—retailers' inspecting cost per unit.
e—reverse cost of defective products per unit which includes manufacturer price and inspection cost.
Q—retailers' order quantity.
Q^*—retailers' optimal order quantity in the decentralized system.
X—random external demand with probability distribution $f(x)$.
$\widetilde{X}$—fuzzy random external demand corresponding to X and expressed as the triangular fuzzy number.
$\widetilde{X}$ = (X-δ_1, X, X+δ_2), where δ_1 and δ_2 are determined by managers depending on their experiences and reflect a kind of fuzzy appreciation from their intrinsic understanding.

Assumptions:

1. Demand follows lognormal distribution.
2. Only one item is considered with decentralized decision-making system.
3. Lifetime of items is very short.

2.2 Mathematical Model

Supply chain is the movement or of transferring manufactured item to the customer. In this process, several entities are included to shift products from manufacturer to the customer. While shifting manufactured product to the retailer, the entities are not cooperating to each other. When the entities are not cooperating to each other, then that supply chain is called as decentralized supply chain. In decentralized system, manufacturer and retailers take their decision independently. Retailers place the order to the manufacturer. Manufacturer fulfills the order (Q) of retailers and tries to satisfy all the requirements of the retailer. The main objective of the supply chain is to satisfy the customer demand and optimize the profit of the supply chain. As soon as the retailer inspect all the items received from manufacturer and send back the defective items to the manufacturer. The total rQe cost of defective items is debited to manufacturer.

Now, there will be $Q(1-r)$ perfect items. Here, demand is fuzzy random variable. The retailer has to bear entire loss due to unsold items. At the end of day, all unsold items are rotted. Hence, the retailer fuzzy random profit becomes:

$$\widetilde{P_1}(Q) = q\ min(\widetilde{X}, Q(1-r)) - (n+s)(Q(1-r) - \widetilde{X})^+$$

$$-m(\widetilde{X} - Q(1-r))^+ - sQ + rQe - fQ$$

where
$min\{\widetilde{X}, Q(1-r)\} = \widetilde{X} - (\widetilde{X} - Q(1-r))^+$,
$(Q(1-r) - \widetilde{X})^+ = Q(1-r) - \widetilde{X} + (\widetilde{X} - Q(1-r))^+$,
and
$(\widetilde{X} - Q(1-r))^+ = Max\big(0, (\widetilde{X} - Q(1-r))\big)$.
By using above results, the retailer fuzzy random profit becomes

$$\widetilde{P_1}(Q) = (q+n+s)\widetilde{X} - (q+n+m+s)(\widetilde{X} - Q(1-r))^+ - (n+s)Q(1-r) - sQ + rQe - fQ$$

The fuzzy expectation of random profit can be obtained by using definition (2).

$$E(\widetilde{P_1}(Q) = (q+n+s)E(\widetilde{X}) - (q+n+m+s)E(\widetilde{X} - Q(1-r))^+)$$

$$-(n+s)Q(1-r) - sQ + rQe - fQ$$

By using definition 3, the defuzzification of expectation is as follows:

$$d(E(\widetilde{P_1}(Q), 0)) = (q+n+s)d(E(\widetilde{X}), 0) - (q+n+m+s)d(E(\widetilde{X} - Q(1-r))^+, 0))$$

$$-(n+s)Q(1-r) - sQ + rQe - fQ$$

where

$$d(E(\widetilde{X}), 0) = \frac{1}{2}\int_0^1 \left[E(\widetilde{X}_\alpha^-) + E(\widetilde{X}_\alpha^+)\right] d\alpha$$

$$d(E(\widetilde{X}), 0) = E(X) + \frac{\delta_2 - \delta_1}{4}$$

$$d(E(\widetilde{X} - Q(1-r))^{+}, 0) = \frac{1}{2}\int_0^1 \Big[E\left[((\widetilde{X} - Q(1-r))^{+}, 0)_{\alpha}^{-}\right]$$

$$+E\left[((\widetilde{X} - Q(1-r))^{+}, 0)_{\alpha}^{+}\right]\Big] d\alpha$$

Using these values in the above equation, we get:

$$d(E(\widetilde{P}_1(Q)), 0) = (q+n+s)\left[E(X) + \frac{\delta_2 - \delta_1}{4}\right]$$

$$-\frac{q+n+m+s}{2}\int_0^1 \Big[E((\widetilde{X} - Q(1-r))^{+})_{\alpha}^{-})$$

$$+[E((X - Q(1-r))^{+})_{\alpha}^{+}\Big] d\alpha - (n+s)Q(1-r) - sQ + rQe - fQ$$

Fuzzy random demand can be written as triangular fuzzy numbers.

$$(\widetilde{X} - \delta_1,\ \widetilde{X},\ \widetilde{X} + \delta_2)$$

The shortage can also be expressed as triangular fuzzy numbers.

$$\left[(\widetilde{X} - Q(1-r))^{+} - \delta_1, (\widetilde{X} - Q(1-r))^{+}, (\widetilde{X} - Q(1-r))^{+} + \delta_2\right].$$

Using α-cuts, the triangular fuzzy number can be expressed as $\left[(\widetilde{X} - Q(1-r))^{+} - \delta_1 + \alpha\delta 1,\ \ (\widetilde{X} - Q(1-r))^{+} + \delta_2 - \alpha\delta_2\right]$. The range of fuzzy demand with shortages for left α-cut of the above interval is $(Q(1-r) + \delta_1 - \alpha\delta_1)$ to ∞. The range of fuzzy demand with shortages for right α-cut of the above interval is $(Q(1-r) - \delta_2 + \alpha\delta_2)$ to ∞. The retailer profit becomes:

$$d(E(\widetilde{P}_1(Q), 0)) = (q+n+s)\left[E(X) + \frac{\delta_2 - \delta_1}{4}\right] - (n+s)Q(1-r) - sQ + rQe - fQ$$

$$-\frac{q+n+m+s}{2}\Big[\int_0^1 \int_{Q(1-r)+\delta_1-\alpha\delta_1}^{\infty} (x - \delta_1 + \alpha\delta_1 - Q(1-r)) f(x) dx d\alpha$$

$$+ \int_0^1 \int_{Q(1-r)-\delta_2+\alpha\delta_2}^{\infty} (x + \delta_2 - \alpha\delta_2 - Q(1-r)) f(x) dx d\alpha\Big] \quad (3)$$

Differentiate Equation (3) with respect to Q and equate to zero to evaluate the optimal order quantity Q

$$\frac{d[d(E(\widetilde{P}_1(Q)),0)]}{dQ} = -(n+s)(1-r) - s + re - f + \frac{q+n+m+s}{2}(1-r)$$

$$\left[2 - \int_0^1 F(Q(1-r) + \delta_1 - \alpha\delta_1) + F(Q(1-r) - \delta_2 + \alpha\delta_2)d\alpha\right] = 0 \quad (4)$$

$$\frac{1}{2}\int_0^1 \left[F(Q^*(1-r) - \delta_2 + \alpha\delta_2) + F(Q^*(1-r) + \delta_1 - \alpha\delta_1)d\alpha\right]$$

$$= \frac{(q+m)(1-r) - s + re - f}{(q+n+m+s)(1-r)} \quad (5)$$

Again, differentiate Equation (4) with respect to Q

$$\frac{d^2[d(E(\widetilde{P}_1(Q)),0)]}{dQ^2} = -\frac{(q+n+m+s)(1-r)^2}{2}\int_0^1 [f(Q(1-r) - \delta_2 + \alpha\delta_2)$$

$$+ f(Q(1-r) + \delta_1 - \alpha\delta_1)]d\alpha$$

For given $\alpha \in [0, 1]$ $\frac{d^2[d(E(\widetilde{P}_1(Q),0)]}{dQ^2} < 0$
Since, $f(Q(1-r) - \delta_2 + \alpha\delta_2) \geq 0$ and $f(Q(1-r) + \delta_1 - \alpha\delta_1) > 0$
That is, the term $d(E(\widetilde{P}_1(Q),0)$ is strictly concave function with respect Q.
After simplifying Equation (3) and by using integration by parts rule (Rudin, 1976) and Leibnitz theorem (Murray, 1985),

$$d(E(\widetilde{P}_1(Q)),0) = (q+n+s)\left[e^{\mu+\frac{\sigma^2}{2}} + \frac{\delta_2-\delta_1}{4}\right] - (n+s)Q(1-r) - sQ + rQe - fQ$$

$$-\frac{q+n+m+s}{2}\left[e^{\mu+\frac{\sigma^2}{2}}\int_0^1\left(\left[2 - \phi\left[\frac{(ln(Q(1-b) + \delta_1 - \alpha\delta_1) - \mu}{\sigma} - \sigma\right]\right]\right.\right.$$

$$\left.-\phi\left[\frac{ln(Q(1-r)) - \delta_2 + \alpha\delta_2) - \mu}{\sigma} - \sigma\right]\right)d\alpha$$

$$-\int_0^1\left[(Q(1-r) - \alpha\delta_1 + \delta_1)\left[1 - \phi\left[\frac{ln(Q(1-r) + \delta_1 - \alpha\delta_1) - \mu}{\sigma}\right]\right]\right]d\alpha$$

$$\left.+\int_0^1\left[(\delta_2 - \alpha\delta_2 - Q(1-r))\left[1 - \phi\left[\frac{ln(Q(1-r) - \delta_2 + \alpha\delta_2) - \mu}{\sigma}\right]\right]\right]d\alpha\right] \quad (6)$$

The optimal order quantity Q^* can be obtained by using Eq. (5) and integration by parts (Rudin, [1976]).

$$2\phi\left[\frac{ln(Q(1-r)-\mu}{\sigma}\right]+\left[\frac{\sigma}{\delta_1^2}-\frac{\sigma}{\delta_2^2}\right]e^{2\mu+2\sigma^2}\phi\left[-2\sigma+\frac{ln(Q(1-r))-\mu}{\sigma}\right]$$

$$+\sigma e^{2\mu+2\sigma^2}\left[\frac{1}{\delta_2^2}\phi\left[-2\sigma+\frac{ln(Q(1-r)-\delta_2)-\mu}{\sigma}\right]-\frac{1}{\delta_1^2}\phi\left[-2\sigma+\frac{ln(Q(1-r)+\delta_1)-\mu}{\sigma}\right]\right]$$

$$+\sigma e^{\mu+\frac{\sigma^2}{2}}\left[\frac{Q(1-r)-\delta_2}{\delta_2^2}-\frac{Q(1-r)+\delta_1}{\delta_1^2}\right]\phi\left[-\sigma+\frac{ln(Q(1-r))-\mu}{\sigma}\right]$$

$$+\sigma e^{\mu+\frac{\sigma^2}{2}}\left[\frac{Q(1-r)+\delta_1}{\delta_1^2}\phi\left[-\sigma+\frac{ln(Q(1-r)+\delta_1)-\mu}{\sigma}\right]\right]$$

$$-\sigma e^{\mu+\frac{\sigma^2}{2}}\left[\frac{Q(1-r)-\delta_2}{\delta_2^2}\phi\left[-\sigma+\frac{ln(Q(1-r)-\delta_2)-\mu}{\sigma}\right]\right]$$

$$=\frac{2\left[(q+m)(1-r)-s+re-f\right]}{(q+n+m+s)(1-r)} \tag{7}$$

where ϕ is the cumulative distribution function of the standard normal distribution. It is already presented by Latpate and Bhosale (2020a). We can find the optimal order quantity Q^* in decentralized system by using Eq. (7) and Newton–Raphson method. If Q^* is the optimal order quantity, then profit for the manufacturer is:

$$M_1(Q^*)=(s-p)Q^*-rQ^*e$$

In decentralized system, the whole supply chain profit is:

$$T_1(Q^*)=d(E(\widetilde{P}_1(Q),0))+M_1(Q^*) \tag{8}$$

3 Numerical Case Study

The bakery product cake has short lifetime. It is produced from Millennium Bakery Koregaon Park Pune, India. The data is collected from Koregaon Park area store at Pune, India. The daily sales and demand are recorded for the whole year 2017. The demand distribution is fitted. It follows lognormal distribution with parameter μ=6.214 and $\sigma=0.09631$. Manufacturer wholesale price is $s=Rs.200$ per kg. The retail price per unit is $q=Rs.300$ per kg, retailer holding cost per unit is $n=Rs.5$, and retailer shortage cost per unit is $m=Rs.50$. The value of defective products per unit is $e=Rs.175$, the production cost is $p=Rs.150$, and inspection cost is $f=$

Table 1 Sensitivity analysis of decentralized supply chain model

δ_1	δ_2	r	Q^*	$dEP_1(Q)$	$M_1(Q^*)$	$T_1(Q^*)$
160	40	0.02	458.6687	27960.87	21328.09	49288.96
160	40	0.04	468.1022	27680.67	20128.39	47809.06
160	40	0.06	477.9312	27388.72	18878.28	46267
160	40	0.08	488.1809	27084.25	17574.51	44658.76
120	80	0.02	472.6324	30836.98	21977.41	52814.39
120	80	0.04	482.2693	30552.02	20737.58	51289.6
120	80	0.06	492.3064	30255.13	19446.1	49701.23
120	80	0.08	502.7693	29945.55	18099.69	48045.24
80	120	0.02	502.7693	32369.71	23378.77	55748.48
80	120	0.04	502.7693	32830.57	21619.08	54449.65
80	120	0.06	502.7693	33020.44	19859.39	52879.83
80	120	0.08	502.7693	32943.93	18099.69	51043.62
40	160	0.02	502.7693	35663.83	23378.77	59042.6
40	160	0.04	502.7693	35685.62	21619.08	57304.7
40	160	0.06	502.7693	35418.83	19859.39	55278.22
40	160	0.08	502.7693	34875.57	18099.69	52975.26

$Rs.5$. The defective products are disposed, and its value is debited to manufacturer. The defective rate changes from $r = 0.02$ to $r = 0.08$.

The fuzzy random demand is expressed as triangular numbers $[\widetilde{X} - \delta_1, \ \widetilde{X}, \ \widetilde{X} + \delta_2]$. The aspiration-level values are δ_1 and δ_2. In decentralized supply chain system, the retailer has to bear the loss due to unsold items. Also, retailer has to bear loss due to shortages. These two events are mutually exclusive. Using above parameters of model, the optimal order quantity Q^* is obtained by using Eq. (7). These results are substituted in Eq. (6) to get the retailer profit. Also, manufacturer and total supply chain profit is obtained by using Eq. (8) and results are presented in Table 1.

4 Results and Discussion

The sensitivity analysis is carried out by changing the different parameters. As defective rate increased, the total supply chain profit decreased as shown in Fig. 1. Since the defective items are large, it leads to loss of total inventory. As defective rate increased, the manufacturer profit decreased which is shown in Table 1 and Fig. 2, because the manufacturer has to bear the whole loss incurred due to defective production. Also, as defective rate increased, the optimal order quantity increased in the supply chain system which is shown in Table 1, whenever there is large number of defective items in consignment. Retailer has no option to increase the order quantity to meet the market demand. As the aspiration level δ_1 increases, the manufacturing

Fig. 1 Graph of defective rate versus total supply chain profit

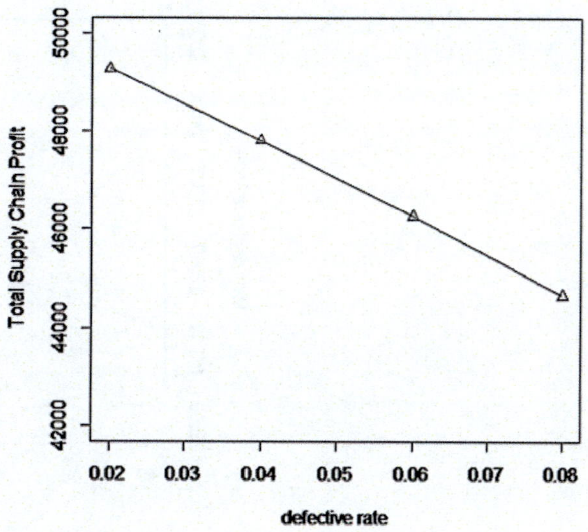

Fig. 2 Graph of δ_1 versus manufacturer profit

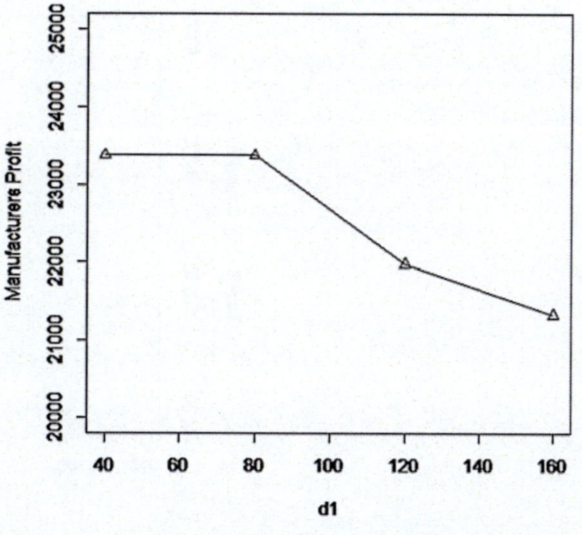

profit decreases. As the aspiration level δ_1 increases, the total optimal order quantity and total supply chain profit decrease which is shown in Fig. 5 and Table 1. As δ_2 increases, the optimal order quantity and total supply chain profit increase as shown in Figs. 3 and 6 and Table 1. It is obvious that the impact of δ_1 causes least optimal order quantity. Hence, there will be manufacturer profit and total supply chain profit decreases.

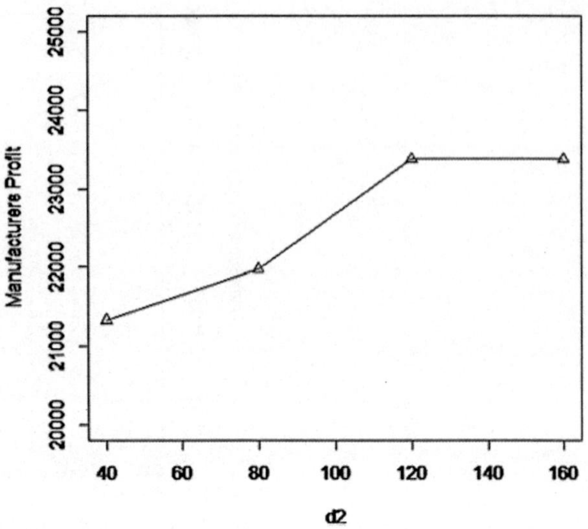

Fig. 3 Graph of δ_2 versus manufacturer profit

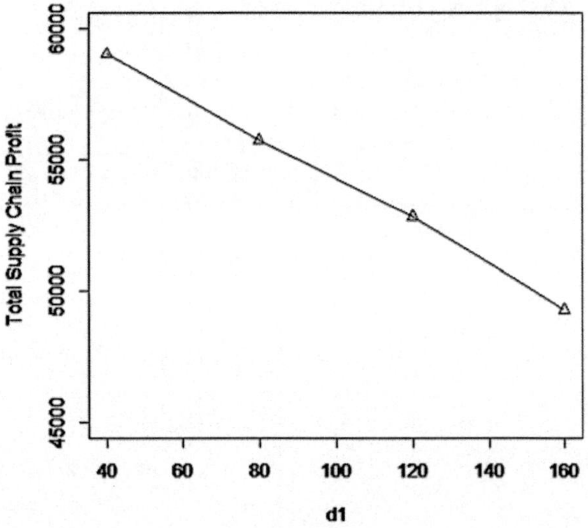

Fig. 4 Graph of δ_1 versus total supply chain profit

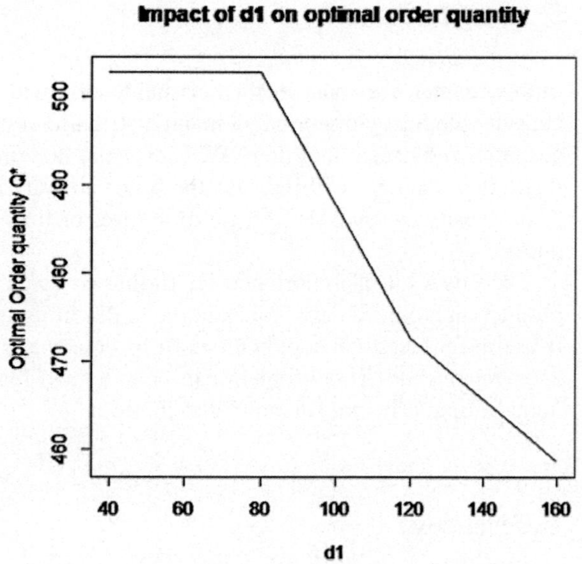

Fig. 5 Impact of δ_1 on optimal order quantity

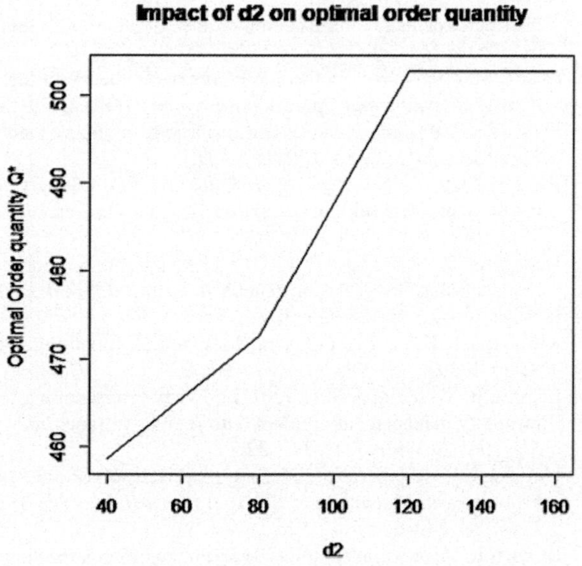

Fig. 6 Impact of δ_2 on optimal order quantity

5 Conclusions

In this chapter, we consider the perishable item with fluctuating demand. Here, we consider the fuzzy lognormal demand distribution fitted for the bakery product with decentralized decision system. Whenever the large numbers of defective items are there, then retailers will increase the order quantity to meet the customer demand. Also, sensitivity analysis of aspiration level of fuzzy parameters δ_1 and δ_2 is conducted.

There is a lot of importance for further research in milk products. Majority of products are utilized daily, for example, milk, fruits and vegetables. Also, this model is useful for electrical supply chain. It has many applications in logistics and management science. This problem can be extended for multi-product and multistage items with multivariate demand distribution.

References

Bhosale, M. R., & Latpate, R. V. (2019). Single stage fuzzy supply chain model with weibull distributed demand for milk commodities. *Granular Computing*. https://doi.org/10.1007/s41066-019-00186-2.11

Chang, H. (2004). An application of fuzzy set theory to the EOQ model with imperfect quality items. *Computers and Operation Research, 31*(12), 2079–2092.

Feng, Y. (2001). The variance and covariance of fuzzy random variables and their applications. *Fuzzy Sets and Systems, 120*, 487–497.

Hu, J., Zhang, H., Xu, R., Ji, Y., & Guo, C. (2010). Supply chain coordination for fuzzy random newsboy problem with imperfect quality. *International Journal of Approximate Reasoning, 51*, 771–784.

Kurade, S. S., & Latpate, R. (2021). Demand and deterioration of items per unit time inventory models with shortages using genetic algorithm. *Journal of Management Analytics 8*(3), 502–529. https://doi.org/10.1080/23270012.2020.1829113

Kwakernaak, H. (1978). Fuzzy random variables: definition and Theorem. *Information Sciences, 15*(1), 1–29.

Latpate, R. V., & Bajaj, V. H. (2011a). Fuzzy programming for multi-objective transportation and inventory management problem with retailer storage. *International Journal of Agricultural and Statistical Sciences 7*(1), 377–326.

Latpate, R. V., Bajaj, V. H. (2011b). Fuzzy Multi-objective, Multi-product, Production Distribution Problem with Manufacturer Storage. *Proceedings of International Congress on PQROM, New Delhi*, pp. 340–355.

Latpate, R. V., & Kurade, S. S. (2017). Fuzzy MOGA for Supply chain model with Pareto decision space at different alpha-cuts. *International Journal of Advanced Manufacturing Technology, 91*, 3861–3876. https://doi.org/10.1007/s00170-016-9966-5

Latpate, R. V., & Bhosale, M. R. (2020). Single cycle supply chain coordination model for fuzzy stochastic demand of perishable items. *Iranian Journal of Fuzzy Systems, 17*(2), 39–48.

Latpate, R., & Kurade, S.S. (2022). Multi-objective multi-index transportation model for crude oil using fuzzy NSGA-II. *IEEE Transactions on Intelligent Transportation Systems 23*(2), 1347–1356. https://doi.org/10.1109/TITS.2020.3024693

Li, X., & Zhao, S. (2006). Supply chain real time coordination method based on buyback contract with fuzzy random demand. In: *IEEE Proceeding of International Conference on fuzzy sets, Shreton Vancouver Wall Center Hotel*, pp. 1543–1549.

Murray, H. P., & Charles, B. M. (1985). *Intermediate Calculus*, 2nd edn. Berlin, Heidelberg: Springer Verlag.

Petrovic, D., Petrovic, R., & Vujosevic, M. (1996). Fuzzy models for newsboy problem. *International Journal of Production Economics, 45*, 435–441.

Petrovic, D., & Petrovic, R. (1999). Supply chain modeling using fuzzy sets. *Journal of Production Economy, 59*, 443–445.

Petrovic, D., Petrovic, R., & Vujosevic, M.(2008). Fuzzy models for the newsboy problem. *International Journal of Production and Economics, 116*(1), 115–128.

Puri, M. L., & Ralescu, D. A. (1986). Fuzzy random Variables. *Journal of Applied Mathematics, 114*, 409–422.

Rudin W. (1976). *Principles of Mathematical Analysis*, 3rd edn. New York: McGraw-Hill Inc.

Walker, J. (1992). The single period inventory problem with uniform demand. *International Journal of Operations and Production Management, 12*, 79–84.

Wang, J., Zhao, R., & Tang, W. (2008). Supply chain coordination by revenue sharing contract with fuzzy demand. *Journal of Intelligent & Fuzzy Systems, 19*(6), 409–420.

Wang, X. B., & Wang, S. H. (2009). Supply chain coordination based on buyback contract with fuzzy demand. In: *IEEE Proceedings on International Conference of Machine Learning and Cybernetics* Vol. 2, pp. 839–843. Heibei, China.

Xu, Y., & Hu, J. (2012). Random Fuzzy demand newsboy problem. *Physics Procedia, 25*, 924–931.

Xu, R., & Zhai, X. (2008). Optimal models for single period supply chain problem with fuzzy demand. *Information Sciences, 178*, 3374–3381.

Xu, R., & Zhai, X. (2010). Analysis of supply chain coordination under fuzzy demand in a two stage supply chain. *Applied Mathematical Modeling, 34*(1), 129–139.

Zadeh, L. A. (1965). Fuzzy sets. *Information and Control, 3*, 338–353.

Zhang, B., Lu, S., Zhang, D., & Wen, K. (2014). Supply chain coordination based on a buyback contract under fuzzy random variable demand. *Fuzzy Sets and Systems, 255*, 1–16.

Probabilistic Supply Chain Models with Partial Backlogging for Deteriorating Items

Sandesh Kurade, Raosaheb Latpate, and David Hanagal

Abstract A supply chain (SC) is a network comprising suppliers, producers, manufacturers, distributors and retailers. Generally, it is represented as single tier, two tier and multi-tier according to its various independent nodes, i.e., entities in the SC. For smooth flow of SC, an inventory is maintained and optimized. At supplier, the raw material inventory is required to dispatch at various producers. The finished product inventory with the help of raw material received from suppliers is produced at producer level. An assembled inventory from finished product is produced at manufacturer level. Various distributors transport this assembled inventory using various modes of transportation to retailers. At final node of SC, the customer will purchase the product. Thus, inventory management is a crucial task in a SC. In probabilistic inventory models, using suitable probability distribution for demand rate, an inventory can be optimized. Here, we develop the inventory models by assuming various probability distributions for demand and deterioration rate under shortages. For modeling, we consider probabilistic demand per unit time as well as the probabilistic deterioration rates. Under these assumptions, probabilistic economic order quantity (EOQ) models are developed under partial backlogging. Classical methods are unable to solve these situations by these assumptions. Therefore, the proposed genetic algorithm is useful to solve the EOQ models. Numerical case study is presented and solved by using non-traditional method, i.e., genetic algorithm. Sensitivity analysis of various parameters is also presented.

Keywords Genetic algorithm · Partial backlogging · Probabilistic inventory model · Multi-echelon supply chain · Variable deterioration

---

S. Kurade
Department of Statistics, MES Abasaheb Garware College, Pune 411004, India

R. Latpate (✉) · D. Hanagal
Department of Statistics, Savitribai Phule Pune University, Pune 411007, India
e-mail: rvl@unipune.ac.in

D. D. Hanagal et al. (eds.), *Applied Statistical Methods*, Springer Proceedings in Mathematics & Statistics 380, https://doi.org/10.1007/978-981-16-7932-2_6

1 Introduction

Supply chain (SC) consists of several stakeholders like suppliers of raw materials, producers of unique product, manufacturers for assembling the produced product, warehouse for storage purpose, retailers for distribution purpose and transporters for shipping the manufactured product from each node of SC to other node. Each stakeholder in the SC has a role to optimize the SC. The major objective of SC is to satisfy or to meet the customer demand in such a way that the profit or cost of maintaining the SC would be maximum or minimum, respectively. All the above stakeholders in the SC operate independently for generating their profits. The SC connects each stakeholder to the end customer. It consists of suppliers, manufacturers, transporters, warehouses and customers. Every entity in the SC has to satisfy the customer demand and to generate profit for itself, whereas customers are the integral part of it. The term SC conjures up supply of product moving from suppliers to manufacturers, manufactures to distributors, distributors to retailers and retailers to customers in a SC. The detailed working procedure of SC network as, producers receive raw materials from various outside suppliers which produces number of units of the perishable product. After production of units of perishable item, it transports to the manufacturers site for packaging or assembling the final product, then manufacturer transports the final product to several independent warehouses for storage purpose. Lastly, produced perishable product transports to various independent retailers for distributing to customers. Here, we have developed the new probabilistic economic order quantity (EOQ) inventory models for multi-tier SC; see Fig. 1. In the literature, several authors developed EOQ inventory models with stock-dependent, replenishment-dependent, ramp-type function of demand, etc. Here, we assumed that the market demand is uncertain and follow a certain probability distribution. The answers of basic questions in inventory like how much to store and optimum order quantity can be obtained from the developed models. Also, a new solution methodology based on evolutionary algorithm is developed. This solution methodology can be applicable to all types of optimization problems involved in inventory management.

Beamon (1998) highlighted the two basic random processes occurring in the SC: (i) production planning and inventory process and (ii) distribution (i.e., transportation) and logistics process. The inventory process deals with the manufacturing and storage problems in SC, whereas production planning deals with the cooperation between each and every manufacturing process. The distribution process deals with how products are transported from suppliers to producers, from producers to manufacturers, from manufacturers to retailers. This process includes procurement of inventory and transportation of raw material as well as finished product. Each stage of SC is connected through the flow of products, information and funds. These flows are in both ways. For effectively managing the SC, a manager has to decide the location, capacity and type of plants, warehouses and retailers to establish the SCN. The SCN problem covers wide range of formulations such as simple single product

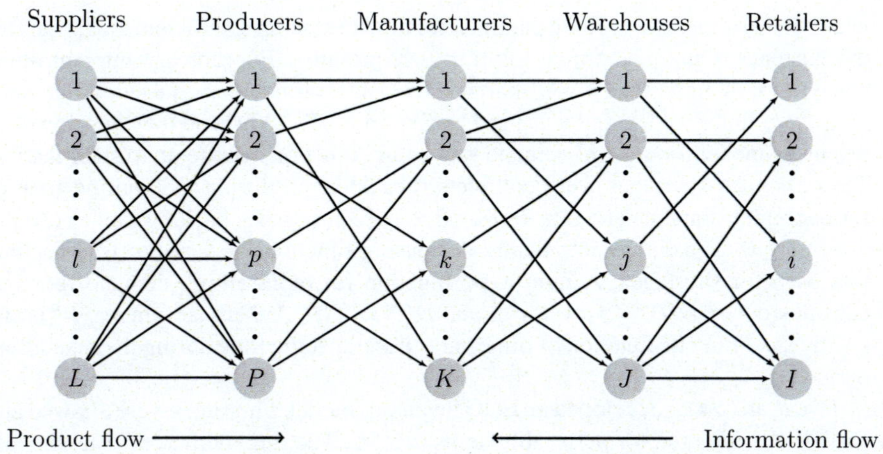

Fig. 1 Supply chain network (SCN)

type to complex multi-product one and from linear deterministic models to complex nonlinear ones. SC connects each stakeholder to the end customer.

EOQ inventory model was firstly introduced by Ghare and Schrader (1963) with constant deterioration; i.e., it follows an exponential distribution $g(t) = \theta > 0$ where $g(t)$ is the rate of deterioration. Later, this model was extended by Covert and Philip (1973), under which they assumed variable deterioration and it follows the Weibull distribution $g(t) = \gamma\beta\, t^{\beta-1}, \quad 0 \leq t \leq T$ where T is the cycle time and $\alpha > 0$ and $\beta > 0$. Afterward, Philip (1974) formulated more general EOQ inventory model with Weibull distributed deterioration. Padmanabhan and Vrat (1995) developed three EOQ inventory models by assuming no, complete and partial backlogging. Under continuous review policy, the inventory models were proposed in which stock-dependent demand rate, i.e., $D(t) = \alpha + \beta\, I(t), \alpha, \beta > 0$ per unit time t, was assumed.

Wee (1995) developed the replenishment strategy for a perishable product under complete as well as partial backlogging with different backlog rates. Bose et al. (1995) formulated an EOQ inventory model for deteriorating items with linear time-dependent demand rate per unit time $D(t) = a + b \cdot t, a > 0, b > 0$. Such model can be applicable to highly deteriorated products like fruits, milk products, etc. Bhunia and Maiti (1997) formulated two deterministic EOQ inventory models under variable production. They modeled the level of inventory at time t during the production period at a finite replenishment rate, i.e., $R(t) = \alpha - \beta\, I(t)$ and $R(t) = \alpha + \beta\, D(t)$. In both the models, demand rate was assumed to be a linearly increasing function of time t. Later, Bhunia and Maiti (1998) extended earlier developed EOQ inventory models under complete backlogging by considering finite rate of replenishment.

Wu (2001) developed an EOQ inventory model by considering ramp-type demand and stochastic deterioration rate. Shortages were allowed in the developed model, and they were partially backlogged with backlogging rate $1/[1 + \delta(T - t)]$ per unit time where $\delta > 0$ is the backlogging parameter. The necessary and sufficient conditions for the existence of unique optimal solution were provided. For representing ramp-type behavior of demand rate, a well-known Heaviside's function was used. Later, such ramp-type function demand was used by Skouri et al. (2009) and they developed an inventory model for general demand rate as any function of time up to stabilization. They assumed shortages were completely backlogged during the waiting time of further replenishments at a rate $\delta(t) \in (0, 1)$ which satisfies backlog rate $= (\delta(t) + T\delta^{'}(t)) \geq 0$. Thus, a monotonically decreasing function $\delta^{'}(t) \leq 0$ of waiting time was used for shortages. A ramp-type function for representing demand rate was considered, i.e., $D(t) = f(t), t < \mu$ and $D(t) = f(\mu)$, otherwise where $f(t)$ is any positive, continuous function of time and μ is the specific time during the scheduling period.

Dye et al. (2005) developed an EOQ inventory model. Shortages were allowed and partially backlogged for perishable items in a SC. The time-dependent backlog rate was assumed in the developed model, i.e., backlog rate $= 1/(1 + \delta[T - t]), \delta > 0$. Later, Eroglu and Ozdemir (2007) presented a deterministic EOQ inventory model for a manufacturer with few defective products in a lot. The proposed model can be applicable to a SC consisting of only a manufacturer and a retailer under shortages. Uniform defective rate of products was considered in the developed model, i.e., $f(p) \sim U(0, 0.1)$ where p is the proportion of defectives in the lot. Raosaheb and Bajaj (2011) formulated transportation and inventory model with retailer storage under uncertain environment. In the same year, Latpate and Bajaj (2011) developed multi-objective production distribution SC model for manufacturer storage under uncertain environment. Later, Kurade and Latpate (2021) proposed different EOQ inventory models under no, complete and partial backlogging. The time-dependent demand and deterioration rates were assumed in the developed model. In the same year, Latpate and Bhosale (2020) proposed SC coordination model with stochastic market demand. Bhosale and Latpate (2019) formulated a fuzzy SC model with Weibull distributed demand for dairy product.

Remainder of the chapter is organized as follows: In Sect. 2, preliminary concepts of demand and deterioration variation of the inventory model are stated. Also, this section is dedicated to the formulation of probabilistic inventory model with assumptions. Subsequently, Sect. 3 develops binary coded genetic algorithm approach to solve the formulated probabilistic inventory model under various demand distributions. The developed probabilistic EOQ inventory model is illustrated with the hypothetical data for various demand distributions in Sect. 4 with results discussed in the same section. Finally, sensitivity analysis of various parameters is discussed in Sect. 5. The managerial implications are added in Sect. 6. Concluding remarks with future scope are given in Sect. 7. Last, an exhaustive list of references is provided.

2 Mathematical Model

The main working behavior of the SC is displayed in Fig. 1. This figure shows the general network of SC, in which product flows from suppliers to retailers through various stages. Thus, from supplier, producer, manufacturer and warehouses the finished product will reach to the customer. The customer is always an integral part of a SC network. In supply chain network (SCN), generally information and funds flow from customers to suppliers and units of product flow from suppliers to customers. Multi-echelon SCN provides an unique optimal way for efficiently and effectively managing SC. It manages product and information flows both in and between several linked but independent stakeholders.

Here, we consider a SCN in Fig. 1 in which the inventory is stored at different independent nodes of SC. At supplier, the inventory of a raw material is stored for the purpose of production of a finished product at several producers. The finished product inventory is stored at producer level. The assembly of a finished product is done at manufacturer level. Thus, the inventory of finished product is stored at manufacturer level. At warehouses, the transported finished product is stored for the purpose of distribution to several retailers. Thus, at each point of a SC different kinds of inventory are stored. For managing this inventory, we have proposed a probabilistic inventory model, in which the demand of a product from the market is assumed to be probabilistic. The goal is to determine an EOQ of the product in the scheduling period, $t \in [0, T]$.

2.1 Preliminaries

2.1.1 Demand Variation

EOQ inventory models are developed by maximizing profit in which demand follows a probability distribution per unit time with known parameters. Demand rate follows an uniform and normal distribution.

Uniform Distribution

An uniform distribution with parameters a and b is denoted by $U(a, b)$, and its probability density function is

$$f(t) = \frac{1}{b-a}; \qquad a < t < b, \qquad a, b \in \mathbf{R}. \tag{1}$$

Normal Distribution

A normal distribution with mean μ and variance σ^2 is denoted by $N(\mu, \sigma^2)$, and its probability density function is

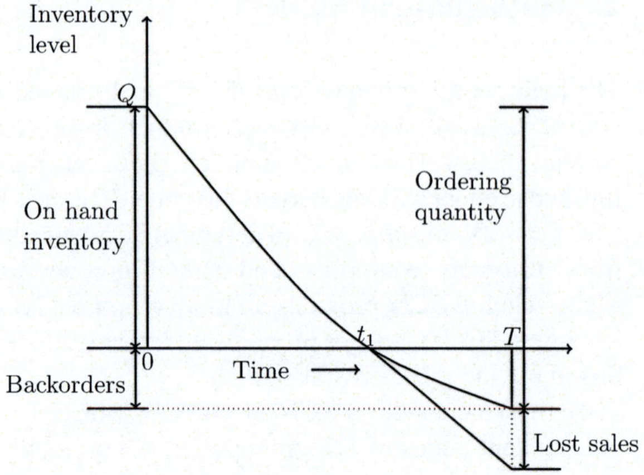

Fig. 2 Graphical representation of inventory system in partial backlogging

$$f(t) = \frac{1}{\sigma\sqrt{2\pi}} e^{-(t-\mu)^2/2\sigma^2}; \qquad t \in \mathbf{R}, \qquad \sigma > 0, \qquad \mu \in \mathbf{R}. \tag{2}$$

2.1.2 Deterioration Variation

During the normal storage period, the deterioration may occur in several perishable products. Deterioration includes vaporization, drying, decay, damage or spoilage such that the product cannot be used for its intended application. For representing the deterioration of perishable product, we use Weibull distribution.

Weibull Distribution: It includes all types of deterioration such as constant, increasing and decreasing. It is defined as

$$g(t) = \gamma\beta t^{\beta-1}; \qquad t > 0, \qquad \gamma > 0, \qquad \beta > 0. \tag{3}$$

Note: If $\beta > 1$, it shows increasing deterioration; if $\beta < 1$, it shows decreasing deterioration; and if $\beta = 1$, it shows constant deterioration.

During the period $t \in [t_1, T]$ (see Fig. 2), it is generally assumed that customers are impatient in nature and do not wish to wait for replenishment. Thus, only a fraction of backlogged demand is considered and backlogging rate is taken to be variable. Backlogging rate depends on the length of time, for the customer waits before receiving the product. Here, backlogging rate is considered as a decreasing exponential function of waiting time (Abad, 2001; Chang & Dye, Chang and Dye (1999); Dye et al., 2007).

$$\therefore \text{Backlogging rate} = \frac{1}{1+\delta(T-t)}; \quad t_1 \le t \le T.$$

Here, we consider the expected demand per unit time of a product which is obtained as:

$$\begin{aligned}\text{Expected demand in } (t,\ t+\Delta t) &= D_0 \cdot P[\text{demand of a product in interval } (t,\ t+\Delta t)]\\ &= D_0 \cdot \lim_{\Delta t \to 0}\left[\frac{F(t+\Delta t)-F(t)}{\Delta t}\right]\\ &= D_0 \cdot f(t).\end{aligned}$$

Assumptions: The assumptions considered in the problem are:

1. Replenishment rate is infinite with negligible lead time.
2. Time periods between two successive demands of an unique perishable product are independent and identically distributed random variables.
3. Continuous review policy of inventory model for single perishable product is considered.
4. The demand and deterioration rate are probabilistic in nature.
5. Storage facility is available at each node of SC.

Notations: These are listed below:

- $I_1(t)$ = positive inventory in a cycle of length T.
- $I_2(t)$ = negative inventory in a cycle of length T.
- D_0 = total demand in the inventory cycle.
- $D(t)$ = demand rate.
- Q = order quantity (per cycle).
- B = maximum inventory level (per cycle).
- $g(t)$ = deterioration rate.
- γ and β = deterioration parameters.
- δ = backlogging parameter in the backlogging period.
- $f(t)$ = probability density function.
- $F(t)$ = cumulative distribution function.
- $P(T, t_1)$ = profit per unit time for partial backlogging model.

Costs:

The notations of various costs involved in the inventory model are listed below:

- C = the purchase cost (per unit).
- C_2 = a finite shortage cost (per unit).
- S = the selling price (per unit), where $S > C$.
- A = the ordering cost (per order).
- R = the cost of lost sales (i.e., opportunity cost) (per unit).
- r = the inventory carrying cost as a fraction (per unit per unit time).

Decision Variables:

The notations of various decision variables involved in the problem are listed below:

- t_1 = length of the duration over which inventory level is positive in a cycle.
- T = length of the scheduling period.

The inventory level decreases in satisfying the market demand as well as due to the deterioration during the period $[0, t_1]$ (see Fig. 2). Thus, the differential equations considering the partial backlogging during the cycle $[0, T]$ are given as:

$$\frac{dI_1(t)}{dt} + g(t)I(t) = -D(t), \qquad 0 \le t \le t_1. \tag{4}$$

$$\frac{dI_2(t)}{dt} = \frac{-D(t)}{[1+\delta(T-t)]}, \qquad t_1 \le t \le T. \tag{5}$$

The solutions of Eqs. (4) and (5), for the boundary condition $I(t_1) = 0$, are

$$I_1(t) = D_0 e^{-\int g(t)dt} \int_t^{t_1} f(x) e^{\int g(x)dx} dx, \qquad 0 \le t \le t_1. \tag{6}$$

$$I_2(t) = -D_0 \int_{t_1}^{t} f(x) \left[\frac{1}{[1+\delta(T-t)]}\right] dx, \qquad t_1 \le t \le T. \tag{7}$$

∴ inventory level at the beginning of the cycle (maximum inventory level) is

$$B = D_0 \int_0^{t_1} f(x) e^{\int g(x)dx} dx. \tag{8}$$

Thus,

$$\begin{aligned}
\text{Sales revenue} &= S\left\{\int_0^{t_1} D(x)dx - I_2(t)\right\} \\
&= S\left\{\int_0^{t_1} D_0 f(x)dx + \int_{t_1}^{T} D_0 f(x)\left[\frac{1}{[1+\delta(T-x)]}\right]dx\right\} \\
&= SD_0\left\{F(t_1) + \int_{t_1}^{T} f(x)\left[\frac{1}{[1+\delta(T-x)]}\right]dx\right\}.
\end{aligned} \tag{9}$$

$$\text{Carrying cost} = rC\int_0^{t_1} I_1(x)dx = rCD_0 \int_0^{t_1} e^{-\int g(t)dt}\left\{\int_t^{t_1} f(x)e^{\int g(x)dx}dx\right\}dt. \tag{10}$$

$$\text{Shortage cost} = C_2\left\{\int_{t_1}^{T}[-I_2(x)]dx\right\}$$
$$= -C_2\left\{\int_{t_1}^{T}\left[-D_0\int_{t_1}^{t} f(x)\left[\frac{1}{[1+\delta(T-x)]}\right]dx\right]dt\right\}$$
$$= C_2 D_0\left\{\int_{t_1}^{T}\left[\int_{t_1}^{t} f(x)\left[\frac{1}{[1+\delta(T-x)]}\right]dx\right]dt\right\}. \quad (11)$$

$$\text{Material cost} = C\left\{\int_{0}^{t_1} I_1(x)dx - \int_{t_1}^{T} I_2(x)dx\right\}$$
$$= \left\{CD_0\int_0^{t_1} f(x)e^{\int g(x)dx}dx\right\} + \left\{\int_{t_1}^{T}\left[\frac{CD_0 f(x)}{[1+\delta(T-x)]}\right]dx\right\}. \quad (12)$$

$$\text{Opportunity cost} = R\int_{t_1}^{T} D(x)\left[1-\frac{1}{[1+\delta(T-x)]}\right]dx$$
$$= R\ D_0\int_{t_1}^{T} f(x)\left[1-\frac{1}{[1+\delta(T-x)]}\right]dx$$
$$= R\ D_0\int_{t_1}^{T} f(x)\left[\frac{\delta(T-x)}{[1+\delta(T-x)]}\right]dx. \quad (13)$$

Therefore from above defined expressions, the profit per unit time is

$$\text{Profit} = \frac{1}{T}(\text{Sales revenue} - \text{order cost} - \text{carrying cost} - \text{shortage cost}$$
$$- \text{material cost} - \text{opportunity cost})$$

$$\therefore P(T, t_1) = \frac{1}{T}\left(SD_0\left\{F(t_1) + \int_{t_1}^{T}\left[\frac{f(x)}{[1+\delta(T-x)]}\right]dx\right\} - CD_0\int_0^{t_1} f(x)e^{\int g(x)dx}dx\right.$$
$$- rCD_0\int_0^{t_1} e^{-\int g(t)dt}\left\{\int_t^{t_1} f(x)e^{\int g(x)dx}dx\right\}dt$$
$$- C_2D_0\left\{\int_{t_1}^{T}\left[\int_{t_1}^{t} f(x)\left[\frac{1}{[1+\delta(T-x)]}\right]dx\right]dt\right\} - A$$
$$\left.- \int_{t_1}^{T}\left[\frac{CD_0 f(x)}{[1+\delta(T-x)]}\right]dx - R\ D_0\int_{t_1}^{T} f(x)\left[\frac{\delta(T-x)}{[1+\delta(T-x)]}\right]dx\right). \quad (14)$$

The optimum values of T and t_1 can be obtained by solving the above nonlinear expression using GA. From this, the optimum order quantity is

$$Q = I_1(0) - I_2(T) = D_0 \left[\int_0^{t_1} f(x) e^{\int g(x)dx} dx + \int_{t_1}^{T} f(x) \left[\frac{1}{[1+\delta(T-t)]} \right] dx \right]. \tag{15}$$

Particular cases: Case 1: Demand rate follows an uniform distribution (Eq. 1) and deterioration rate follows a Weibull distribution (Eq. 3), i.e., $f(t) \sim U(a,b)$ and $g(t) \sim W(\gamma, \beta)$.
Thus Equation 14 becomes,

$$\begin{aligned} P(T,t_1) = \frac{1}{T} \Bigg(& SD_0 \left\{ \frac{t_1 - a}{b-a} + \int_{t_1}^{T} \frac{1}{b-a} \left[\frac{1}{[1+\delta(T-x)]} \right] dx \right\} \\ & - CD_0 \int_0^{t_1} \frac{e^{\gamma x^{\beta}}}{b-a} dx - rCD_0 \int_0^{t_1} e^{-\gamma t^{\beta}} \left\{ \int_t^{t_1} \frac{e^{\gamma x^{\beta}}}{b-a} dx \right\} dt \\ & - C_2 D_0 \left\{ \int_{t_1}^{T} \left[\int_{t_1}^{t} \frac{1}{b-a} \left[\frac{1}{[1+\delta(T-x)]} \right] dx \right] dt \right\} - A \\ & - \int_{t_1}^{T} \frac{1}{b-a} \left[\frac{CD_0}{[1+\delta(T-x)]} \right] dx - R\, D_0 \int_{t_1}^{T} \frac{1}{b-a} \left[\frac{\delta(T-x)}{[1+\delta(T-x)]} \right] dx \Bigg). \end{aligned} \tag{16}$$

Case 2: Demand rate follows a normal distribution (Eq. 2) and deterioration rate follows a Weibull distribution (Eq. 3), i.e., $f(t) \sim N(\mu, \sigma^2)$ and $g(t) \sim W(\gamma, \beta)$.
Thus Equation 14 becomes,

$$\begin{aligned} P(T,t_1) = \frac{1}{T} \Bigg(& SD_0 \left\{ \int_{-\infty}^{t_1} \frac{e^{-\frac{(x-\mu)^2}{2\sigma^2}}}{\sigma\sqrt{2\pi}} + \int_{t_1}^{T} \frac{e^{-\frac{(x-\mu)^2}{2\sigma^2}}}{\sigma\sqrt{2\pi}} \left[\frac{1}{[1+\delta(T-x)]} \right] dx \right\} \\ & - CD_0 \int_0^{t_1} \frac{e^{-\frac{(x-\mu)^2}{2\sigma^2}}}{\sigma\sqrt{2\pi}} e^{\gamma x^{\beta}} dx - rCD_0 \int_0^{t_1} e^{-\gamma t^{\beta}} \left\{ \int_t^{t_1} \frac{e^{-\frac{(x-\mu)^2}{2\sigma^2}}}{\sigma\sqrt{2\pi}} e^{\gamma x^{\beta}} dx \right\} dt \\ & - C_2 D_0 \left\{ \int_{t_1}^{T} \left[\int_{t_1}^{t} \frac{e^{-\frac{(x-\mu)^2}{2\sigma^2}}}{\sigma\sqrt{2\pi}} \left[\frac{1}{[1+\delta(T-x)]} \right] dx \right] dt \right\} - A \\ & - \int_{t_1}^{T} \frac{e^{-\frac{(x-\mu)^2}{2\sigma^2}}}{\sigma\sqrt{2\pi}} \left[\frac{CD_0}{[1+\delta(T-x)]} \right] dx - R\, D_0 \int_{t_1}^{T} \frac{e^{-\frac{(x-\mu)^2}{2\sigma^2}}}{\sigma\sqrt{2\pi}} \left[\frac{\delta(T-x)}{[1+\delta(T-x)]} \right] dx \Bigg). \end{aligned} \tag{17}$$

3 Genetic Algorithm

Genetic algorithm (GA) is an optimization technique, which achieves better optimization of the problem through random search. It is a population-based random search algorithm. Holland (1992) was the main founder of GA. Initially, he introduced this for solving the problems of natural system. Later, it has been widely applied by several researchers for solving their optimization problems. During those days, his *Schema theorem* was gained much attention by several researchers.

At initial stage, the research work about GA was found in proceedings of international conferences. It is a biologically inspired search and stochastic algorithm which works using genetic operators Deb (2005), namely reproduction/selection, crossover and mutation. Mainly, it is inspired by Darwin's theory of evolution which is one of the competitive intelligent algorithms used for optimization. Its advantage is that researchers require minimum problem information about various parameters of GA. Its main parameters are population size N, crossover probability P_{Cross} and mutation probability P_{Mut}. According to the initialization of population, there are two types of GA, binary coded (BCGA) and real coded (RCGA). If initial population is generated using binary number, then the resultant GA is called as BCGA; otherwise, it is called as RCGA.

It always deals with the coding of the problem, and it requires only information about objective functions for computing fitness function. In single objective optimization problem, fitness function is nothing but simply the value of objective function. But in case of multi-objective optimization problems, it is a suitably well-defined function by considering all objectives. The solutions obtained from GA are always efficient and robust since it works with a set of feasible points instead of single point in the search space. Generally, multi-objective optimization problems are handled by two different techniques. One is to concatenate all the objectives to get the single objective with feasible constraints. The second one is to determine the Pareto optimal solution set. Pareto optimal solutions are the non-dominated solutions. Generally, if solution of the problem is strictly better than at least in one objective function, then it is considered as a non-dominated solution.

Several researchers contributed for the development of GAs like Latpate and Kurade (2017) formulated a fuzzy multiple objective genetic algorithm (fuzzy–MOGA). The performance of this algorithm was analyzed using hypothetical case study for a SC network. In this network, a manufacturing company having multiple plants in different geographical regions was assumed. It consists of five raw material suppliers and four manufacturing plants which produce single type of product, for distributing six warehouses and eight retailers. Pareto decision space for various uncertainty levels in demand and cost parameters was provided. Later, Latpate and Kurade (2020) developed new fuzzy non-dominated sorting GA (fuzzy–NSGA II) for optimizing crude oil SC of India. The formulated transportation model was suitable for deciding optimum routes and modes of shipping. The hybridization of ant colony optimization (ACO) and GA was proposed by Maiti (2020). In this hybridization, the initial population of candidate solutions was generated by ACO. Efficiency of the algorithm was tested for different test functions. Maity et al. (2017) used MOGA for solving their proposed multi−item inventory model in which demand was stock dependent. GA has a major drawback like it requires more computational complexity and its convergence performance. For convergence, it requires more simulation runs. To overcome this, a compound mutation strategy in intelligent bionic genetic algorithm (IB−GA) and multi-stage composite genetic algorithm (MSC−GA) was proposed by Li et al. (2011). The latter one has better convergence with high accuracy. Using Markov chain theory, they studied the global convergence under the elitist preserving strategy.

Here, we have proposed binary coded GA (BCGA) for solving the formulated probabilistic EOQ inventory models with roulette wheel selection, single-point crossover and bitwise mutation for development of solution methodology; see Algorithm 1.

Genetic Operators

(1) **Roulette wheel selection**: There are various types of selection techniques like roulette wheel selection, tournament selection, crowded tournament selection, etc. The roulette wheel selection obtains duplicate copies of best chromosomes and eliminates worst from the population, keeping its size fixed. In the proposed BCGA, initial population is randomly generated from a continuous uniform distribution. Each randomly generated individual chromosome in the initial population is a candidate solution to the problem. In this selection mechanism, chromosomes are assigned a probability of being selected, based on their fitness values.

(2) **Single-point crossover**: It is used for exchanging information between randomly selected parent chromosomes by recombining parts of their genetic materials. This operation performed probabilistically combines parts of two parent chromosomes to generate offspring. Its step-by-step procedure is explained below:

   (a) It works using crossover probability say P_{cross}. Thus, only $(M \cdot P_{cross})$ chromosomes in the population go for crossover where M is the population size.
   (b) Randomly select any two parent chromosomes from the population of mating pool. It is generated, when a selection operator is applied on the population. Mating pool has size M.
   Let $X_1 = \{X_{11}, X_{12}, \cdots, X_{1(k-1)}, X_{1k}, X_{1(k+1)}, \cdots, X_{1N}\}$ and $X_2 = \{X_{21}, X_{22}, \cdots, X_{2(k-1)}, X_{2k}, X_{2(k+1)}, \cdots, X_{2N}\}$ be the two parent chromosomes selected for crossover operation.
   (c) Draw a random number in continuous uniform distribution from 1 to N, i.e., $U(1, N)$. Let it be $k \in [1, N]$.
   (d) Then, the resulting offspring becomes $X_1{}' = \{X_{11}, X_{12}, \cdots, X_{1(k-1)}, X_{2k}, X_{2(k+1)}, \cdots, X_{2N}\}$ and $X_2{}' = \{X_{21}, X_{22}, \cdots, X_{2(k-1)}, X_{1k}, X_{1(k+1)}, \cdots, X_{1N}\}$.

(3) **Bitwise mutation**: It is applied to forbid the premature convergence, and it has the ability to explore the new solution space. Mutation is the process in which

the genetic structure of a chromosome is randomly altered. It leads to genetic diversity in a population. It is a step-by-step working procedure explained below.

(a) It works using mutation probability say P_{mut}. Thus, only $(M \cdot P_{mut})$ genes in the population go for mutation.
(b) Draw a random number from continuous uniform distribution, i.e., $j \in [1, M]$.
(c) Let a chromosome $X_j = \{X_{j1}, X_{j2}, \cdots, X_{j(k-1)}, X_{jk}, X_{j(k+1)}, \cdots, X_{jN}\}$ of length N be randomly selected for mutation.
(d) Again draw two points from continuous uniform distribution, i.e., $r_1, r_2 \in [1, N]$.
(e) Let $r_1 = 1$ and $r_2 = k$. Then, 1st and kth genes are selected for the mutation, i.e., X_{j1} and X_{jk}.
(f) Let $X_{j1} = 1$ and $X_{jk} = 0$, then the new chromosome becomes

$X_j{}' = \{X_{j1}{}', X_{j2}, \ldots, X_{j(k-1)}, X_{jk}{}', X_{j(k+1)}, \ldots, X_{jN}\}$ where $X_{j1}{}' = 0$ and $X_{jk}{}' = 1$.

Algorithm 1

1: Start with a random initial population P_0 which consists of offspring whose values are 1 and 0. Set $gen = 0$. Let M be initial population size.
2: Using following mapping function to compute the time point T_i for ith string,

$$T_i = T_i^{min} + \frac{T_i^{max} - T_i^{min}}{2^{l_i} - 1} * DV(S_i)$$

where T_i^{min} is the possible minimum value of time, T_i^{max} is the possible maximum value of time, l_i is the string length and $DV(S_i)$ is the decoded value of the ith string.
3: Using following defined mapping function, compute time point $t_i \in (0, T_i]$ for i^{th} string,

$$t_i = t_i^{min} + \frac{t_i^{max} - t_i^{min}}{2^{l_i} - 1} * DV(S_i)$$

where t_i^{min} is the possible minimum value of time and $t_i^{max} = T_i$.
4: Calculate fitness function using the objective function for the initial population.
5: Thereafter, the roulette wheel selection method is applied on the initial population to select parents for the mating pool.
6: Then, single-point crossover and bitwise mutation operators are applied on the mating pool until offspring population Q_{gen} of size M is filled.
7: Set $iter = gen + 1$, and $P_{(iter)} = Q_{gen}$; go to Step 2.

4 Numerical Example

The SC described in earlier section is demonstrated using a hypothetical example. Let demand per unit time follow uniform distribution, i.e., $f(t) \sim U(a = 0, b = 1)$, and

Table 1 Optimum solutions for uniform distributed demand

γ	Carrying charge r					
		0.12	0.13	0.15	0.17	0.20
0.05	Q	63.22	61.39	61.10	56.82	52.11
	B	62.79	62.99	56.55	55.32	53.05
	P	669.77	668.42	667.08	665.24	662.93
	t	0.623	0.602	0.564	0.545	0.519
	T	0.859	0.989	0.976	0.999	0.753
0.10	Q	57.54	57.99	53.88	51.04	49.07
	B	54.98	54.98	52.88	50.39	47.22
	P	666.94	665.65	664.54	662.85	660.71
	t	0.546	0.557	0.514	0.513	0.472
	T	0.878	0.990	0.830	0.752	0.921
0.20	Q	50.34	48.44	48.07	45.86	44.71
	B	49.07	47.09	46.61	44.64	43.22
	P	662.39	661.36	660.34	658.99	657.08
	t	0.474	0.461	0.451	0.434	0.414
	T	0.998	0.987	0.884	0.926	0.980

normal distribution, i.e., $f(t) \sim N(\mu = 1, \sigma^2 = 0.04)$. Other parameter values for the model are: $D_0 = 100, A = 10, r = 0.15, C = 3, S = 10, \beta = 1.5$ and $\delta = 2$. We have used GA with roulette wheel selection, single-point crossover, bitwise mutation, number of iterations 100, crossover probability 0.8, mutation probability 0.2, string length 40 and population size 20. Our aim is to determine the values of t_1 and T which maximize $P(T, t_1)$. Using Algorithm 1, EOQ (Q) and profit (P) are evaluated by using Eqs. 16 and 17. The effect of parameter γ with respect to carrying charge (r) is shown in Tables 1 and 2.

From Table 1, we conclude that as r increases with fixed γ the optimum profit and EOQ both decrease for uniform distributed demand. Similar results are seen for normal distributed demand; see Table 2. Also, as γ increases with fixed r, EOQ and optimum profit both decrease. From Tables 1 and 2, we see that the profit in uniform distribution is more than normal distribution. The codes are written by R software and run on i3–3110M, CPU @ 2.40 GHz and 4 GB RAM.

5 Sensitivity Analysis

The effect of various parameters, viz., total order quantity D_0, order cost A, per unit purchase cost C and selling cost S on the optimality of solution, is studied through the sensitivity analysis. For fixed $\gamma = 0.2$ and $\beta = 1.5$, the effect of 50% over- and under-estimation of these parameters on EOQ (Q) and optimum profit (P) has been

Table 2 Optimum solutions for normal distributed demand

γ	Carrying charge r					
		0.12	0.13	0.15	0.17	0.20
0.05	Q	51.59	51.59	51.58	51.59	51.37
	B	46.64	47.70	46.43	47.10	45.37
	P	319.07	317.57	315.64	313.09	309.37
	t	0.974	0.979	0.973	0.976	0.968
	T	1.00	0.999	1.00	0.999	0.999
0.10	Q	53.34	53.45	53.11	52.95	53.24
	B	48.45	49.60	46.65	46.65	47.79
	P	312.91	311.40	309.35	306.31	303.33
	t	0.975	0.979	0.966	0.966	0.972
	T	0.999	0.999	0.999	0.999	0.999
0.20	Q	56.03	56.61	57.02	56.16	55.946
	B	47.23	49.94	51.70	47.68	46.99
	P	300.75	299.33	296.51	295.05	291.34
	t	0.953	0.965	0.972	0.9552	0.952
	T	0.999	0.999	0.999	0.999	0.999

examined. That means the sensitivity analysis is performed by changing each of the parameters by −50%, −30%, +20% and +50% taking one parameter at a time and keeping the remaining parameters unchanged. The obtained results for uniform and normal distributed demand rate are shown in Table 3.

On the basis of the results of Table 3, the following observation can be made:

(1) All parameters except order cost effects equally on P.
(2) Total demand during the scheduling period, purchase cost and selling cost for uniform distribution effects mostly on P.
(3) Order cost has low effect on P for uniform distributed demand rate.
(4) Effect of almost all parameters except order cost for normal distribution is much high.
(5) The parameters except selling cost affects high on backlogging time for uniform distribution.
(6) For normal distribution, all parameters have low effect on backlogging and cycle time.

6 Managerial Implications

According to the demand of deteriorated item, the manager of a company finds the suitable probability distribution for the deteriorated item. Also, by comparing various probability distributions manager can decide the unique probability distribution

Table 3 Effect of changes in the parameters of the probabilistic demand inventory model

	% change in	Percentage of over- and under-estimation of parameter							
		Uniform distribution				Normal distribution			
		−50%	−30%	+20%	+50%	−50%	−30%	+20%	+50%
D_0	Q	−31.20	−17.24	11.51	26.66	−49.22	−30.01	20.22	51.08
	P	−51.42	−30.93	20.70	51.81	−51.75	−31.07	20.49	51.77
	t	35.90	15.61	−7.06	−14.90	1.75	0.09	0.40	0.73
	T	−11.67	−16.05	4.04	1.27	−0.02	−0.03	−0.01	0.01
A	Q	−25.18	−10.08	11.38	22.96	−0.94	−2.76	−2.00	0.07
	P	1.91	1.07	−0.65	−1.53	1.89	1.03	−0.49	−1.60
	t	−23.37	−15.55	8.16	19.20	−1.05	−2.86	−2.15	−0.02
	T	8.07	3.22	8.98	0.35	0.01	−0.02	−0.00	0.03
C	Q	35.74	15.09	−7.32	−15.87	0.12	0.60	−2.92	−4.66
	P	24.25	14.48	−9.57	−23.84	32.11	19.06	−12.63	−30.56
	t	34.88	14.37	−8.84	−17.38	0.17	0.84	−2.63	−4.84
	T	9.55	−8.82	8.02	−7.32	−0.01	−0.04	−0.14	0.01
S	Q	0.75	1.91	−3.12	−0.2	−1.78	−1.51	0.39	3.15
	P	−75.67	−45.41	30.28	75.7	−83.19	−50.08	33.36	83.21
	t	−1.05	−4.23	−1.81	−3.56	−1.86	−1.39	0.46	3.50
	T	0.48	−2.72	−5.48	−2.74	0.02	−0.05	0.01	0.02

which maximizes the profit of a company. The proposed models help the manager to understand the uncertainty in the market. Also, these models can assist the manager in accurately determining the optimal order quantity and profit. Before applying the model and proposed algorithm, manager has to collect necessary information about SC of the company. Such types of probabilistic EOQ inventory models can be useful in manufacturing and distributing industries. Moreover, these models can be used in inventory control of certain deteriorating items such as food items, electronic components, fashionable commodities and others.

7 Conclusions

In this study, EOQ inventory models under probabilistic market demand for a multi-echelon SC have been proposed for items with Weibull distributed deterioration. In the developed EOQ inventory models, shortages were allowed and they were partial backlogged. The backlogging rate is a variable, and it depends on the length of time for the customer waits before receiving the item. Thus, it is considered as a decreasing exponential function of waiting time. In the literature, deterministic inventory models were developed by several researchers by assuming market demand of a deteriorating item depends on stock, on time, on replenishment, etc. But here, we have developed probabilistic EOQ inventory models, which can be helpful where demand of the deteriorating item is uncertain in nature. Also, we are providing a novel solution methodology for solving proposed inventory models using binary coded GA. This methodology can be helpful for solving deterministic as well as probabilistic inventory models.

In a future study, it is hoped to further incorporate the proposed model into more realistic assumptions, such as lead time as a decision variable and a finite rate of replenishment. Also, formulated models can be solved by using particle swarm optimization, ant colony optimization, etc.

References

Abad, P. L. (2001). Optimal price and order size for a reseller under partial backlogging. *Computers and Operations Research, 28*(1), 53–65.

Beamon, B. M. (1998). Supply chain design and analysis: Models and methods. *International Journal of Production Economics, 55*(3), 281–294.

Bhosale, M. R., & Latpate, R. V. (2019). Single stage fuzzy supply chain model with Weibull distributed demand for milk commodities. *Granular Computing*, 1–12.

Bhunia, A. K., & Maiti, M. (1998). Deterministic inventory model for deteriorating items with finite rate of replenishment dependent on inventory level. *Computers and Operations Research, 25*(11), 997–1006.

Bhunia, A. K., & Maiti, M. (1997). Deterministic inventory models for variable production. *Journal of the Operational Research Society, 48*(2), 221–224.

Bose, S., Goswami, A., & Chaudhuri, K. S. (1995). An EOQ model for deteriorating items with linear time-dependent demand rate and shortages under inflation and time discounting. *Journal of the Operational Research Society, 46*(6), 771–782.

Chang, H. J., & Dye, C. Y. (1999). An EOQ model for deteriorating items with time varying demand and partial backlogging. *Journal of the Operational Research Society, 50*(11), 1176–1182.

Covert, R. P., & Philip, G. C. (1973). An EOQ model for items with Weibull distribution deterioration. *AIIE Transactions, 5*(4), 323–326.

Deb, K. (2005). *Multi-objective optimization using evolutionary algorithms* (Vol. 16). Wiley.

Dye, C. Y., Hsieh, T. P., & Ouyang, L. Y. (2007). Determining optimal selling price and lot size with a varying rate of deterioration and exponential partial backlogging. *European Journal of Operational Research, 181*(2), 668–678.

Dye, C. Y., & Ouyang, L. Y. (2005). An EOQ model for perishable items under stock-dependent selling rate and time-dependent partial backlogging. *European Journal of Operational Research, 163*(3), 776–783.

Eroglu, A., & Ozdemir, G. (2007). An economic order quantity model with defective items and shortages. *International Journal of Production Economics, 106*(2), 544–549.

Ghare, P. M., & Schrader, G. F. (1963). A model for exponentially decaying inventory. *Journal of Industrial Engineering, 14*(5), 238–243.

Holland, J. H. (1992). *Adaptation in natural and artificial systems: An introductory analysis with applications to biology, control, and artificial intelligence*. MIT Press.

Kurade, S. S., & Latpate, R. (2021). Demand and deterioration of items per unit time inventory models with shortages using genetic algorithm. *Journal of Management Analytics, 8*(3), 502–529. https://doi.org/10.1080/23270012.2020.1829113

Latpate, R. V., & Kurade, S. S. (2017). Fuzzy MOGA for supply chain models with Pareto decision space at different α-cuts. *The International Journal of Advanced Manufacturing Technology, 91*(9), 3861–3876.

Latpate, R., & Kurade, S. S. (2022). Multi-objective multi-index transportation model for crude oil using fuzzy NSGA-II. *IEEE Transactions on Intelligent Transportation Systems, 23*(2), 1347–1356. https://doi.org/10.1109/TITS.2020.3024693.

Latpate, R. V., & Bhosale, M. R. (2020). Single cycle supply chain coordination model for fuzzy stochastic demand of perishable items. *Iranian Journal of Fuzzy Systems, 17*(2), 39–48.

Latpate, R. V., & Bajaj, V. H. (2011). Fuzzy multi-objective, multi-product, production distribution problem with manufacturer storage. In *Proceedings of International Congress on PQROM* (pp. 340–355).

Li, F., Da, Xu., Jin, L., & C. and Wang, Hong. (2011). Structure of multi-stage composite genetic algorithm (MSC-GA) and its performance. *Expert Systems with Applications,38*(7), 8929–8937.

Maiti, A. K. (2020). Multi-item fuzzy inventory model for deteriorating items in multi-outlet under single management. *Journal of Management Analytics, 7*(1), 44–68.

Maity, S., Roy, A., & Maiti, M. (2017). An intelligent hybrid algorithm for 4-dimensional TSP. *Journal of Industrial Information Integration, 5,* 39–50.

Padmanabhan, G., & Vrat, P. (1995). EOQ models for perishable items under stock dependent selling rate. *European Journal of Operational Research, 86*(2), 281–292.

Philip, G. C. (1974). A Generalized EOQ model for items with Weibull distribution deterioration. *AIIE Transactions, 6*(2), 159–162.

Raosaheb, L., & Bajaj, V. H. (2011). Fuzzy programming for multi-objective transportation and inventory management problem with retailer storage. *International Journal of Agricultural and Statistical Sciences, 7*(1), 317–326.

Skouri, K., Konstantaras, I., Papachristos, S., & Ganas, I. (2009). Inventory models with ramp type demand rate, partial backlogging and Weibull deterioration rate. *European Journal of Operational Research, 192*(1), 79–92.

Wee, H. M. (1995). A deterministic lot-size inventory model for deteriorating items with shortages and a declining market. *Computers and Operations Research, 22*(3), 345–356.

Wu, K. S. (2001). An EOQ inventory model for items with Weibull distribution deterioration, ramp type demand rate and partial backlogging. *Production Planning & Control, 12*(8), 787–793.

The Evolution of Dynamic Gaussian Process Model with Applications to Malaria Vaccine Coverage Prediction

Pritam Ranjan and M. Harshvardhan

Abstract Gaussian process (GP)-based statistical surrogates are popular, inexpensive substitutes for emulating the outputs of expensive computer models that simulate real-world phenomena or complex systems. Here, we discuss the evolution of dynamic GP model—a computationally efficient statistical surrogate for a computer simulator with time-series outputs. The main idea is to use a convolution of standard GP models, where the weights are guided by a singular value decomposition (SVD) of the response matrix over the time component. The dynamic GP model also adopts a localized modelling approach for building a statistical model for large datasets. In this chapter, we use several popular test function-based computer simulators to illustrate the evolution of dynamic GP models. We also use this model for predicting the coverage of Malaria vaccine worldwide. Malaria is still affecting more than eighty countries concentrated in the tropical belt. In 2019 alone, it was the cause of more than 435,000 deaths worldwide. The malice is easy to cure if diagnosed in time, but the common symptoms make it difficult. We focus on a recently discovered reliable vaccine called Mosquirix (RTS,S) which is currently going under human trials. With the help of publicly available data on dosages, efficacy, disease incidence and communicability of other vaccines obtained from the World Health Organization, we predict vaccine coverage for 78 Malaria-prone countries.

Keywords Gaussian process (GP) model · Big data · Dynamic GP · Time-series valued process

1 Introduction

Computer simulators are widely used to understand complex physical systems in many areas such as aerospace, renewable energy, climate modelling and manufacturing. For example, Greenberg (1979) developed a finite volume community ocean model (FVCOM) for simulating the flow of water in the Bay of Fundy; Bower et al.

P. Ranjan (✉) · M. Harshvardhan
Indian Institute of Management Indore, Indore, Madhya Pradesh, India
e-mail: pritamr@iimidr.ac.in

D. D. Hanagal et al. (eds.), *Applied Statistical Methods*, Springer Proceedings in Mathematics & Statistics 380, https://doi.org/10.1007/978-981-16-7932-2_7

(2006) discussed the formation of galaxies using a simulator called GALFORM; Bayarri et al. (2009) used a simulator called TITAN2D for modelling the maximum volcanic eruption flow height; Zhang et al. (2018) used a TDB simulator to model the population growth of European red mites. Realistic computer simulators can also be computationally expensive to run, and thus statistical surrogates used as an inexpensive substitute for a deeper understanding of the underlying phenomena. Sacks et al. (1989) proposed using a realization of the Gaussian process (GP) model as a surrogate for such simulator outputs.

The types of simulator output structures dealt with are as varied as the applications. One is faced with scalar, multivariate, functional, time-series and spatial–temporal data, to name a few. In this chapter, we discuss the evolution of GP-based surrogate models for computer simulators with time-series outputs, which we refer to as *dynamic computer simulators*. Such simulators arise in various applications, for example, rainfall–runoff model (Conti et al., 2009), vehicle suspension system (Bayarri et al., 2007) and TDB model (Zhang et al., 2018).

The emulation of dynamic computer simulators has been considered by many (Kennedy & O'Hagan, 2001; Stein, 2005; Bayarri et al., 2007; Higdon et al., 2008; Conti et al., 2009; Liu & West, 2009; Farah et al., 2014; Hung et al., 2015). In this chapter, we highlight the singular value decomposition (SVD)-based GP models, which was originally introduced by Higdon et al. (2008) for computer model calibration with high-dimensional outputs. However, Zhang et al. (2018) generalized it further for time-series responses and developed the empirical Bayesian inference for large-scale computer simulators.

Fitting GP models requires the inversion of $N \times N$ spatial correlation matrices, which gets prohibitive if N (the sample size) becomes large. In other words, fitting GP models over the entire training set can often be computationally infeasible for large-scale dynamic computer experiments involving thousands of training points. A naive popular approach is to build localized models for prediction in the big data context. To search for the most relevant data for local neighbourhood in a more intelligent way, Emery (2009) built a local neighbourhood by sequentially including data that make the kriging variance decrease more. Gramacy and Apley (2015) improved the prediction accuracy by using a sequential greedy algorithm and an optimality criterion for finding a non-trivial local neighbourhood set, and Zhang et al. (2018) further extended the idea for dynamic simulator outputs.

In this chapter, we illustrate the implementation of dynamic SVD-based GP model for several test function-based simulator outputs and a real-life modelling problem where the objective is to predict the usage of a new Malaria vaccine. Malaria is a mosquito-borne disease caused by a *Plasmodium*, a malarial parasite. Although Malaria is not life-threatening by its nature, if left untreated, it can cause severe illness and prove to be fatal. The disease was eliminated from American and European continents by first half of twentieth century but is still very common in South Asia and Sub-Saharan Africa. In 2017 alone, there were more than 219 million cases of Malaria which resulted in deaths of more than 435,000 people worldwide (World Health Organization, 2019).

In February 2019, a new Malaria vaccine `RTS,S`—known by the trade name *Mosquirix*—was approved for human trials in three countries—Ghana, Malawi and Kenya—coordinated by WHO. The study is expected to get over by December 2022. However, in last few months, several pharmaceutical majors have begun showing interest in the vaccine's mass production, and the investors want to estimate the coverage ratio—defined by *the vaccine population count divided by the total population*.

The chapter is outlined as follows. In Sect. 2, we start with the standard GP model for scalar-valued response and present the dynamic SVD-based GP model. Further, we discuss the localized dynamic GP model for handling big data. Section 3 explains how dynamic GP model is used for predicting vaccination coverage, with model inputs and built-in R packages. They are illustrated with model outputs on a world map. Finally, concluding remarks and recommendations are suggested in Sect. 4.

2 Evolution of Dynamic Gaussian Process Model

In this section, we present a sequence of statistical surrogate models starting from the most basic GP model which emulates deterministic computer simulators returning scalar outputs, to dynamic GP model that acts as a surrogate to time-series valued simulators. The models are supported by a brief explanation of their theoretical foundations, an associated example and R implementation.

2.1 Basic Gaussian Process Model

Gaussian process models are immensely popular in computer experiment literature for emulating computer simulator outputs. In one of the pioneering researches, Sacks et al. (1989) suggested using realizations of Gaussian stochastic process to model deterministic scalar-valued simulator outputs. However, the notion of such statistical models originates from the kriging literature in geostatistics.

Let the training data consist of d-dimensional input and 1-dimensional output of the computer simulator, denoted by $x_i = (x_{i1}, x_{i2}, \dots, x_{id})$ and $y_i = y(x_i)$, respectively. Then, the GP model is written as

$$y_i = \mu + z(x_i), \quad i = 1, 2, \dots, n, \tag{1}$$

where μ is the overall mean and $\{z(x), x \in [0,1]^d\} \sim GP(0, \sigma_z^2 R(,))$ with $E(z(x)) = 0$, $Var(z(x)) = \sigma_z^2$, and $Cov(z(x_i), z(x_j)) = \sigma_z^2 R(x_i, x_j)$ where $R(,)$ is a positive definite correlation function. Then, any finite subset of variables $\{z(x_1), z(x_2), \dots, z(x_n)\}$, for $n \geq 1$, will jointly follow multivariate normal distribution. That is, $Y = (y_1, y_2, \dots, y_n)' \sim MVN(\mu 1_n, \sigma_z^2 R_n)$, where 1_n is an $n \times 1$ vector of all 1's and R_n is an $n \times n$ correlation matrix with (i, j)th element given by $R(x_i, x_j)$

(see Sacks et al. (1989), Santner et al. (2003), Rasmussen and Williams (2006) for more details).

The model described by (1) is typically fitted by either maximizing the likelihood or via Bayesian algorithms like Markov chain Monte Carlo (MCMC). As a result, the predicted response $\hat{y}(x_0)$ for an arbitrary input x_0 can be obtained as a conditional expectation from the following $(n+1)$-dimensional multivariate normal distribution:

$$\begin{pmatrix} y(x_0) \\ Y \end{pmatrix} = N\left(\begin{pmatrix} \mu \\ \mu 1_n \end{pmatrix}, \begin{pmatrix} \sigma_z^2 & \sigma_z^2 r'(x_0) \\ \sigma_z^2 r(x_0) & \sigma_z^2 R_n \end{pmatrix} \right), \tag{2}$$

where $r(x_0) = [corr(x_1, x_0), \ldots, corr(x_n, x_0)]'$. The predicted response $\hat{y}(x_0)$ is the same as the conditional mean:

$$E(y(x_0)|Y) = \mu + r(x_0)' R_n^{-1} (Y - 1_n \mu), \tag{3}$$

and the associate prediction uncertainty estimate (denoted by $s^2(x_0)$) can be quantified by the conditional variance:

$$Var(y(x_0)|Y) = \sigma_z^2 (1 - r'(x_0) R_n^{-1} r(x_0)). \tag{4}$$

The most crucial component of such a GP model is the spatial correlation structure, $R(,)$, which dictates the 'smoothness' of the interpolator that passes through the observations. By definition, any positive definite correlation structure would suffice, but the most popular choice is the power exponential correlation family given by

$$R(x_i, x_j) = \prod_{k=1}^{d} \exp\{-\theta_k |x_{ik} - x_{jk}|^{p_k}\}, \tag{5}$$

where θ_k and p_k control the wobbliness of the surrogate in the kth coordinate. A special case with $p_k = 2$ for all $k = 1, 2, \ldots, d$, represents the most popular Gaussian correlation also known as radial basis kernel in machine learning literature. Figure 1 demonstrates the significance of p_k in the smoothness of the mean prediction.

Example 1 Suppose the simulator output is generated by a one-dimensional test function $f(x) = \ln(x + 0.1) + \sin(5\pi x)$, and $X = \{x_1, \ldots, x_7\}$ is a randomly generated training set as per the space-filling Latin hypercube design (McKay et al., 1979). We use an R library called **GPfit** (MacDonald et al., 2015) for fitting the model via maximum likelihood approach. Figure 1 shows the fitted surrogate along with the true simulator response curves.

Clearly, the choice of p_k in (5) plays an important role in determining the smoothness of the predictor. It can be noticed from Fig. 1 that $p_k = 2$ versus $p_k = 1.95$ does not make visible difference in terms of smoothness. However, it turns out that by

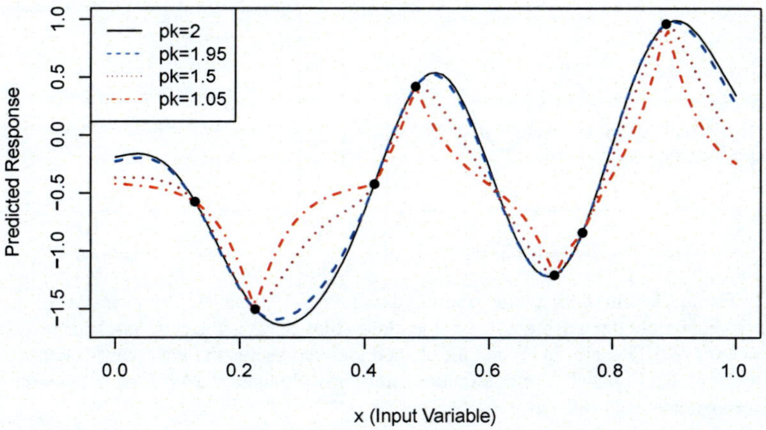

Fig. 1 Mean predictions obtained using GPfit, when the true simulator response is generated using $f(x) = \ln(x + 0.1) + \sin(5\pi x)$

changing the power from $p_k = 2$ to $p_k = 1.95$, the numerical stability of the correlation matrix inversion can be immensely increased.

Depending upon the parameter estimation approach used (i.e. maximum likelihood method, empirical Bayesian or full Bayesian), the prediction uncertainty estimate may vary. For instance, in empirical Bayesian approach, the parameters μ, σ and θ in R_n are replaced by their maximum a posteriori (MAP) estimates. On the other hand, the MLE-based approach starts by maximizing the likelihood with respect to μ and σ_z^2, giving closed form expressions as

$$\hat{\mu} = (1_n' R_n^{-1} 1_n)^{-1} (1_n' R_n^{-1} Y), \tag{6}$$

and

$$\hat{\sigma}_z^2 = \frac{(Y - 1_n\mu)' R_n^{-1} (Y - 1_n\mu)}{n}, \tag{7}$$

conditional on the value of θ in R_n. The hyper-parameter θ is further estimated by maximizing the profile likelihood, which is typically an intensive optimization problem. Sacks et al. (1989) report the prediction uncertainty estimate as

$$s^2(x_0) = \sigma_z^2 \left(1 - r'(x_0) R_n^{-1} r(x_0) + \frac{(1 - \mathbf{1_n}' R_n^{-1} r(x_0))^2}{\mathbf{1_n}' R_n^{-1} \mathbf{1_n}'} \right), \tag{8}$$

which accounts for additional uncertainty due to the prediction of unknown constant mean μ. Of course, the difference between (8) and (4) can be somewhat substantial. See Example 2 for an illustration using a test function-based computer simulator.

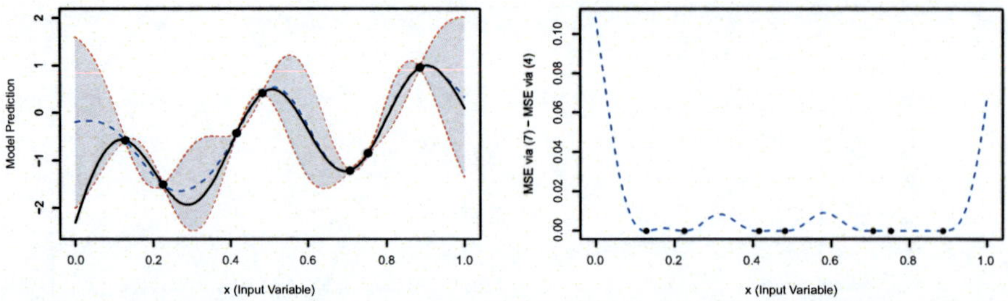

Fig. 2 The black solid dots are the training data points. Left panel: The blue dashed curve is the mean prediction obtained using `GPfit`, the black solid curve is the true simulator response curve $f(x) = ln(x + 0.1) + \sin(5\pi x)$, and the shaded area represents the uncertainty quantification via $\hat{y}(x) \pm 2s(x)$. Right panel: The prediction uncertainty obtained via MLE as in (8)—the posterior variance estimate as in (4)

Example 2 Considering the same set-up as in Example 1, Fig. 2 shows the fitted surrogate along with the prediction uncertainty estimates obtained via GPfit and the two formulations (4) and (8).

From the right panel of Fig. 2, it is clear that the third term in the prediction uncertainty estimate (in (8)) is relatively large in the unexplored input regions. As a result, it is recommended to account for uncertainty quantification due to the estimation of unknown model parameters.

Several additional theoretical and numerical issues on GP models require more careful understanding. See Santner et al. (2003), Rasmussen and Williams (2006) and Harshvardhan and Ranjan (2019), for more details on optimization of likelihood, near-singularity of correlation matrices, choice of correlation kernel, parametrization of hyper-parameters and the choice of mean function.

2.2 *Dynamic Gaussian Process Model*

Experimentation via dynamic computer simulators arises in various applications, for example, rainfall–runoff model (Conti et al., 2009), vehicle suspension system (Bayarri et al., 2007) and population growth model for European red mites (Zhang et al., 2018). The real-life application presented in this chapter comes from the pharmaceutical industry, where the investors want to predict the coverage of a particular Malaria vaccine called RTS,S/AS01 (Mosquirix) around the globe over a 20-year window.

The time-series dependence in the simulator response makes the statistical emulation substantially more challenging as compared to the standard GP model presented in the previous section. Recently, a few attempts have been made in this regard. For example, Conti et al. (2009) constructed dynamic emulators by using a one-step transition function of state vectors to emulate the computer model movement from one time step to the next. Liu and West (2009) proposed time varying autoregression

(TVAR) models with GP residuals. Farah et al. (2014) extend the TVAR models in Liu and West (2009) by including the input-dependent dynamic regression term. Another clever approach is to represent the time-series outputs as linear combinations of a fixed set of basis such as singular vectors (Higdon et al., 2008) or wavelet basis (Bayarri et al., 2007) and impose GP models on the linear coefficients. Zhang et al. (2018) further extended the singular value decomposition (SVD)-based approach for large-scale data. Next, we discuss the basic version of SVD-based GP model developed by Higdon et al. (2008).

Suppose the computer simulator outputs have been collected at N design points and stored in the $N \times q$ design matrix $\boldsymbol{X} = [\boldsymbol{x}_1, \ldots, \boldsymbol{x}_N]^T$, and $\boldsymbol{Y} = [\boldsymbol{y}(\boldsymbol{x}_1), \ldots, \boldsymbol{y}(\boldsymbol{x}_N)]$ is the corresponding $L \times N$ matrix of time-series responses. Then, the SVD on $\boldsymbol{Y}$ gives

$$\boldsymbol{Y} = \boldsymbol{U}\boldsymbol{D}\boldsymbol{V}^T,$$

where $\boldsymbol{U} = [\boldsymbol{u}_1, \ldots, \boldsymbol{u}_k]$ is an $L \times k$ column-orthogonal matrix, $\boldsymbol{D} = \mathrm{diag}(d_1, \ldots, d_k)$ is a $k \times k$ diagonal matrix of singular values sorted in decreasing order, $\boldsymbol{V}$ is an $N \times k$ column-orthogonal matrix of right singular vectors and $k = \min\{N, L\}$. Higdon et al. (2008) suggested modelling the simulator response as

$$\boldsymbol{y}(\boldsymbol{x}) = \sum_{i=1}^{p} c_i(\boldsymbol{x})\boldsymbol{b}_i + \boldsymbol{\epsilon}, \tag{9}$$

where $\boldsymbol{x} \in \mathbb{R}^q$ and $\boldsymbol{b}_i = d_i\boldsymbol{u}_i \in \mathbb{R}^L$, for $i = 1, \ldots, p$ represent the orthogonal basis. The coefficients c_i's in (9) are assumed to be independent Gaussian processes, i.e. $c_i \sim \mathcal{GP}(0, \sigma_i^2 K_i(\cdot, \cdot; \boldsymbol{\theta}_i))$ for $i = 1, \ldots, p$, where K_i's are correlation functions. We use the popular anisotropic Gaussian correlation, $K(\boldsymbol{x}_1, \boldsymbol{x}_2; \boldsymbol{\theta}_i) = \exp\{-\sum_{j=1}^{q} \theta_{ij}(x_{1j} - x_{2j})^2\}$. The residual term $\boldsymbol{\epsilon}$ in (9) is assumed to be independent $\mathcal{N}(0, \sigma^2 \boldsymbol{I}_L)$. The number of significant singular values, p, in (9), is determined empirically by the cumulative percentage criterion $p = \min\{m : (\sum_{i=1}^{m} d_i)/(\sum_{i=1}^{k} d_i) > \gamma\}$, where γ is a threshold of the explained variation.

In this chapter, we discuss the implementation of this so-called svdGP model by Zhang et al. (2018). R library called *DynamicGP* (Zhang et al., 2020) provides user-friendly functions for quick usage. The most important function is `svdGP`, and its usage is illustrated as follows:

```
svdGP(design, resp, frac=0.95, nthread=1, clutype="PSOCK", ...)
```

where `design` is the input design matrix, `resp` is the output response matrix, `frac` specifies $\gamma = 95\%$, and `nthread` and `clutype` controls the parallelization of the implementation. There are a few additional arguments of `svdGP()` that accounts for other nuances of the model fitting process.

For all the model parameters in (9), Zhang et al. (2018, 2020) used the maximum a posteriori (MAP) values as the plug-in estimates. To obtain the MAP estimates of process and noise variance parameters, σ_i^2 and σ^2, inverse Gamma priors were used, i.e.

$$[\sigma_i^2] \sim \mathrm{IG}\left(\frac{\alpha_i}{2}, \frac{\beta_i}{2}\right), i = 1, \ldots, p, \qquad [\sigma^2] \sim \mathrm{IG}\left(\frac{\alpha}{2}, \frac{\beta}{2}\right),$$

and Gamma prior was used for the hyper-parameter $1/\theta_{ij}$ of the correlation function.

Zhang et al. (2018) show that the approximate predictive distribution for an arbitrary untried $\boldsymbol{x}_0 \in \mathbb{R}^q$ is obtained by

$$\pi(\boldsymbol{y}(\boldsymbol{x}_0)|\boldsymbol{Y}) \approx \pi(\boldsymbol{y}(\boldsymbol{x}_0)|\boldsymbol{V}^*, \hat{\boldsymbol{\Theta}}, \hat{\sigma}^2) \approx \mathcal{N}\big(\boldsymbol{B}\hat{\boldsymbol{c}}(\boldsymbol{x}_0|\boldsymbol{V}^*, \hat{\boldsymbol{\Theta}}), \boldsymbol{B}\boldsymbol{\Lambda}(\boldsymbol{V}^*, \hat{\boldsymbol{\Theta}})\boldsymbol{B}^T + \hat{\sigma}^2\boldsymbol{I}_L\big), \tag{10}$$

where $\boldsymbol{B} = [d_1\boldsymbol{u}_1, \ldots, d_p\boldsymbol{u}_p] = \boldsymbol{U}^*\boldsymbol{D}^*$, with $\boldsymbol{U}^* = [\boldsymbol{u}_1, \ldots, \boldsymbol{u}_p]$, $\boldsymbol{D}^* = \mathrm{diag}(d_1, \ldots, d_p)$ and $\boldsymbol{V}^* = [\boldsymbol{v}_1, \ldots, \boldsymbol{v}_p]^T$, and $\hat{\boldsymbol{\Theta}} = \{\hat{\boldsymbol{\theta}}_1, \ldots, \hat{\boldsymbol{\theta}}_p\}$ and $\hat{\sigma}^2$ are the MAP estimates of the correlation parameters and noise variance σ^2, respectively. As shown in Zhang et al. (2018),

$$\begin{aligned} \hat{\boldsymbol{\theta}}_i &= \underset{\boldsymbol{\theta}_i \in \mathbb{R}^q}{\mathrm{argmax}}\ |\boldsymbol{K}_i|^{-1/2}\left(\frac{\beta_i + \psi_i}{2}\right)^{-(\alpha_i+N)/2} \pi(\boldsymbol{\theta}_i), \text{ and} \\ \hat{\sigma}^2 &= \frac{1}{NL + \alpha + 2}\left(\boldsymbol{r}^T\boldsymbol{r} + \beta\right), \end{aligned} \tag{11}$$

where $\boldsymbol{K}_i$ is the $N \times N$ correlation matrix on the design matrix $\boldsymbol{X}$ with the (j, l)th entry being $K(\boldsymbol{x}_j, \boldsymbol{x}_l; \hat{\boldsymbol{\theta}}_i)$ for $i = 1, \ldots, p$ and $j, l = 1, \ldots, N$, $\psi_i = \boldsymbol{v}_i^T\boldsymbol{K}_i^{-1}\boldsymbol{v}_i$, $\pi(\boldsymbol{\theta}_i)$ is the prior distribution of $\boldsymbol{\theta}_i$ and $\boldsymbol{r} = \mathrm{vec}(\boldsymbol{Y}) - (I_N \otimes \boldsymbol{B})\mathrm{vec}(\boldsymbol{V}^{*T})$ with $\mathrm{vec}(\cdot)$ and $\otimes$ being the vectorization operator and the Kronecker product for matrices, respectively.

The vector of predictive mean of the coefficients at $\boldsymbol{x}_0$ is

$$\begin{aligned} \hat{\boldsymbol{c}}(\boldsymbol{x}_0, |\boldsymbol{V}^*, \hat{\boldsymbol{\Theta}}) &= [\hat{c}_1(\boldsymbol{x}_0|\boldsymbol{v}_1, \hat{\boldsymbol{\theta}}_1), \ldots, \hat{c}_p(\boldsymbol{x}_0|\boldsymbol{v}_p, \hat{\boldsymbol{\theta}}_p)]^T \\ &= [\boldsymbol{k}_1^T(\boldsymbol{x}_0)\boldsymbol{K}_1^{-1}\boldsymbol{v}_1, \ldots, \boldsymbol{k}_p^T(\boldsymbol{x}_0)\boldsymbol{K}_p^{-1}\boldsymbol{v}_p]^T, \end{aligned} \tag{12}$$

where $\boldsymbol{k}_i(\boldsymbol{x}_0) = [K(\boldsymbol{x}_0, \boldsymbol{x}_1; \hat{\boldsymbol{\theta}}_i), \ldots, K(\boldsymbol{x}_0, \boldsymbol{x}_N; \hat{\boldsymbol{\theta}}_i)]^T$. The predictive variance $\boldsymbol{\Lambda}(\boldsymbol{V}^*, \hat{\boldsymbol{\Theta}})$ of the coefficients at $\boldsymbol{x}_0$ is a $p \times p$ diagonal matrix with the ith diagonal entry being

$$\hat{\sigma}_i^2(\boldsymbol{x}_0|\boldsymbol{v}_i, \hat{\boldsymbol{\theta}}_i) = \frac{(\beta_i + \boldsymbol{v}_i^T\boldsymbol{K}_i^{-1}\boldsymbol{v}_i)\left(1 - \boldsymbol{k}_i^T(\boldsymbol{x}_0)\boldsymbol{K}_i^{-1}\boldsymbol{k}_i(\boldsymbol{x}_0)\right)}{\alpha_i + N}. \tag{13}$$

Example 3 illustrates the implementation of svdGP model for a test function-based computer simulator model via the R library DynamicGP (Zhang et al., 2020).

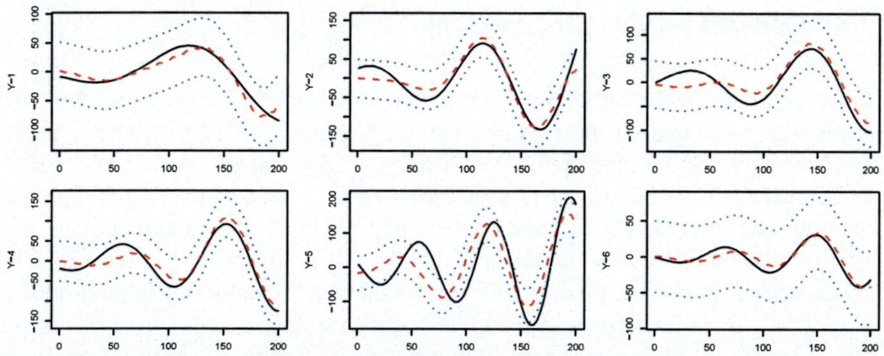

Fig. 3 Model prediction for six randomly chosen inputs. Each panel shows the true simulator response (black solid curve), the mean predicted svdGP fit (dashed red curve) and the uncertainty bounds (blue dotted curves)

Example 3 Suppose the time-series valued response is generated using the following test function (Forrester et al., 2008) which takes 3-dimensional inputs,

$$f(\boldsymbol{x}, t) = (x_1 t - 2)^2 \sin(x_2 t - x_3), \tag{14}$$

where $\boldsymbol{x} = (x_1, x_2, x_3)^T \in [4, 10] \times [4, 20] \times [1, 7]$ and $t \in [1, 2]$ is on a 200-point equidistant time grid. We used `svdGP()` function in the R library *DynamicGP* for easy implementation. Figure 3 illustrates the implementation, by first fitting the svdGP model to a training set of 20 input points randomly generated via maximin Latin hypercube design in the three-dimensional hyper-rectangle $[4, 10] \times [4, 20] \times [1, 7]$ and then predicting the time-series valued simulator output using `svdGP()` function.

From Fig. 3, it is clear that the fitted surrogate model predictions are reasonable approximations of the simulator outputs at the design points. We fitted svdGP model using the default settings of *DynamicGP* package. Of course, one can play around with other arguments to obtain better (more accurate) predictions.

Both, the basic GP models (in Sect. 2.1) which emulate scalar-valued simulator outputs and the svdGP models (in Sect. 2.2) used for emulating time-series valued dynamic simulator responses, require numerous inversions of $n \times n$ correlation matrices—this is computationally intensive and prohibitive if N (the sample size) is large. For instance, in our motivating application where the training data size is $N = 146$ (see Sect. 3), model fitting via either likelihood method or a Bayesian approach would be computationally burdensome unless the codes are parallelized on heavy computing clusters. The next section briefly reviews GP-based models for large data.

2.3 Generalizations for Big Data

Thus far, several techniques have been proposed to account for the large size of the data while building a GP-based surrogate; see Santner et al. (2003), Harshvardhan and Ranjan (2019) for quick reference. A naive yet popular approach is to fit several local inexpensive (somewhat less accurate) models instead of one big (supposedly more precise) model. The method of searching for local neighbourhood can be as simple as finding the *k-nearest neighbour* (k-NN) at the point of prediction. For scalar-valued simulators, Emery (2009) built a more efficient local neighbourhood by sequentially including data that make the kriging variance decrease more. Gramacy and Apley (2015) improved the prediction accuracy by using a greedy algorithm and an optimality criterion for finding a non-trivial local neighbourhood set. Zhang et al. (2018) extended this approach further for the svdGP model.

Assume the total training data size is N, and we wish to predict the simulator response at x_0. Then, the main idea behind this greedy approach in Gramacy and Apley (2015), Zhang et al. (2018) is to first use k-NN approach for finding n_0 neighbours from the training data and then sequentially obtain the remaining $n - n_0$ points by using an optimality criterion. This proposed greedy sequential method known as lasvdGP (locally approximate svdGP) is computationally very efficient as compared to the full scalar GP/svdGP and much more accurate than the naive k-NN-based svdGP model (referred to as knnsvdGP).

The following functions in the R library *DynamicGP* can be used for easy implementation:

```
knnsvdGP(design,resp, nn=20, ..., nthread = 1, clutype="PSOCK")
lasvdGP(design, resp, n0=10, nn=20, ..., nthread = 1, clutype="PSOCK")
```

where `design`, `resp`, `nthread` and `clutype` are the same as in `svdGP()`, and the important additional parameters are `nn`—the size of the local neighbourhood set (on which the local GP models have to be built), and `n0`—size of the local neighbourhood set to be found via k-nearest neighbours which will serve as the starting point of the greedy sequential approach for building the local neighbourhood set.

Example 4 Suppose the simulator response is generated using the same test function as in Example 3, but the training data are obtained on a $N = 500$-point random Latin hypercube design in the input space: $[4, 10] \times [4, 20] \times [1, 7]$. In such a case, fitting a full svdGP is certainly infeasible on a regular laptop or desktop. Thus, we rely on fitting the localized surrogate models like knnsvdGP and lasvdGP. Figure 4 shows the surrogate fits with $n_0 = 20$ and $nn = 30$ local neighbourhood point sets for lasvdGP model.

From Fig. 4, it is clear that the surrogate fits are much better approximations of the underlying truth (as compared to the illustration in Example 3), which is however expected as the training size is 500 (much bigger than 20-point design in Example 3).

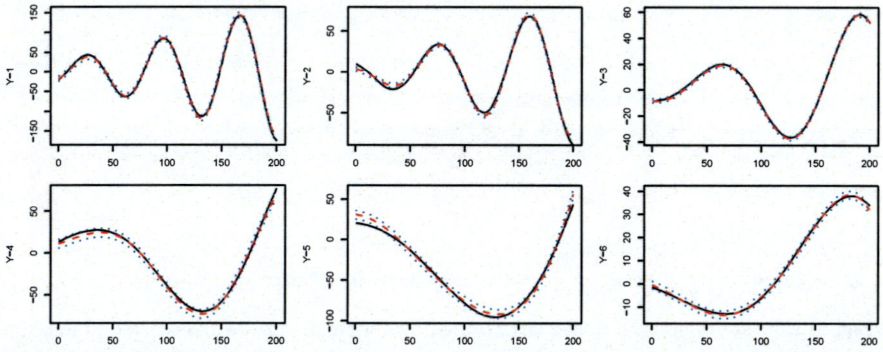

Fig. 4 Model prediction for six randomly chosen inputs. Each panel shows the true simulator response (black solid curve), the mean predicted lasvdGP fit (dashed red curve) and the uncertainty bounds (blue dotted curves)

Interestingly, the error bounds around the predicted mean response are too narrow and sometimes do not cover the true simulator output. It can perhaps be attributed to the fact that the R library *DynamicGP* uses MAP estimators and not the full Bayesian approach. It is often believed that the latter approach accounts for more uncertainty in the model fitting process.

3 Application: Malaria Vaccination Coverage

In this section, we use the historical data on worldwide vaccination coverage for several diseases to predict the coverage ratio of a new Malaria vaccine. The diseased typically experience fevers, chills and flu like illnesses (Centers for Disease Control, 2019) with the symptoms varying in their severity on a case-by-case basis. This can be lethal if not treated properly, and a 2002 study by Greenwood and Mutabingwa (2002) tells us the serious state of Malaria in the world (see World Health Organisation (2019) for latest detailed report).

Recently, a new Malaria vaccine `RTS,S` (also known as *Mosquirix*) has been showing promising results for human trials in Ghana, Malawi and Kenya. Malaria Vaccine Implementation Programme (MVIP), coordinated by WHO, is being funded by a global fund comprising (1) Gavi—The Vaccine Alliance, (2) UNITAID and (3) PATH. As of now, no results have been made public, and the study is expected to get over by December 2022. However, in last few months, several pharmaceutical majors have begun showing interest in the vaccine's mass production.

Major limitations in the success of a Malaria vaccine are technical and economic feasibility (Moorthy et al., 2004). With the current human trials underway, the former is largely solved; however, the latter remains. A study on predicting coverage ratios would immensely benefit to attract global monies—by corporates and philanthropist funds—to the cause. Recall that the coverage ratio is defined by *the vaccine pop-*

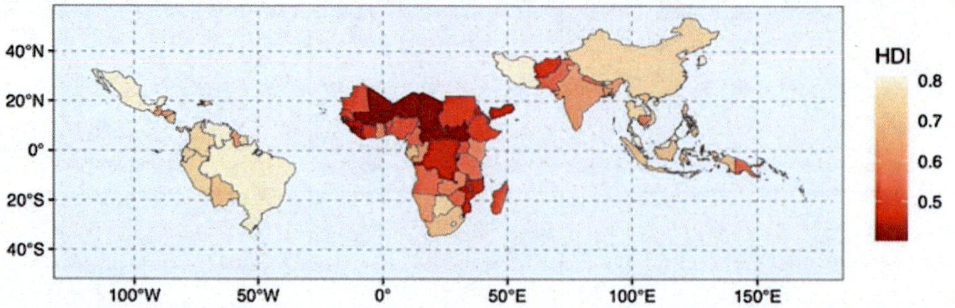

Fig. 5 Human development index (HDI) value for different countries considered in this study around the globe

ulation count divided by the total population. Thus, our objective is to predict the coverage ratio for this Malaria vaccine, using the available data on the coverage ratio of other vaccines. Based on earlier studies on vaccines, the following variables have been identified as predictors:

- *Dosage number* (X_1): The value is k, if k doses of the vaccine have already been given. Luman et al. (2005) suggested higher the number of dosages, lower the chance of completing the entire treatment.
- *Dosage time* (X_2): Number of months after birth when the first dosage is taken; 0 represents 'at birth'. Luman et al. (2005) found vaccines which were given at birth had higher coverage as there is no extra effort needed to come to health centre.
- *Efficacy* (X_3): Recorded in percentage—ability of the vaccine to actually prevent the disease (see McLean, 1995). Vaccination does not guarantee prevention, assuming if chances of prevention are better, more people will be vaccinated.
- *Incidence per lac* (X_4): It is more likely that the parents will give the vaccine to their children if the occurrence of the disease is high. When incidences are high, the population is more careful about prevention.
- *Communicable* (X_5): Binary (0: non-communicable, 1: communicable)—assuming that the fear of contagion may drive the vaccination.
- *Years active* (X_6): How long has the vaccine been around for public use (in years).

We used data for several vaccines (e.g. tuberculosis, diphtheria, hepatitis B, polio, Japanese encephalitis, measles, NTetanus, rubella and yellow fever) collected on aforementioned variables from 78 countries. We pooled the countries using human development index (HDI) values into 8 groups of size 8 and 2 groups of size 7 each. Figure 5 depicts the HDI value of different countries. In total, the data consist of 146 observations—the coverage ratio of different vaccines for 10 country groups observed over 38 year period (from 1980 to 2017), i.e. $y_t(x_i)$, for $t = 1, \ldots, 38$ and the corresponding input $x_i = (x_{i1}, x_{i2}, \ldots, x_{i7})$, where $i = 1, 2, \ldots, 146$ represent the observation number, $X_1, \ldots, X_6$ are predictor variables described above and the seventh input (X_7) is the average HDI value of the country group.

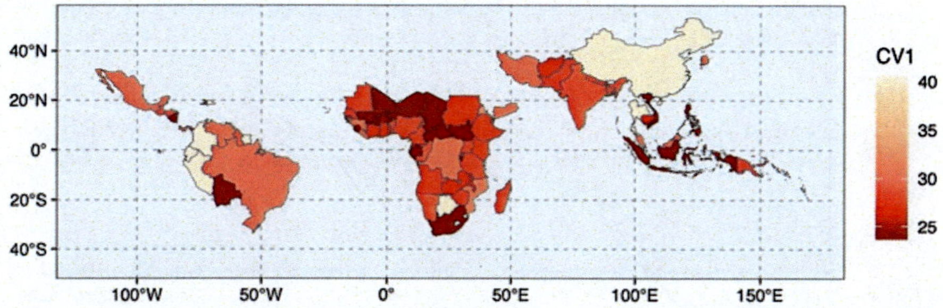

Fig. 6 Prediction of coverage ratio for Mosquirix at $t = 0$ for different country groups using lasvdGP model with $nn = 50$ and $n_0 = 30$ points

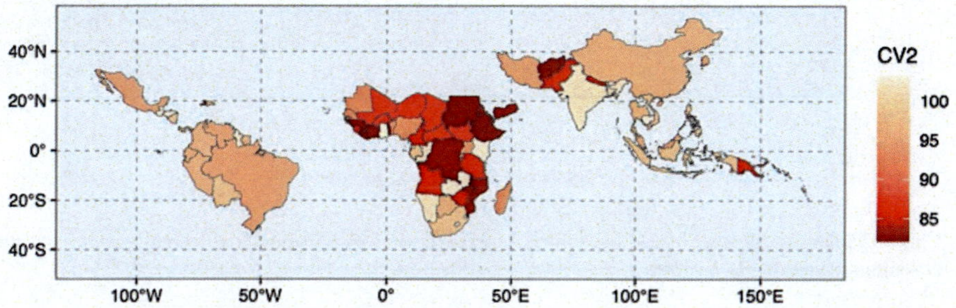

Fig. 7 Prediction of coverage ratio for Mosquirix at $t = 38$ for different country groups using lasvdGP model with $nn = 50$ and $n_0 = 30$ points

Since the training data size is too big to fit a full svdGP model on a standard laptop, we implement the localized model (i.e. lasvdGP model) developed by Zhang et al. (2018) for the model fitting. For a quick illustration, we predict the coverage ratio of the proposed Malaria vaccine *Mosquirix* for the first dose ($X_1 = 0$) given to a 6-month-old child ($X_2 = 6$), assuming the disease is not communicable ($X_5 = 0$) and the vaccine has been around since 1980 (the study period). We run the model with the average observed value of the incidence ($X_4 = 60$) and a conservative efficacy ($X_3 = 70$) as compared to other vaccines. We vary the value of X_7 for predicting the coverage ratio of Mosquirix at $t = 0$ and $t = 38$ for different country groups; see Figs. 6 and 7, respectively.

Note that the development of an accurate model for predicting the coverage ratio is beyond the scope of this chapter. Our main objective is to illustrate the usage of lasvdGP model in a complex real-life statistical problem. Although the overall pattern between Figs. 5, 6 and 7 shows positive association among HDI value and coverage ratio, more conclusive remarks require extensive modelling and analysis. One should also look at the dependence with respect to other predictor variables.

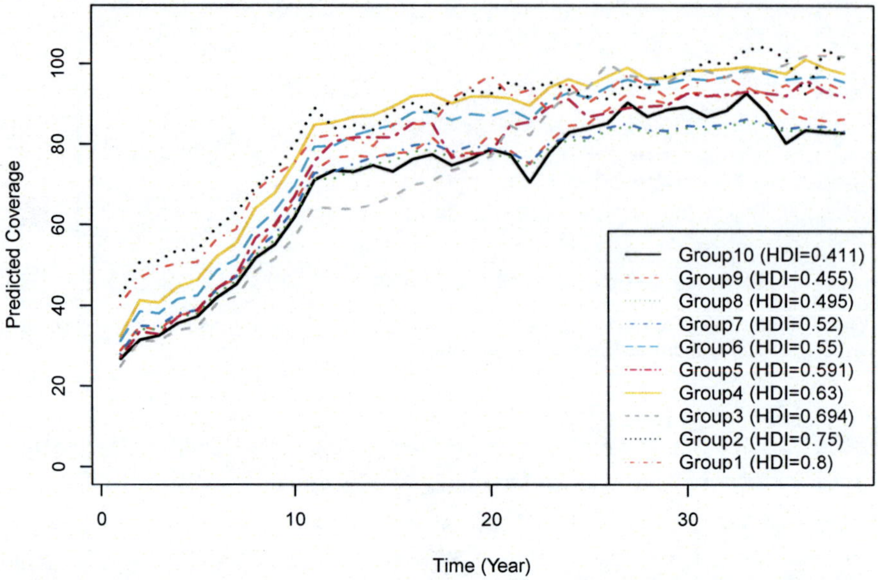

Fig. 8 Mean predictions of the Mosquirix coverage ratio over time for different country groups classified based on HDI values

Figure 8 shows the predicted coverage ratio over time—the direct output of lasvdGP model for different country groups.

Clearly, the coverage ratio increases to 100%. This is expected from this model, but an in-depth analysis is required for more meaningful inference.

4 Concluding Remarks

In this chapter, we talked about the popular Gaussian process models and its importance in computer-aided experiments for emulating real-world phenomena. We discussed various fundamental concepts that drive Gaussian process models, and the statistical interpretations and usages. These models, however, suffer from computational instability due to a variety of reasons, major ones being related to the near-singularity and the cost of inverting correlation matrices. Due to the computational overload, the process is expensive for numerous evaluations, which are needed for parameter estimation. Under the umbrella of big data, we present efficient localized GP models for emulating dynamic (time-series valued) computer simulators.

The concepts and R implementations are illustrated via several test functions. Finally, we presented an elaborate case study of how a new Malaria vaccine coverage can be predicted using the dynamic SVD-based GP model. Of course, this is just an illustration and not an attempt to accurately solve the case study. An elaborated second-level modelling and analysis is required to understand how and why the coverage ratios of Mosquirix would vary for different countries.

One could consider alternative approaches in predicting the coverage ratios, for example, clustering techniques to distribute the countries through their holistic characteristics instead of artificially binning into groups using HDI. One could also simply use a time-series modelling through AR, MA, ARIMA, etc., to predict coverage ratios.

Acknowledgements We would like to thank Pradeep Charan, an IPM student at IIM Indore, for brainstorming on the applications of dynamic GP and helping in collecting country-level data. We would also like to thank Aspect Ratio (http://aspectratio.in) for stimulating and inspiring this research. The authors also thank the editor and two reviewers for their helpful comments which led to significant revision of the chapter.

Appendix: R Codes

The following R code generates the prediction curves in Fig. 1 of Example 1. One can change the 'power' argument in `GP_fit` and 'predict' to fit GP model with different power exponential correlation structures.

```
#-------------------------------------------
n = 7; d = 1;
computer_simulator <- function(x) {
   y <-  log(x+0.1)+sin(5*pi*x)
   return(y)
}

set.seed(1)
library(lhs)
library(GPfit)

x = maximinLHS(n,d)
y = computer_simulator(x)

xpred = seq(0,1,length=100)
ytrue = computer_simulator(xpred)

GPmodel = GP_fit(x,y, corr = list(type="exponential", power=1.95))
pred=predict(GPmodel,xnew=xpred, corr = list(type="exponential", power=1.95))
yhat = pred$Y_hat
#-------------------------------------------
```

The following R code generates the prediction curves in Fig. 3 of Example 3. `ret$pmean[,i]` contains the predicted mean response for the ith input, and `ret$ps2[,i]` contains the corresponding mean square error estimates.

```
#-------------------------------------------
set.seed(1234568)
```

```
library("lhs")
library(DynamicGP)

forretal <- function(x,t,shift=1)
{
    par1 <- x[1]*6+4
    par2 <- x[2]*16+4
    par3 <- x[3]*6+1
    t <- t+shift
    y <- (par1*t-2)^2*sin(par2*t-par3)
}
timepoints <- seq(0,1,len=200)

train <- maximinLHS(20,3)
resp <- apply(train,1,forretal,timepoints)
test <- randomLHS(50,3)

ret <- svdGP(train,resp,test,nstarts=5)
#-------------------------------------------
```

For generating the predictions in Fig. 4 of Example 4, we only need to replace the last line of the previous code ('`ret <- svdGP(...)`') with the following code.

```
#-------------------------------------------
retl <- lasvdGP(atrain,resp,atrain,nn=30,n0=20,nstarts=5)
#-------------------------------------------
```

References

Bayarri, M., Berger, J., Cafeo, J., Garcia-Donato, G., Liu, F., Palomo, J., et al. (2007). Computer model validation with functional output. *The Annals of Statistics, 35*(5), 1874–1906.

Bayarri, M. J., Berger, J. O., Calder, E. S., Dalbey, K., Lunagomez, S., Patra, A. K., et al. (2009). Using statistical and computer models to quantify volcanic hazards. *Technometrics, 51*(4), 402–413.

Bower, R. G., Benson, A., Malbon, R., Helly, J., Frenk, C., Baugh, C., et al. (2006). Breaking the hierarchy of galaxy formation. *Monthly Notices of the Royal Astronomical Society, 370*(2), 645–655.

Centers for Disease Control. (2019). CDC - Parasites - Malaria. Retrieved August 26, 2019, from https://www.cdc.gov/parasites/malaria/index.html.

Conti, S., Gosling, J. P., Oakley, J. E., & O'Hagan, A. (2009). Gaussian process emulation of dynamic computer codes. *Biometrika, 96*(3), 663–676.

Emery, X. (2009). The kriging update equations and their application to the selection of neighboring data. *Computational Geosciences, 13*(3), 269–280.

Farah, M., Birrell, P., Conti, S., & Angelis, D. D. (2014). Bayesian emulation and calibration of a dynamic epidemic model for A/H1N1 influenza. *Journal of the American Statistical Association, 109*(508), 1398–1411.

Forrester, A., Sobester, A., & Keane, A. (2008). *Engineering design via surrogate modelling: A practical guide*. Wiley.

Gramacy, R. B., & Apley, D. W. (2015). Local Gaussian process approximation for large computer experiments. *Journal of Computational and Graphical Statistics, 24*(2), 561–578.

Greenberg, D. A. (1979). A numerical model investigation of tidal phenomena in the Bay of Fundy and Gulf of Maine. *Marine Geodesy, 2*(2), 161–187.

Greenwood, B., & Mutabingwa, T. (2002). Malaria in 2002.
Harshvardhan, M., & Ranjan, P. (2019). Statistical modelling and analysis of the computer-simulated datasets. In *Handbook of research on cloud computing and big data applications in IoT* (pp. 202–228). IGI Global.
Higdon, D., Gattiker, J., Williams, B., & Rightley, M. (2008). Computer model calibration using high-dimensional output. *Journal of the American Statistical Association, 103*(482), 570–583.
Hung, Y., Joseph, V. R., & Melkote, S. N. (2015). Analysis of computer experiments with functional response. *Technometrics, 57*(1), 35–44.
Kennedy, M. C., & O'Hagan, A. (2001). Bayesian calibration of computer models. *Journal of the Royal Statistical Society: Series B (Statistical Methodology), 63*(3), 425–464.
Liu, F., & West, M. (2009). A dynamic modelling strategy for Bayesian computer model emulation. *Bayesian Analysis, 4*(2), 393–411.
Luman, E. T., Barker, L. E., Shaw, K. M., McCauley, M. M., Buehler, J. W., & Pickering, L. K. (2005). Timeliness of childhood vaccinations in the united states: Days undervaccinated and number of vaccines delayed. *JAMA, 293*(10), 1204–1211.
MacDonald, B., Ranjan, P., & Chipman, H. (2015). GPfit: An R package for fitting a Gaussian process model to deterministic simulator outputs. *Journal of Statistical Software, 64*(12), 1–23.
McKay, M. D., Beckman, R. J., & Conover, W. J. (1979). A comparison of three methods for selecting values of input variables in the analysis of output from a computer code. *Technometrics, 42*(1), 55–61.
McLean, A. R. (1995). Vaccination, evolution and changes in the efficacy of vaccines: A theoretical framework. *Proceedings of the Royal Society of London. Series B: Biological Sciences, 261*(1362), 389–393.
Moorthy, V. S., Good, M. F., & Hill, A. V. (2004). Malaria vaccine developments. *The Lancet, 363*(9403), 150–156.
Rasmussen, C. E., & Williams, C. K. (2006). *Gaussian processes for machine learning* (Vol. 38, pp. 715–719). The MIT Press.
Sacks, J., Welch, W. J., Mitchell, T. J., & Wynn, H. P. (1989). Design and analysis of computer experiments. *Statistical Science, 4*(4), 409–423.
Santner, T. J., Williams, B. J., & Notz, W. I. (2003). Space-filling designs for computer experiments. In *The design and analysis of computer experiments* (pp. 121–161). Springer.
Stein, M. L. (2005). Space-time covariance functions. *Journal of the American Statistical Association, 100*(469), 310–321.
World Health Organisation. (2019). Malaria. Retrieved August 26, 2019, from https://www.who.int/news-room/fact-sheets/detail/malaria.
Zhang, R., Lin, C. D., & Ranjan, P. (2018). Local Gaussian process model for large-scale dynamic computer experiments. *Journal of Computational and Graphical Statistics, 27*(4), 1–10.
Zhang, R., Lin, C. D., & Ranjan, P. (2020). DynamicGP: Modelling and analysis of dynamic computer experiments. R Package Version 1.1-6.

Grey Relational Analysis for the Selection of Potential Isolates of *Alternaria Alternata* of Poplar

Kartik Uniyal, Girish Chandra, R. U. Khan, and Y. P. Singh

Abstract Mapping the variability of the pathogenic population and selection of potential pathogenic isolates remains a challenge. The chapter deals with the grey relational analysis (GRA), a method for multiple attribute decision-making situations, to select the potent isolates of *Alternaria alternata* causing leaf spot on poplar. Earlier, three methods, viz. rough gauging, equal class interval and unequal class interval, were attempted. Poplar is an important tree in the agri-silvicultural system. The year-round availability of the poplar in field encourages the survival and persistence of the pathogen with high parasitic fitness. A survey was conducted in poplar fields at different geographical locations. Altogether, 72 isolates of *A. alternata* were collected from four commercial poplar clones. Fifteen *A. alternata* isolates were selected based on growth rate, sporulation and spore size (maximum length and breadth). All the-larger-the-better attributes were the choice for isolate selection in GRA. For choosing top fifteen isolates, equal and variable weights to all the four attributes were used. Distinguishing coefficients of 0.1–0.9 with step-up size 0.2 were taken for the analysis. GRA normalizes the measured values of all attributes making the isolates comparable. Moreover, the output of isolates was not much influenced by the variation in distinguishing coefficient. It was observed that changing weights influenced the selection of isolates substantially and growth played a significant role in selection of the isolates followed by the sporulation.

Keywords Distinguishing coefficient · Forest pathogen · Grey relational grade · Grey relational generating

K. Uniyal (✉) · Y. P. Singh
Forest Pathology Division, Forest Research Institute, Dehradun 240006, India

G. Chandra
Division of Forest Statistics, Indian Council of Forestry Research and Education, Dehradun 248006, India

R. U. Khan
Faculty of Agricultural Sciences, Department of Plant Protection, Aligarh Muslim University, Aligarh 202002, India

D. D. Hanagal et al. (eds.), *Applied Statistical Methods*, Springer Proceedings in Mathematics & Statistics 380, https://doi.org/10.1007/978-981-16-7932-2_8

1 Introduction

The idea of grey relational theory was initially proposed by Deng (1982) with the aim to cope with the uncertainty of a system (so-called grey system). The grey system provides multidisciplinary approaches for analysis and abstract modelling of systems for which the information is limited, incomplete and characterized by random uncertainty (Sifen & Forrest, 2007). GRA, a part of grey theory, is one of the methods solving the problems on complicated interrelationships between multiple alternatives and attributes. GRA solves multi-attribute decision-making problems by combining the entire range of performance attribute values being considered for every alternative into one single value. This reduces the original problem to a single-objective decision-making problem. It uses information from the grey system to dynamically compare each factor quantitatively, based on the level of similarity and variability among factors to establish their relation.

GRA has been successfully applied on cluster and prediction analysis, robot path planning, project selection, performance and factor effect evaluation and multiple criteria decision (Chang & Yeh, 2005; Lu & Yeh, 2002; Yeh & Lu, 2000). GRA has been used in various fields, for example, the restoration planning for power distribution systems (Chen, 2005), the detection of silicon wafer slicing defects (Lin et al., 2006), financial performance of venture capital (Kung & Wen, 2007), in agriculture (Xie et al., 2009), determining the optimum process parameters for open-end spinning yarn (Hasani et al., 2012), etc.

In forestry research, the use of GRA in assessing and optimizing small biomass boilers has been attempted by Morán et al. (2006). Optimization of drilling parameters for medium-density fibreboard (MDF) panel with multiple performance characteristics using GRA was carried out by Prakash et al. (2015). The best possible combination of various parameters of self-healing glass fibre reinforced plastic to optimize the mechanical properties was obtained by Mercy et al. (2017). GRA method was used to individually study grey correlations among six factors (rainfall duration and intensity, antecedent soil moisture, vegetation cover and type, and slope gradient) and soil cumulative infiltration by Juan et al. (2013). The relational degree between Yanbian's forestry industry and three forestry industries was systematically analysed by Xinning and Yufen (2017). Recently, Bisht et al. (2020) performed GRA for prioritization of the essential oil samples of *Santalum album* L. using the two important attributes α- and β-santalols. Saha et al. (2020, 2021) have successfully used GRA in creating a comparability sequence to obtain the rank of the various root quality parameters of eucalyptus clones at Forest Research Institute, Dehradun. Other possible areas of GRA in forestry and environmental science may be chosen from Chandra et al. (2020).

Poplar occupies a unique and important position in rural economy of India. There are 35 species of poplar currently recognized in the world. *Populus deltoides* is the most popular exotic species in India. Poplar suffers from several diseases as they are being raised as single clone monocultures and are, thus, prone to disease outbreaks (Singh et al., 2012). While the potential for rapid growth exists, poplar productivity is frequently reduced by diseases (Broderick et al., 2010; Widin & Schipper, 1981). Diseases from nursery to plantations have been reported and studied extensively by various workers (Singh et al., 2012). For development of an effective breeding programme for disease resistance, a comprehensive understanding about morphological, physiological and pathogenic variations of causal organism is essential. In this background, the variation in the population of a potent pathogen of poplar *A. alternata* was assessed.

Seventy-two isolates of the pathogen were collected from four commercial clones, viz. G48, Udai, WSL22 and WSL39, from three distinct geographical locations of poplar culture. Four attributes, i.e. growth rate, sporulation and spore size (length maximum (L_M) and breadth maximum (B_M)), of the isolates were studied. In forest pathology, one confronts for the selection of potent isolates from the collected population as performance of each isolate depends upon multiple attributes. Therefore, mapping the variability of the pathogenic population and selection of potential isolate remains a challenge. It was attempted and resolved through a variety of statistical methods (Uniyal et al., 2018). Further, in this chapter, GRA was used for the same purpose.

2 Material and Methods

2.1 Survey and Collection of Alternaria Isolates

The surveys for collection of fungal isolates were conducted at three nurseries of Wimco Seedlings Private Limited, India, during the period 2007 to 2011 as detailed in Table 1. Leaves infected with *A. alternata* like symptoms were collected from each clone. In all, 72 isolates were obtained, and their morphological attributes were studied.

Table 1 Details of sampled nurseries and *P. deltoides* clones

S. no	Name of the nursery	Geographical location	Clone of *P. deltoides*
1	Bagwala, Udham Singh Nagar, Uttarakhand	29° 30′ N and 79° 28′ E	G48, Udai, WSL22 and WSL39
2	Maheshwari and Paniyala, Roorkee, Uttarakhand	29° 52′ N and 77° 53′ E	G48, Udai, WSL22 and WSL39
3	Thana Chappar, Yamuna Nagar, Haryana	30° 7′ N and 77° 18′ E	G48, WSL22 and WSL39

2.2 GRA for Selection of Potent Isolates

Suppose N number (72 in the present case) of isolates of *A. alternata* were collected during the survey from the study area. These isolates depend on p growth attributes (four, viz. growth rate, sporulation, L_M and B_M which were measured on the ratio scale). Based on isolate attributes, the value of p recorded varied and random. The objective of the study was to select n number of isolates (15) out of the population having higher combine values of the attributes. The flow diagram of GRA procedure is presented in Fig. 1.

The essential four steps in obtaining top n isolates through GRA method are explained as under:

Step 1. Grey Relational Generating: This step, also called the data processing to supplement information, is aimed to process complicate and tedious attributes at one place for a bird's eye view. When the units in which performance is measured are different for different attributes, for example, growth in cm and spore size in micrometre (μm), the influence of some attributes may be underestimated. It may

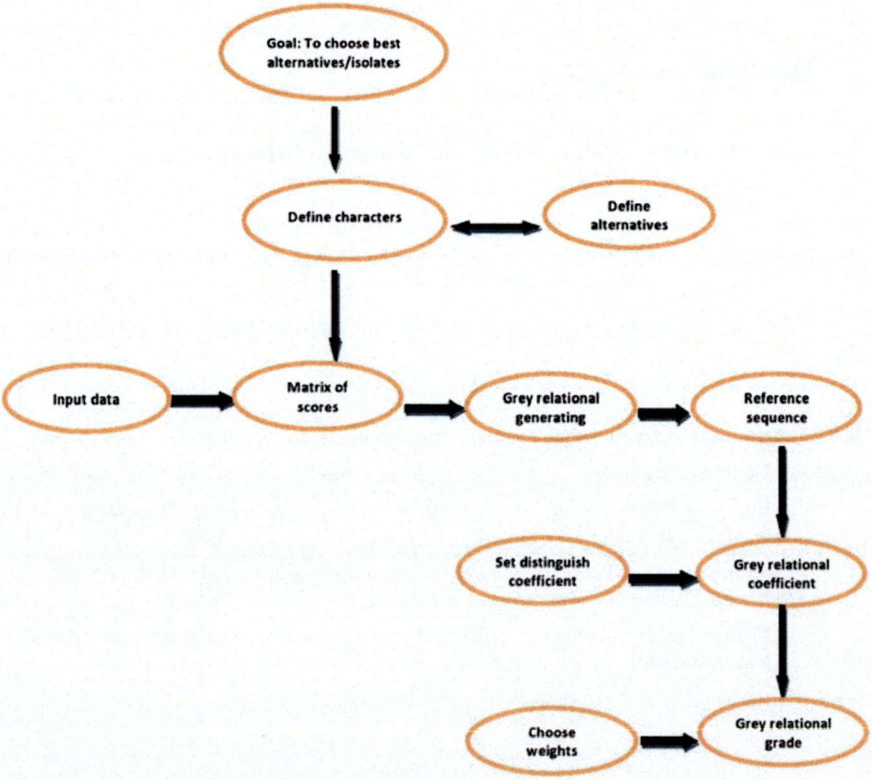

Fig. 1 Flow diagram showing the steps of GRA

further happen if performance of some attributes has a very large range, for example, sporulation in the present case (133.33–1,181.67), it may neglect the influence of those attributes. Also, in the sequence, if the attributes' goals and directions are different, the other relational analysis might also produce incorrect results. Therefore, preprocessing of the data is necessary. It could be done by transformation of all performance values for each isolate into a comparability sequence, a process analogous to normalization. This process is called grey relational generating in GRA.

From the calculation point of view, *i*th isolate, $i = 1, 2, \ldots, 72$ may be denoted as

$Y_i = (y_{i1}, y_{i2}, y_{i3}, y_{i4})$, where y_{ij} is the performance value of *j*th attribute of *i*th isolate. Now, Y_i can be transformed into the comparability sequence, $X_i = (x_{i1}, x_{i2}, x_{i3}, x_{i4})$ using Eq. (2.1).

$$x_{ij} = \frac{y_{ij} - Min\{y_{ij}, i = 1, 2, \ldots, 72\}}{Max\{y_{ij}, i = 1, 2, \ldots, 72\} - Min\{y_{ij}, i = 1, 2, \ldots, 72,\}} \quad i = 1, 2, \ldots, 72;\ j = 1, 2, 3, 4 \tag{2.1}$$

It should be noted that the above equation is used when the-larger-the-better attributes are the choice for selection. The present case pertains to this category.

Step 2. Defining Reference Sequence (Ideal Target Sequence): In this step, the values of x_{ij}'s are scaled in [0, 1]. The scaled value of a particular attribute which is closer to 1 shall be the better one for that attribute. Therefore, an isolate will be the best choice if all its performance values are closest to or equal to 1. However, this kind of isolates, generally, does not exist in nature. In such situation, a reference sequence may be defined for the best isolate as $X_0 = \{x_{01}, x_{02}, x_{03}, x_{04}\} = \{1, 1, 1, 1\}$. The aim shall be to find the isolates whose comparability sequence is the closest to the reference sequence.

Step 3. Determination of Grey Relational Coefficient: Determination of the grey relational coefficient between all comparability sequences and the reference sequence is needed to see the closeness of x_{ij} with x_{0j}, $j = 1, 2, 3, 4$. The larger the grey relational coefficient means there is more closeness of x_{ij} with x_{0j}. The grey relational coefficient, $\delta(x_{ij}, x_{0j})$ can be determined as:

$$\delta(x_{ij}, x_{0j}) = \frac{\Delta_{\min} + d\Delta_{\min}}{\Delta_{ij} + d\Delta_{\max}},\ i = 1, 2, \ldots, 72;\ j = 1, 2, 3, 4 \tag{2.2}$$

where $\Delta_{ij} = |x_{ij} - x_{0j}|$, $\Delta_{\min} = Min(\Delta_{ij})$, $\Delta_{\max} = Max(\Delta_{ij})$ and d is a distinguishing coefficient with $d \in [0, 1]$.

The purpose of the distinguishing coefficient is to expand or compress the range of the grey relational coefficient. When the values of x_{ij}'s are very small, the value of distinguish coefficient may be adjusted nearer to 1 or vice versa. The different distinguishing coefficients shall produce different values of grey relational grades (in Step 4) with change in the difference of values between the graded but usually the output, i.e. the rank of isolates shall remain same. In this article, the results are based upon the distinguishing coefficient $d = 0.5$, while other values of d were also tested to see the effect on the grey relational grades.

Step 4. Calculation of Grey Relational Grades: Finally, based on the grey relational coefficients, the grey relational grade, $\Gamma(X_0, X_i)$, between the reference sequence (X_0) and every comparability sequence (X_i, $i = 1, 2, \ldots, 72$) was calculated. If a comparability sequence translated from an isolate has the highest grey relational grade between the reference sequence and itself, that isolate will be the best choice. The grey relational grade also indicates the degree of similarity between the comparability and the reference sequence (Fung, 2003). It is calculated as:

$$\Gamma(X_0, X_i) = \sum_{j=1}^{4} w_j \delta(x_{ij}, x_{0j}),\ i = 1, 2, \ldots, 72 \quad (2.3)$$

where w_j denotes the weight assigned to the jth attribute by the decision maker with $\sum_{j=1}^{4} w_j = 1$. The weights are totally depending on researcher's perception about the importance of the attribute.

3 Results

3.1 *Growth Attributes of* A. Alternata

The fungal attribute values of *A. alternata* are given in Table 2 which has been used for the computation of grey relational generating.

3.2 *Grey Relational Generating, Coefficients and Grades*

The calculated values of grey relational generating were obtained by Eq. (2.1). For example, in case of growth attribute, the maximum and minimum values were 7.00 and 3.82 cm, respectively. The transformed value using Eq. (2.1) for isolate A1 is $x_{11} = \frac{5.92-3.82}{7.00-3.82} = 0.66$ and so on.

The grey relational coefficients $\delta(x_{ij}, x_{0j})$ with $d = 0.5$ were calculated using Eq. (2.2) and given in Table 3.

The values of grey relational grades, $\Gamma(X_0, X_i)$, $i = 1, 2, \ldots, 72$ for equal (0.25) and unequal values of w_j, $j = 1, 2, 3, 4$, with $d = 0.5$ are shown in Table 4. In unequal case, based upon personal preference, weight of 0.5 was assigned to growth, 0.3 to sporulation and equal weight of 0.1 to both L_M and B_M.

It was observed that barring four isolates, viz. isolates A19, A45, A63 and A67, all the isolates were common in both the situations. In comparison, the isolates A63 and A67 were not found in the unequal case, while A19 and A45 were absent in

Table 2 Values of the growth attributes of *A. alternata* isolates

Isolate no	Growth (cm)	Sporulation (no. × 10^4/ml)	L_M (μm)	B_M (μm)	Isolate no	Growth (cm)	Sporulation (no. × 10^4/ml)	L_M (μm)	B_M (μm)
A1	5.92	643.33	32.83	15.43	A37	6.02	598.33	25.30	10.47
A2	5.85	520.00	36.83	14.63	A38	5.15	878.33	29.30	12.50
A3	5.97	693.33	31.57	15.90	A39	4.43	1,101.67	27.13	12.80
A4	6.02	761.67	34.40	12.00	A40	7.00	981.67	28.20	13.57
A5	5.97	1,020.00	27.93	13.23	A41	7.00	796.67	27.90	14.03
A6	5.50	741.67	30.87	15.27	A42	5.65	500.00	32.47	19.80
A7	6.52	1,051.67	34.93	15.10	A43	6.50	645.00	26.67	13.70
A8	4.92	746.67	32.07	15.40	A44	4.48	765.00	29.20	13.30
A9	5.85	1,075.00	25.73	13.03	A45	6.97	648.33	28.97	12.17
A10	5.73	1,033.33	31.40	17.20	A46	5.45	520.00	33.43	12.00
A11	5.03	1,180.00	29.50	14.67	A47	7.00	983.33	36.27	16.60
A12	6.73	1,181.67	36.20	14.37	A48	6.05	331.67	29.70	13.27
A13	6.75	1,173.33	31.57	15.03	A49	6.15	498.33	33.47	14.13
A14	5.63	880.00	37.40	13.77	A50	5.98	553.33	31.00	13.03
A15	6.57	1,026.67	40.30	15.63	A51	6.45	646.67	34.77	12.77
A16	6.60	1,161.67	28.07	14.97	A52	6.65	653.33	37.60	12.33
A17	6.07	776.67	22.20	13.73	A53	4.02	226.67	44.30	15.77
A18	6.10	695.00	33.40	13.17	A54	5.02	148.33	36.67	16.90
A19	6.77	941.67	29.60	12.13	A55	6.55	418.33	26.57	13.50
A20	6.45	741.67	22.03	11.63	A56	4.47	133.33	41.23	18.20
A21	6.43	646.67	32.07	17.23	A57	4.48	175.00	36.10	17.43

(continued)

Table 2 (continued)

Isolate no	Growth (cm)	Sporulation (no. × 10^4/ml)	L_M (μm)	B_M (μm)	Isolate no	Growth (cm)	Sporulation (no. × 10^4/ml)	L_M (μm)	B_M (μm)
A22	5.93	531.67	32.17	10.97	A58	3.82	578.33	28.10	14.00
A23	5.88	638.33	30.70	14.43	A59	5.08	706.67	36.50	12.87
A24	6.67	916.67	33.17	11.83	A60	5.53	410.00	34.40	11.77
A25	7.00	956.67	30.17	13.10	A61	5.05	938.33	29.43	12.37
A26	6.62	763.33	36.50	14.60	A62	5.05	870.00	35.10	13.97
A27	5.45	600.00	28.57	12.70	A63	5.58	903.33	48.80	14.30
A28	6.32	496.67	31.40	11.87	A64	6.98	1,043.33	35.83	16.77
A29	6.57	575.00	31.07	11.97	A65	7.00	1,108.33	38.73	13.27
A30	6.07	441.67	24.60	12.03	A66	4.80	618.33	41.57	16.03
A31	4.15	490.00	40.60	13.93	A67	5.02	871.67	53.77	14.97
A32	7.00	895.00	32.00	13.17	A68	5.07	865.00	36.73	13.00
A33	4.47	588.33	33.00	14.33	A69	4.75	670.00	38.23	13.97
A34	6.28	351.67	31.50	14.37	A70	6.43	975.00	26.03	12.23
A35	6.35	615.00	35.53	12.03	A71	6.48	926.67	45.57	14.73
A36	6.38	551.67	30.03	15.07	A72	6.57	823.33	28.27	13.37

Table 3 Grey relational coefficients for growth attributes of *A. alternata* isolates ($d = 0.5$)

Isolate no	Growth	Sporulation	L_M	B_M	Isolate no	Growth	Sporulation	L_M	B_M
A1	0.60	0.49	0.43	0.52	A37	0.62	0.47	0.36	0.33
A2	0.58	0.44	0.48	0.47	A38	0.46	0.63	0.39	0.39
A3	0.61	0.52	0.42	0.54	A39	0.38	0.87	0.37	0.40
A4	0.62	0.56	0.45	0.37	A40	1.00	0.72	0.38	0.43
A5	0.61	0.76	0.38	0.42	A41	1.00	0.58	0.38	0.45
A6	0.51	0.54	0.41	0.51	A42	0.54	0.43	0.43	1.00
A7	0.77	0.80	0.46	0.50	A43	0.76	0.49	0.37	0.43
A8	0.43	0.55	0.42	0.51	A44	0.39	0.56	0.39	0.42
A9	0.58	0.83	0.36	0.41	A45	0.98	0.50	0.39	0.38
A10	0.56	0.78	0.41	0.64	A46	0.51	0.44	0.44	0.37
A11	0.45	1.00	0.40	0.48	A47	1.00	0.73	0.48	0.59
A12	0.86	1.00	0.47	0.46	A48	0.63	0.38	0.40	0.42
A13	0.86	0.98	0.42	0.49	A49	0.65	0.43	0.44	0.45
A14	0.54	0.63	0.49	0.44	A50	0.61	0.45	0.41	0.41
A15	0.79	0.77	0.54	0.53	A51	0.74	0.49	0.46	0.40
A16	0.80	0.96	0.38	0.49	A52	0.82	0.50	0.50	0.38
A17	0.63	0.56	0.33	0.43	A53	0.35	0.35	0.63	0.54
A18	0.64	0.52	0.44	0.41	A54	0.45	0.34	0.48	0.62
A19	0.87	0.69	0.40	0.38	A55	0.78	0.41	0.37	0.43
A20	0.74	0.54	0.33	0.36	A56	0.39	0.33	0.56	0.74
A21	0.74	0.49	0.42	0.65	A57	0.39	0.34	0.47	0.66
A22	0.60	0.45	0.42	0.35	A58	0.33	0.46	0.38	0.45
A23	0.59	0.49	0.41	0.47	A59	0.45	0.52	0.48	0.40
A24	0.83	0.66	0.44	0.37	A60	0.52	0.40	0.45	0.37
A25	1.00	0.70	0.40	0.41	A61	0.45	0.68	0.39	0.39
A26	0.81	0.56	0.48	0.47	A62	0.45	0.63	0.46	0.44
A27	0.51	0.47	0.39	0.40	A63	0.53	0.65	0.76	0.46
A28	0.70	0.43	0.41	0.37	A64	0.99	0.79	0.47	0.61
A29	0.79	0.46	0.41	0.37	A65	1.00	0.88	0.51	0.42
A30	0.63	0.41	0.35	0.38	A66	0.42	0.48	0.57	0.55
A31	0.36	0.43	0.55	0.44	A67	0.45	0.63	1.00	0.49
A32	1.00	0.65	0.42	0.41	A68	0.45	0.62	0.48	0.41
A33	0.39	0.47	0.43	0.46	A69	0.41	0.51	0.51	0.44
A34	0.69	0.39	0.42	0.46	A70	0.74	0.72	0.36	0.38
A35	0.71	0.48	0.47	0.38	A71	0.75	0.67	0.66	0.48
A36	0.72	0.45	0.40	0.50	A72	0.79	0.59	0.38	0.42

Table 4 Grey relational grades and ranking of selected *A. alternata* isolates at $d = 0.5$ for equal and unequal weights

Rank	$w_1 = w_2 = w_3 = w_4 = 0.25$		$w_1 = 0.5, w_2 = 0.3, w_3 = w_4 = 0.1$	
	Isolate no	GRG	Isolate no	GRG
1	A64	0.7031	A65	0.8562
2	A65	0.7019	A64	0.8397
3	A47	0.6986	A47	0.8245
4	A12	0.6983	A12	0.8219
5	A13	0.6900	A13	0.8186
6	A16	0.6588	A40	0.7983
7	A15	0.6567	A25	0.7912
8	A71	0.6416	A32	0.7774
9	A67	0.6412	A16	0.7758
10	A40	0.6337	A41	0.7557
11	A7	0.6310	A15	0.7315
12	A25	0.6281	A7	0.7195
13	A32	0.6203	A19	0.7193
14	A41	0.6010	A45	0.7154
15	A63	0.6007	A71	0.6932

equal weight situation. Subjectivity of the weight selection of the attributes by the individual may change the ranking of the individual isolate. For example, A64 and A65 had highest ranking of 1 and 2 that mutually changed depending upon the weight selection.

The relation between distinguishing coefficient and grey relational coefficient for first three isolates is given in Fig. 2. It showed that as distinguishing coefficient

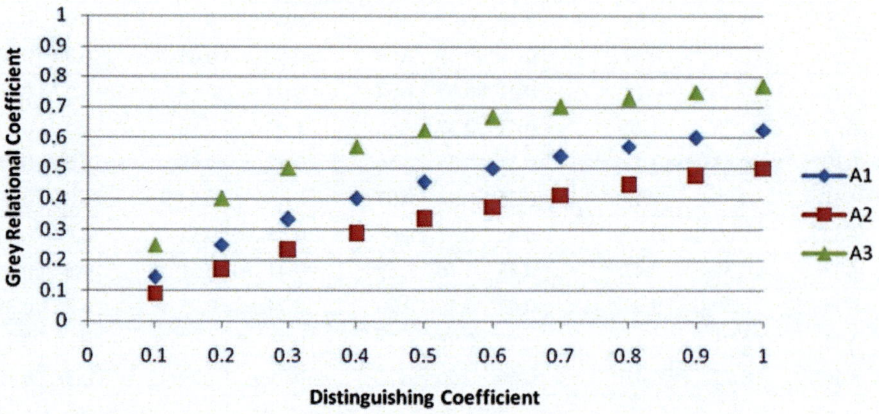

Fig. 2 Relationship between distinguishing coefficient and grey relational coefficient

increased, the value of grey relational coefficients also increased for all three isolates, therefore, to the final grey relational grades were not much affected.

To understand the effect of distinguish coefficient (d) and different weight assignments, five different values of $d = 0.1$ to 0.9 with step-up size 0.2 were taken for comparison. Table 5 gives the effect for the case of both equal and unequal weight selections. For the case of equal weight, with reference to $d = 0.5$, the selection of isolates was not influenced by the variation in d in top five isolates. In selecting total top 15 isolates, 11 isolates were in common for each values of d. With reference to $d = 0.5$, top 3 isolates did not lose their position with changing the value of d at unequal weights. Thereafter, d (particularly small d) influences the ranking of the selected isolates. For the value of $d \geq 0.5$, all the 15 selected isolates were common. With decreasing the value of $d = 0.5$, number of common rank isolates slightly decreased. In all, 11 isolates were found common.

Now, the critical contributing attribute with respect to the allocation of weight was worked out.

3.3 *Performance Evaluation of Selected Isolates*

The performance of highest rank isolate, i.e. A64 for $d = 0.5$ at equal weight, was initially checked. There was a total of ${}^4C_2 = 6$ pair combinations of the four attributes (Table 6). With each pair, the weight of first attributed was changed either with increasing or decreasing value; subsequently, the weight of other attribute automatically adjusted with the rule that the sum of weights of all four attributes remained one. It is to be noted that the weight of other two attributes which were not the part of the pair was fixed with the weights 0.25. Performance in terms of ranks was checked for two decimal places of weights. For example, the first pair $G \times S$ of Table 6, the weights of two attributes L_M and B_M were fixed at 0.25 for both the case of increasing and decreasing weights of G. In the case of increasing weights of G, there was no change in the performance of the isolate. While in the case of decreasing weights of G, the performance was decreasing. In all the combinations, where either G or S was first partner, if the weight gets increased of the first partner, no change was registered in rank of the isolate. However, the performance gets decreased when the weight of first partner decreased. In the pair of $L_M \times B_M$, the trend remained opposite. When the weight of L_M increased, the performance was decreased; however, no changes in the performance when the weight of L_M decreased.

To understand the contribution of a combination of attributes in the performance/ranking of an isolate in a population, the values of sum of squares of the residuals (SSR) for each pair of attributes were calculated using the formula.

$SSR = \sum_c (R_c - R)^2$, where R_c and R denotes the values of changed rank and the original rank (in our case $R = 1$).

In the present case, the combination $G \times L_M$ had the highest SSR among all six possible combinations reflecting the importance of G and L_M in the performance of an isolate or its ranking (Table 7).

Table 5 Comparison of ranking of selected *A. alternata* isolates with $d = 0.1(0.2)0.9$ of equal and unequal weights

Isolate no	Ranking at equal weight					Isolate no	Ranking at unequal weight				
	$d = 0.1$	$d = 0.3$	$d = 0.5$	$d = 0.7$	$d = 0.9$		$d = 0.1$	$d = 0.3$	$d = 0.5$	$d = 0.7$	$d = 0.9$
A64	4	1	1	1	1	A65	1	1	1	1	1
A65	1	2	2	3	3	A64	2	2	2	2	2
A47	5	5	3	2	2	A47	3	3	3	3	3
A12	2	3	4	4	4	A12	7	4	4	4	4
A13	3	4	5	5	5	A13	9	5	5	5	5
A16	8	6	6	7	8	A40	4	6	6	6	6
A15	15	9	7	6	6	A25	5	7	7	7	7
A71	17	12	8	8	7	A32	6	8	8	9	9
A67	9	7	9	9	10	A16	11	9	9	8	8
A40	6	8	10	11	11	A41	8	10	10	10	11
A7	16	13	11	10	9	A15	14	13	11	11	10
A25	7	10	12	12	12	A7	15	14	12	12	12
A32	10	11	13	13	14	A19	12	12	13	13	13
A41	11	14	14	17	17	A45	10	11	14	14	15
A63	21	17	15	14	13	A71	17	16	15	15	14

Table 6 Effect of weight on ranks of the *A. alternata* isolate A64 with all attribute combinations

Combination of attributes	Type	Weight				Rank
		Growth (G)	Sporulation (S)	L_M	B_M	
$G \times S$	Increasing G	No change				
	Decreasing G	0.18	0.32	0.25	0.25	2
		0.17	0.33	0.25	0.25	4
		0.12	0.38	0.25	0.25	5
$G \times L_M$	Increasing G	No change				
	Decreasing G	0.22	0.25	0.28	0.25	3
		0.21	0.25	0.29	0.25	4
		0.2	0.25	0.3	0.25	5
		0.14	0.25	0.36	0.25	6
		0.13	0.25	0.37	0.25	7
		0.11	0.25	0.39	0.25	8
		0.1	0.25	0.4	0.25	9
		0.09	0.25	0.41	0.25	11
		0.08	0.25	0.42	0.25	12
		0.06	0.25	0.44	0.25	13
		0.03	0.25	0.47	0.25	16
		0.02	0.25	0.48	0.25	17
$G \times B_M$	Increasing G	No change				
	Decreasing G	0.21	0.25	0.25	0.29	2
		0.2	0.25	0.25	0.3	4
		0.18	0.25	0.25	0.32	5
		0.1	0.25	0.25	0.4	7
		0.06	0.25	0.25	0.44	9
		0.04	0.25	0.25	0.46	10
		0.03	0.25	0.25	0.47	11
		0.02	0.25	0.25	0.48	12
$S \times L_M$	Increasing S	No change				
	Decreasing S	0.25	0.21	0.29	0.25	2
		0.25	0.2	0.3	0.25	4
		0.25	0.17	0.33	0.25	5
		0.25	0.07	0.43	0.25	7
		0.25	0.02	0.48	0.25	9
$S \times B_M$	Increasing S	No change				
	Decreasing S	0.25	0.18	0.25	0.32	2
		0.25	0.16	0.25	0.34	4

(continued)

Table 6 (continued)

Combination of attributes	Type	Weight				Rank
		Growth (G)	Sporulation (S)	L_M	B_M	
		0.25	0.12	0.25	0.38	5
$L_M \times B_M$	Increasing L_M	0.25	0.25	0.34	0.16	2
		0.25	0.25	0.37	0.13	4
		0.25	0.25	0.43	0.07	5
	Decreasing L_M	No change				

Table 7 Values of SSR for each pair of attributes under study

Type	$G \times S$	$G \times L_M$	$G \times B_M$	$S \times L_M$	$S \times B_M$	$L_M \times B_M$
SSR	26	1,049	428	126	26	26

Table 8 Marginal SSR value of all four attributes

Attribute	G	S	L_M	B_M
SSR	1,503	178	1,201	480

Further, the performance or ranking was resolved at individual attribute level. Table 8 gives the marginal values of SSR for each attribute in a pair. The marginal SSR of an attribute was obtained by summing all SSRs in which that particular attribute was present. For example, the marginal SSR of G was obtained by summing the SSR values from all those pair where G was a partner in Table 7. From Table 8, it is observed that the G (growth) had the maximum effect on the weight followed by S, L_M and B_M.

4 Discussion and Conclusion

An attempt has been made to select the most potent isolate from a diverse population of *A. alternata* with the help of grey relational analysis. The characteristics of GRA method lie in making two incomparable attributes (e.g. growth in measurable unit and sporulation in absolute number) comparable and normalizing hugely different values of an attribute. To understand this process, the normal curves of both values, original and transformed (grey relational generating), were drawn and with the help of Kolmogorov statistic, and their p-values were calculated for comparison. It was interesting to note that no change was observed in the shape of the normal curve and p-values. Moreover, for each growth attribute, grey relational generating had same order as that of its original values. Though in all attributes the generating had values between 0 and 1, that range did not provide any basis for the comparison among attributes due to their different unit of measurements and inherent variation within attribute. The basic purpose of the normalization step seems to be (i) transform the original values of the attributes in the range 0 to 1 with 1 as the reference value for the case of larger-the-better and (ii) make the attributes independent from their unit of measurements by converting them into a range from 0 to 1.

The distance between reference and normalized value is not just the absolute difference between them. This distance is obtained through grey relational coefficient comprising of distinguishing coefficient, the absolute difference and its minimum and maximum values. The grey relational coefficient is more precise distance as it includes the variations within normalized values and, therefore, a two-step normalization process. It was also observed that the absolute difference and grey relational coefficients are not always found in the same order, in turn affecting grey relational grades.

GRA makes a multiple attribute problem into a single-attribute decision-making solution as evident with the conclusion that growth out of four attributes, in the present case, influences the choice of an isolate most. It was achieved by calculating sum of squares of residuals which enabled to pinpoint the contribution of an attribute. Besides, sporulation or the number of spores played a role, second to growth, in isolate selection. If one observes the infection process of a fungal pathogen, these two biological attributes, that is, growth and sporulation, represent the crucial input in terms of pathogen proliferation and spread. This linking of biological fact with the mathematical interpretation gives a real meaning to the selection of a potent isolate. Thus, the present method paves the way not only to understand the population structure but also to mark the most important attribute for selection. This method may be adopted for further investigations in forestry pertaining to population studies.

References

Bisht, S. S., Chandra, G., Ravindra, M., & Kumar, R. (2020). Evaluation of chemical and physical properties of commercial sandalwood essential oils and their comparison with essential oil extracted from *Santalum album* L. *Journal of Essential Oil Bearing Plants, 23*(2), 345–355.

Broderick, N. A., Vasquez, E., Handelsman, J., & Raffa, K. F. (2010). Effect of clonal variations among hybrid poplars on susceptibility of gypsy moth (Lepidoptera: Lymantriidae) to *Bacillus thuringiensis* sub sp. kurstaki. *Journal of Economic Entomology, 103*(3), 718–725.

Chandra, G., Nautiyal, R., & Chandra, H. (Eds.). (2020). *Statistical methods and applications in forestry and environmental sciences*. Springer.

Chang, K. C., & Yeh, M.-F. (2005). Grey-relational analysis based approach for data clustering. *IEEE Proceedings – Vision, Image and Signal Processing, 152*(2), 165–172.

Chen, W. H. (2005). Distribution system restoration using the hybrid fuzzy-grey method. *IEEE Transactions on Power Systems, 20*, 199–205.

Deng, J. (1982). Control problems of grey systems. *Systems & Control Letters, 1*, 288–294.

Fung, C. P. (2003). Manufacturing process optimization for wear property of fibre-reinforced polybutylene terephthalate composites with grey relational analysis. *Wear, 254*, 298–306.

Hasani, H., Tabatabaei, S. A., & Amiri, G. (2012). Grey relational analysis to determine the optimum process parameters for open-end spinning yarns. *Journal of Engineered Fibers and Fabrics, 7*(2), 81–86.

Juan, W., Pute, W., & Xining, Z. (2013). Soil infiltration based on bp neural network and grey relational analysis. *Revista Brasileria de Ciência do Solo, 37*(1), 97–105.

Kung, C. F., & Wen, K. L. (2007). Applying GRA and grey decision making to evaluate the relationship between company attributes and its financial performance: A case study of ventures capital in Taiwan. *Decision Support Systems, 43*, 842–852.

Lin, C. T., Chang, C. W., & Chen, C. B. (2006). The worst ill-conditioned silicon wafer machine detected by using grey relational analysis. *International Journal of Advanced Manufacturing Technology, 31*, 388–395.

Lu, H. C., & Yeh, M. F. (2002). Robot path planning based on modified grey analysis. *Cybernetics and Systems, 33*(2), 129–159.

Mercy, J. L., Prakash, S., Krishnamoorthy, A., Ramesh, S., & Anand, D. A. (2017). Multiresponse optimization of mechanical properties in self-healing glass fiber reinforced plastic using grey relational analysis. *Measure, 110*, 344–355.

Morán, J., Granada, E., Míguez, J. L., & Porteriro, J. (2006). Use of grey relational analysis to assess and optimize small biomass boilers. *Fuel Processing Technology, 87*(2), 123–127.

Prakash, S., Mercy, J. L., Salugu, M. K., & Vineeth, K. S. (2015). Optimization of drilling characteristics using grey relational analysis (GRA) in medium density fiber board (MDF). *Materials Today, 2*(4–5), 1541–1551.

Saha, R., Ginwal, H. S., Chandra, G., & Barthwal, S. (2020). Integrated assessment of adventitious rhizogenesis in *Eucalyptus*: Root quality index and rooting dynamics. *Journal of Forestry Research, 31*, 2145–2161.

Saha, R., Ginwal, H. S., Chandra, G., & Barthwal, S. (2021). A comparative study on grey relational analysis and C5.0 classification algorithm on adventitious rhizogenesis of *Eucalyptus. Trees, 35*, 43–52.

Sifen, L., & Forrest, J. (2007). The current developing status on grey system theory. *The Journal of Grey System, 2*, 111–123.

Singh, Y. P., Uniyal, K., Bagwari, A., Kavita, Barthwal, S., Dhiman, R. C., & Gandhi, J. N. (2012). Status of poplar diseases in India. *ENVIS Forest Bulletin, 12*(1), 84–99.

Uniyal, K., Chandra, G., Khan, R. U., & Singh, Y. P. (2018). Selection of potent isolates from a population of *Alternaria alternata*, a leaf spot pathogen of poplar. *American Journal of Applied Mathematics and Statistics, 6*(6), 232–238.

Widin, K. D., & Schipper, A. L. (1981). Effect of Melampsora medusae leaf rust infection on yield of hybrid poplars in the north central United States. *European Journal of Plan Pathology, 11*, 438–448.

Xie, Y., Li, M., Sun, J., & Wang, F. (2009). Application of grey relational analysis to agricultural management decision-making. In *Proceedings of the SPIE 7490, PIAGENG: Intelligent Information, Control, and Communication Technology for Agricultural Engineering*, 74902N. https://doi.org/10.1117/12.836756

Xinning, L., & Yufen, X. (2017). Relational analysis of Yanbian's primary industry of forestry and forestry industry. *Asian Agricultural Research, 9*(2), 10–11.

Yeh, M. F., & Lu, H. C. (2000). Evaluating weapon systems based on grey relational analysis and fuzzy arithmetic operations. *Journal of the Chinese Institute of Engineers, 23*(2), 211–221.

Decision Making for Multi-Items Inventory Models

Nidhi D. Raykundaliya and Dharmesh P. Raykundaliya

Abstract Decision making plays a vital role in any field of science, commerce, engineering, arts and even in our day-to-day life. The main concern of decision making here is to optimize pricing and ordering strategies for the inventory system. Most of the decisions are subject to various constraints, and these constraints emphasize different pricing and ordering policy. In this chapter, we develop multi-items inventory models under declining demand with the Weibull distributed deterioration rate for two different scenarios, namely credit period and quantity discount, to settle the account due against purchases and suggest suitable inventory model, which maximize total profit. For validity of the model and decision making, numerical example and comparison study are also carried out.

Keywords Credit policy · Declining demand · Quantity discount · Weibull deterioration

1 Introduction

The economic order quantity (EOQ) models introduced by Wilson (1934) become foundation for the development of more advanced and complex inventory models. In any system, uncontrolled inventory creates various complex problems; as a result, one cannot survive in competitive environment. In literature, several authors have discussed inventory models to solve the problem of uncontrolled inventory by incorporating various scenarios such as deterioration rate, credit policy and quantity discount for both single and multi-items. Further, most of the inventory models are developed under the assumption that demand is either constant or changes uniformly over time.

N. D. Raykundaliya (✉)
Department of Mathematics, Gujarat Arts and Science College, Ahmedabad, India
e-mail: nidhi.raykaundaliya@gascellisbridge.gujgov.edu.in

D. P. Raykundaliya
Department of Statistics, Sardar Patel University, Vallabh Vidyanagar, Anand, Gujarat, India

D. D. Hanagal et al. (eds.), *Applied Statistical Methods*, Springer Proceedings in Mathematics & Statistics 380, https://doi.org/10.1007/978-981-16-7932-2_9

Deterioration of goods may be in the form of damage or decay with time, and this factor affects behavior of the demand in inventory analysis. Due to deterioration, it loses its marginal value and decreases usefulness from the original one. Medicines, chemicals, food stuff, fashion goods, vegetables and fruits are examples of deteriorating goods. The deterioration rate of goods is either constant, or it will increase with respect to time. Dave and Patel (1981), Hollier and Mak (1983), Sachan (1984), Raafat (1991), Shah and Shah (2000), Goyal and Giri (2001), Chang et al. (2006) Shah and Raykundaliya (2014), Khanna et al. (2016a) and Raykundaliya (2017) derive inventory model for deteriorating items.

In the competitive environment, supplier trade credit is one of the important promotional tools for many enterprises, and due to this, it has impact on finance and accounting of enterprise. It also helps retailer to keep revenue with an interest-bearing account and earn some interest during credit period. On the contrary, retailer does not require to pay any interest to supplier during permissible period while settling account as per mutual contract, but if payment is not made in specified time period as per contract, then after permissible period interest will be charged to retailer. Bringham (1995) defines "net 30" which means a supplier allows 30-days credit period to settle total amount owed to him. In the same line, Jaggi and Aggarwal (1995), Sarker et al. (2000), Chung and Huang (2003), Ouyang et al. (2006, 2009), Shah and Raykundaliya (2009a, b, 2014), Khanna et al. (2016b, 2017) and Jaggi et al. (2018) discuss credit policy scenario for single item. Jiangtao et al. (2014) and Leopoldo et al. (2015) develop model using two-level trade credit policies for multi-items.

In the era of globalization, to survive in competitive environment, only credits policy of the company does not help to attract buyers but quantity discount pricing policy of the company is also another important promotional tool to attract buyers. Wee and Junas (1997) develop lot-for-lot discount pricing policy when units in inventory are subject to deterioration with constant rate. Using the principal of mutual benefit, Chakravarty and Martin (1988), Papachristos and Skouri (2003), Shah and Raykundaliya (2009a, b), Lin and Chia (2011) and Pandey and Vaish (2017) derive inventory models by considering discount on selling price to attract buyers and motivate them for bulk purchasing.

In this chapter, we develop multi-items inventory models for deteriorating items when deterioration rate follows the two parameters Weibull distribution with two scenarios: credit policy and quantity discount under decline demand market. In Sect. 2, necessary notations and assumptions for inventory model are given. Mathematical models are derived for credit policy, and quantity discount is given in Sect. 3. Section 4 discusses numerical example and comparison study for both models developed in the Sect. 3. Concluding remark is given in the Sect. 5.

2 Notations and Assumptions

2.1 Assumptions

1. The inventory system under consideration deals with the multi-items.
2. The planning horizon is infinite.
3. The demand of the product is declining function of time.
4. Shortages are not allowed, and lead time is zero.
5. The units in inventory subject to deteriorate with time follow the two parameters Weibull distribution. The deteriorated units can neither be repaired nor replaced during the cycle time.
6. The retailer can deposit generated sales revenue in an interest-bearing account during permissible credit period. At the end of this period, the retailer settles account for all units sold keeping the difference for day-to-day expenditure and paying the interest charges on the unsold items of stock.

2.2 Notations

$R(t)$	:	$(= a(1 - bt))$ The annual demand, decreasing function of time, where $a > 0$ is constant demand and b denotes the rate of the demand
C	:	Purchase cost per unit of an item
P	:	The unit selling price with (P > C)
h	:	The inventory holding cost per unit per year
A	:	Ordering cost per order
M	:	The permissible credit period offered by the supplier to the retailer for settling an account
Ic	:	The interest charged per monetary unit in stock per annum by the supplier
Ie	:	The interest earned per monetary unit per year. $Ic > Ie$
d_1	:	Percentage discount offered before deterioration per unit
d_2	:	Percentage discount offered after deterioration per unit
f_1	:	$\left(= (1 - d_1)^{-n_1}, n_1 \epsilon R\right)$ is effect of pre-deterioration discount on selling price
f_2.	:	$\left(= (1 - d_2)^{-n_2}, n_2 \epsilon R\right)$ is effect of post-deterioration discount on selling price
γ	:	The time after which deterioration starts
Q	:	The order quantity
$I(t)$	:	The inventory level at any instant of time t, $0 \leq t \leq T$
T	:	The replenishment cycle time (a decision variable)

(continued)

(continued)

		$(= \alpha\beta t^{\beta-1})$, deterioration units in inventory system follows the two parameters Weibull distribution, where $\alpha(0 \le \alpha < 1)$ denotes scale parameter and $\beta(> 1)$ denotes shape parameter
TP_i	:	The total profit per time unit where $i = 1, 2$

3 Mathematical Model

In this section, we develop two different types of deteriorating inventory models with declining demand for multi-items by considering promotional tools credit period (Model 1) and quantity discount (Model 2), respectively, to attract the buyers.

Model 1: When Credit Period is Offered

The inventory level *I(t)* depletes to meet demand and deterioration. The rate of change of an inventory level is governed by following differential equation:

$$\frac{dI(t)}{dt} + \theta(t)I(t) = -R(t), 0 \le t \le T, \tag{1}$$

with the initial condition $I(0) = Q$ and the boundary condition $I(T) = 0$.

Using series expansion of exponential series and under the assumption that $0 \le \alpha < 1$ (neglecting α^2 and its higher-order term), the solution of differential Eq. (1) is

$$\begin{aligned} I(t) = a(T-t) - \frac{ab\left(T^2 - t^2\right)}{2} + \frac{a\alpha\left(T^{\beta+1} - t^{\beta+1}\right)}{\beta+1} - \frac{ab\alpha\left(T^{\beta+2} - t^{\beta+2}\right)}{\beta+2} \\ -a\alpha\left(Tt^{\beta} - t^{\beta+1}\right) + \frac{ab\alpha\left(T^2t^{\beta} - t^{\beta+2}\right)}{2} \quad 0 \le t \le T \end{aligned} \tag{2}$$

and the order quantity

$$Q = aT - \frac{abT^2}{2} + \frac{a\alpha T^{\beta+1}}{\beta+1} - \frac{ab\alpha T^{\beta+2}}{\beta+2} \tag{3}$$

The total profits of the inventory system per unit time consist of followings:
Sales revenue per unit time

$$SR = \frac{PQ}{T} \tag{4}$$

Purchase cost per unit time

$$PC = \frac{CQ}{T} \tag{5}$$

Ordering cost

$$OC = \frac{A}{T} \tag{6}$$

Deterioration cost per unit time

$$DC = \frac{C}{T}\left(Q - \int_0^T R(t)dt\right). \tag{7}$$

Inventory holding cost per unit time

$$IHC = \frac{h}{T}\int_0^T I(t)dt. \tag{8}$$

Regarding interest earned and interest charged two cases arises on the bases of length of T and M but as market scenario suggesting that the case regarding $(T \le M)$ is not favorable so here we develop model regarding the case when $M \le T$.

On unsold item, supplier charges an interest rate and hence interest charged per unit time

$$IC = \frac{CI_C}{T}\int_M^T I(t)dt \tag{9}$$

During credit period, retailer sells product and earned interest, and hence, interest earned per unit time

$$IE = \frac{CI_e}{T}\int_0^M R(t)tdt \tag{10}$$

Hence, total profit per unit time is

$$TP(T) = SR - PC - DC - IHC - OC - IC + IE \tag{11}$$

The optimal value of T can be obtained by solving $\frac{\partial TP(T)}{\partial T} = 0$ and obtained T guarantees maximize profit if $\frac{\partial^2 TP(T)}{\partial T^2} < 0$ hold.

Model 2: When Quantity Discount is Offered

The system starts with Q units, and this Q units deplete due to only demand up to time γ. After time γ, deterioration starts and inventory level changes due to demand

and deterioration units. It is assumed that pre-deterioration discount $d_1\%$(up to time $(\gamma - t_1)$) on unit selling price on fresh item to enhance the demand and then $d_2\%$ on unit selling price is given for the deteriorated items. The inventory level at any instant of time is governed by the following differential equation:

$$\frac{dI(t)}{dt} = \begin{cases} -a(1-bt), 0 \le t \le t_1, \\ -af_1(1-bt), t_1 \le t \le \gamma, \\ -af_2(1-bt) - \theta(t)I(t)\gamma \le t \le T. \end{cases} \tag{12}$$

with the initial condition $I(0) = Q$ and the boundary condition $I(T) = 0$.

Using continuity and the series expansion of exponential series and under the assumption that, $0 \le \alpha < 1$ (neglecting α^2 and its higher power), the solution of differential Eq. (12) is

$$I(t) = \begin{cases} Q - at + \frac{abt^2}{2}; & 0 \le t \le t \\ af_1(\gamma - t) - \frac{abf_1}{2}\left(\gamma^2 - t^2\right) + af_2(T-\gamma) - \frac{abf_2}{2}\left(T^2 - \gamma^2\right) + \frac{a\alpha f_2}{\beta+1}\left(T^{\beta+1} - \gamma^{\beta+1}\right) & t_1 \le t \le \gamma \\ -\frac{ab\alpha f_2}{\beta+2}\left(T^{\beta+2} - \gamma^{\beta+2}\right) - a\alpha f_2\left(T\gamma^{\beta} - \gamma^{\beta+1}\right) + \frac{ab\alpha f_2}{2}\left(T^2\gamma^{\beta} - \gamma^{\beta+2}\right); & \\ af_2(T-t) - \frac{abf_2}{2}\left(T^2 - t^2\right) + \frac{a\alpha f_2}{\beta+1}\left(T^{\beta+1} - t^{\beta+1}\right) - \frac{ab\alpha f_2}{\beta+2}\left(T^{\beta+2} - \gamma^{\beta+2}\right) & \gamma \le t \le T \\ -\ a\alpha f_2\left(Tt^{\beta} - t^{\beta+1}\right) + \frac{ab\alpha f_2}{2}\left(T^2\gamma^{\beta} - \gamma^{\beta+2}\right); & \end{cases} \tag{13}$$

Where

$$Q = at_1 - \frac{abt_1^2}{2} + af_1(\gamma - t_1) - \frac{abf_1}{2}(\gamma^2 - t_1^2) + af_2(T-\gamma) - \frac{abf_2}{2}(T^2 - \gamma^2) - \frac{ab\alpha f_2}{\beta+2}(T^{\beta+2} - \gamma^{\beta+2}) - a\alpha f_2(T\gamma^{\beta} - \gamma^{\beta+1}) + \frac{ab\alpha f_2}{2}T^2\gamma^{\beta} - \gamma^{\beta+2} \tag{14}$$

The total profit of inventory system per unit time consists of the following:

Sales revenue per unit time

$$SR = \frac{P}{T}\left(\int_0^{t_1} R(t)dt + \int_{t_1}^{\gamma} f_1(1-d_1)R(t)dt + \int_{\gamma}^{T} f_2(1-d_2)R(t)dt\right) \tag{15}$$

Purchase cost per unit time

$$PC = \frac{CQ}{T} \tag{16}$$

Ordering cost

$$OC = \frac{A}{T} \tag{17}$$

Inventory holding cost

$$IHC = \frac{h}{T}\left(\int_{0}^{t_1} I(t)t\,dt + \int_{t_1}^{\gamma} I(t)(t - t_1)dt + \int_{\gamma}^{T} I(t)(t - \gamma)dt\right) \tag{18}$$

Hence, total profit per unit time is

$$TP(t_1, T) = SR - PC - IHC - OC \tag{19}$$

The optimal value of t_1 and T can be obtained by solving $\frac{\partial TP(t_1,T)}{\partial t_1} = 0$; $\frac{\partial TP(t_1,T)}{\partial T} = 0$ and obtained solution guarantees maximize profit if $\frac{\partial^2 TP(t_1,T)}{\partial T^2} < 0$ and $\left(\frac{\partial^2 TP(t_1,T)}{\partial T^2}\frac{\partial^2 TP(t_1,T)}{\partial t_1^2} - \frac{\partial^2 TP(t_1,T)}{\partial T \partial t_1}\right) > 0$ hold.

4 Numerical Example and Comparison Study

In this section, we consider one to one correspondence between retailer and supplier, where supplier sells three different products to the retailer. Different parametric values common to both models are given in Table 1. Data required for credit policy and discounting policy are mentioned in Tables 2 and 3, respectively, and results obtained for both policies and their comparison are given in Table 4.

Convexity of Fig. 1 of all examples proved that resulting parameter guarantees maximum profit for both the Model 1 and the Model 2. Further, from Table 4, we can observe that in the cycle time, quantity demanded is less but profit is more in the Model 1 compared to somewhat large quantity with larger cycle time but less profit

Table 1 Common Parametric Values of Model 1 and Model 2

Parameter	Product 1	Product 2	Product 3
a	600	1000	800
b	0.1	0.15	0.12
C	15	20	18
P	40	45	40
A	50	60	45
h	1	1.2	1.1
α	0.1	0.15	0.15
β	1.1	1.2	1.1

Table 2 Specific Parametric Values For Model 1

Parameter	Product 1	Product 2	Product 3
M	60/365	30/365	50/365
I_c	0.15	0.15	0.18
I_e	0.09	0.09	0.05

Table 3 Specific Parametric Values For Model 2

Parameter	Product 1	Product 2	Product 3
γ	3.5	3	3.2
n_1	2	2	2
n_1	5	5	5
d_1	0.15	0.15	0.05
d_2	0.30	0.30	0.2

Table 4 Comparative Study For Model1 and Model 2

Credit policy				Discounted policy			
Parameter	Product 1	Product 2	Product 3	Parameter	Product 1	Product 2	Product 3
T	0.18	0.11	0.14	T	4.216	3.40	3.51
Q	109	111	116	Q	3797	3517	2665
TP	14,778	24,410	17,201	TP	13,003	19,143	13,735

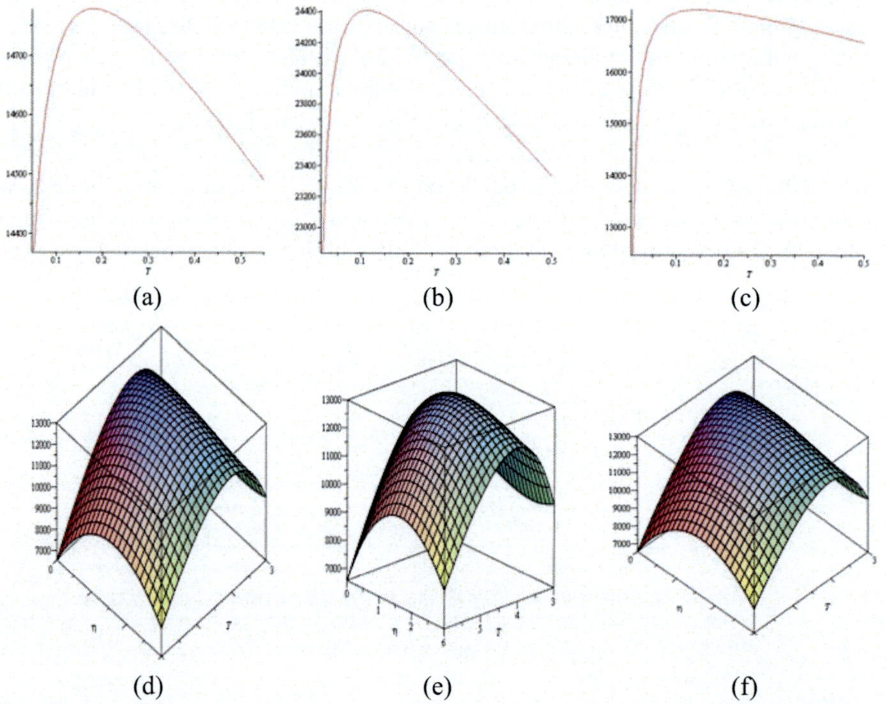

Fig. 1 Total profit (a–c for example 1–3) for model 1 and (d–f for example 1–3) for model 2

in the Model 2 for all three examples. Hence, overall the Model 1 is more preferable over the Model 2 even though quantity discount is more appealing promotional tool.

5 Conclusion

In this chapter, we compared two different deteriorating inventory models by offering either credit policy or quantity discount, in declining demand market. We conclude that though both policies are very promising in nature, one should earn more profit in minimum time period by offering credit period policy. In future, one can also extend the research by combining both policies or by allowing shortages for more relevant result.

References

Bringham, E. F. (1995). *Fundamentals of Financial Management*. The Dryden Press.

Chakravarty, A. K., & Martin, G. E. (1988). An optimal joint buyer-seller discount pricing model. *Computers and Operations Research, 15*, 271–281.

Chang, H. J., Teng, J. T., Ouyang, L. Y., & Dye, C. Y. (2006). Retailer's optimal pricing and lot-sizing policies for deteriorating items with partial backlogging. *European Journal of Operational Research, 168*, 51–64.

Chung, K. J., & Huang, Y. F. (2003). The optimal cycle time for EPQ inventory model under permissible delay in payments. *International Journal of Production Economics, 84*(3), 30–318.

Dave, U., & Patel, L. K. (1981). (T, Si) – Policy inventory model for deteriorating items with time proportional demand. *Journal of the Operational Research Society, 32*, 137–142.

Goyal, S. K., & Giri, B. C. (2001). Recent trends in modeling of deteriorating inventory. *European Journal of Operational Research, 134*, 1–16.

Hollier, R. H., & Mak, K. L. (1983). Inventory replenishment policies for deteriorating items in declining market. *International Journal of Production Research, 21*, 813–826.

Jaggi, C. K., & Aggarwal, S. P. (1995). Ordering policy of deteriorating items under permissible delay in payments. *Journal of the Operational Research Society, 46*, 658–662.

Jaggi, C. K. Gautam, P., & Khanna, A. (2018). Inventory decisions for imperfect quality deteriorating items with exponential declining demand under trade credit and partially backlogged shortages. *In Quality, IT and Business Operations Springer*, Singapore, pp. 213–229.

Jiangtao, M. C., Guimel, F., & Ting, M. H. (2014). Optimal ordering policies for perishable multi-item under stock dependent demand and two-level trade credit. *Applied Mathematical Modelling, 38*, 2522–2532.

Lin, Y.J., & Chia, H. (2011). Integrated inventory model with quantity discount and price sensitive demand. *TOP: An Official Journal of the Spanish Society of Statistics and Operations Research, 18*(1), 177–188.

Khanna, A., Mittal, M., Gautam, P., & Jaggi, C. K. (2016a). Credit financing for deteriorating imperfect quality items with allowable shortages. *Decision Science Letters, 5*(1), 45–60.

Khanna, A. Jaggi, C. K. Gautam, P. (2016b). Coordinating vendor-buyer decisions for imperfect quality items considering trade credit and fully backlogged shortages. *In AIP Conference Proceedings*, AIP Publishing LLC., 1715(1), 020065.

Khanna, A., Gautam, P., & Jaggi, C. K. (2017). Inventory modeling for deteriorating imperfect quality items with selling price dependent demand and shortage backordering under credit financing. *International Journal of Mathematical, Engineering and Management Sciences, 2*(2), 110–124.

Leopoldo, E., Cardenas, B., & ShibSankar, S. (2015) Multi-item EOQ inventory model in a two-layer supply chain while demand varies with promotional effort. *Applied Mathematical Modelling, 39*, 6725-6737.

Ouyang, L. Y., Teng, J. T., & Chen, L. H. (2006). Optimal ordering policy for deteriorating items with partial backlogging under permissible delay in payments. *Journal of Global Optimization, 34*, 245–271.

Ouyang, L. Y., Teng, J. T., Goyal, S. K., & Yang, C. T. (2009). An economic order quantity model for deteriorating items with partially permissible delay in payments linked to order quantity. *European Journal of Operational Research, 194*, 418–431.

Pandey, R. K., & Vaish, B. (2017). Optimal inventory policy foe Deteriorating items with seasonal demand under the effect of price discounting on lost sales. *IOSR Journal of Mathematics, 13*(3), 35–42.

Papachristos, S., & Skouri, K. (2003). An inventory model with deteriorating items, quantity discount, pricing and time- dependent partial backlogging. *International Journal of Production Economics, 83*(3), 247–256.

Raafat, F. (1991). Survey of literature on continuously deteriorating inventory models. *Journal of the Operational Research Society, 40*, 27–37.

Raykundaliya, N. D. (2017). Inventory model for the Rayleigh distributed deterioration under demand inclining market condition. *(PRAJNA) Journal of Pure and Applied Science,* 24–25, 7–10.

Sachan, R. S. (1984). On (T, Si) – policy inventory model for determining itemswith time proportional demand. *Journal of the Operational Research Society, 35*, 1013–1019.

Sarker, B. R., Jamal, A. M. M., & Wang, S. (2000). Supply chain model for perishable products under inflation and permissible delay in payment. *Computers and Operations Research, 27*, 59–75.

Shah, N. H., & Raykundaliya, N. D. (2009a). Optimal Policies for integrated inventory system for deteriorating items using quantity discount in price-sensitive declining market. *Tecnia Journal of Management Studies, 5*(1), 15–25.

Shah, N. H., & Raykundaliya, N. D. (2009b). Optimal inventory policies for Weibull deterioration under trade credit in declining market. *Indus Journal of Management and Social Sciences (IJMSS), 3*(2), 11–20.

Shah, N. H., & Raykundaliya, N. D. (2014). A Collaborative vendor-buyer deteriorating inventory system in declining market when trade credit is offered. *Asian Journal of Management Research, 4*(3), 404–418.

Shah, N. H., & Shah, Y. K. (2000). Literature survey on inventory models for deteriorating items. *Economic Annals, 44*, 221–237.

Wee, H. M., & Jonas, Y. (1997). A deteriorating inventory model with a temporary price discount. *International Journal of Production Economics, 53*(1), 81–90.

Wilson, R. H. (1934). A scientific routine for stock control. *Harvard Business Review, 13*, 116–128.

Modeling Australian Twin Data Using Generalized Lindley Shared Frailty Models

Arvind Pandey, David D. Hanagal, Shikhar Tyagi, and Pragya Gupta

Abstract A new class of shared frailty models based on generalized Lindley distribution is established. We propose shared frailty models based on reversed hazard rate. We estimate the parameters in these frailty models and use the Bayesian paradigm of the Markov chain Monte Carlo technique. Model selection criteria have been performed for the comparison of models. We analyze Australian twin data and suggest a better model.

Keywords Bayesian estimation · Generalized Rayleigh distribution · Left censoring · MCMC · Modified inverse Weibull distribution · Generalized Lindley frailty · Reversed hazard rate

1 Introduction

Vaupel et al. (1979) introduced a random impact which is unobservable risk shared by the subject characterized as frailty. To handle such kind of problems, many models have been derived in survival analysis. Since the establishment of the proportional hazard model given by Cox (1972), survival function has been dominated by hazard rate models. The reason behind the popularity of this model is the significance of known covariates can be tested, and also a relationship between lifetimes and covariates can be incorporated. Augmentation of Cox's proportional hazard model provided away to introduce the unknown covariates,

$$\begin{aligned}\phi(t|\underline{K}) &= \phi_0(t)e^{\underline{K}'\underline{\beta_0}+\underline{V}'\underline{\beta_1}}\\ &= w\phi_0(t)e^{\underline{K}'\underline{\beta_0}}\end{aligned}$$

A. Pandey · S. Tyagi · P. Gupta
Department of Statistics, Central University of Rajasthan, Rajasthan, India

D. D. Hanagal (✉)
Department of Statistics, S. P. Pune University, Pune, India

D. D. Hanagal et al. (eds.), *Applied Statistical Methods*, Springer Proceedings in Mathematics & Statistics 380, https://doi.org/10.1007/978-981-16-7932-2_10

where $\phi(t|\underline{K})$ stands for conditional hazard rate and $\phi_0(t)$ stands for baseline hazard rate. $\underline{K}^{'} = (K_{1j}, K_{2j}, ..., K_{mj})$ and $\underline{V}^{'} = (V_{1j}, V_{2j}, ..., V_{mj})$ are considered as the vector of known and unknown covariates, respectively; $\underline{\beta_0}$ and $\underline{\beta_1}$ are indicated as the vector of corresponding regression coefficients of order m. $w = e^{\underline{V}^{'}\underline{\beta_1}}$ is called as frailty effect.

In the last decade, frailty regression models in mixture distribution have been discussed by Hanagal (2008). Hougaard (2000) had discussed the different aspects of frailty on a broad scale. Hanagal and Dabade (2013, 2015) proposed modeling of the inverse Gaussian frailty model and comparison of different frailty models for analyzing kidney infection data. Modeling kidney infection data for inverse Gaussian shared frailty was done by Hanagal and Pandey (2014a). Gamma frailty models for bivariate survival data were given by Hanagal and Pandey (2015a). Hanagal and Pandey (2017a) used the shared inverse Gaussian frailty models based on additive hazard. For reversed hazard rate setup, Hanagal and Pandey (2014b, 2015b, 2016a, b, 2017b) have contemplated gamma and inverse Gaussian shared frailty models with different baseline distribution functions. Hanagal and Sharma (2013, 2015a, b, c) analyzed acute leukemia data, kidney infection data and diabetic retinopathy data using shared gamma and inverse Gaussian frailty models for the multiplicative model. Compound Poisson frailty was used by Hanagal and Kamble (2015) for Bayesian estimation. Analysis of kidney infection data and Australian twin data was done by Hanagal and Bhambure (2014, 2015, 2016) with different frailty distributions. Hanagal (2011, 2017, 2019) and Wienke (2011) gave extensive literature review on different shared frailty models.

The main aim of this article has three objectives. First, generalized Lindley (GL) shared frailty models for reversed hazard rate with modified inverse Weibull and generalized Rayleigh as baseline distributions have been introduced. Second, Bayesian approach of estimation has been employed to estimate the unknown parameters under random censoring. Third, simulation study and data analysis have been done for the Australian twin data set.

2 Reversed Hazard Rate

There are many situations, when it is more realistic to use reversed hazard rate instead of hazard rate. It is more relevant to analyze the left censored and right truncated survival data. The reversed hazard rate can be given as

$$\psi(t) = \frac{f(t)}{F(t)} \tag{2.1}$$

where $F(t)$ is distribution function and $f(t)$ is probability density function at time t. For the further details in reversed hazard rate, you can see Shaked and Shantikumar (1994) and Block et al. (1998). The reversed hazard rate gives more appropriate result when the failure is quantity of interest in predicting the actual time of failure. In the left censored and right truncated data, the reversed hazard rate (RHR) gives more proper reliability functions. Sankaran and Gleeja (2011) introduced frailty as a common random effect that acts multiplicatively on RHR, which is useful for the analysis of left censored data. Gamma shared frailty model based on reversed hazard rate for bivariate survival data was discussed by the Hanagal and Pandey (2014b). Shared frailty models based on reversed hazard rate for modeling Australian twin data were studied by the Hanagal and Pandey (2015b). Hanagal and Pandey (2016a) proposed the gamma shared frailty models based on reversed hazard rate (RHR). Hanagal and Pandey (2016b) discussed the inverse Gaussian shared frailty model based on the RHR.

3 General Shared Frailty Model

Let the lifetime of an individual be defined as T. For a given frailty W_j=w_j, the conditional reversed hazard rate is

$$\psi(t_j|w_j) = w_j\psi_0(t_j)e^{\underline{K}'\underline{\beta}_0} \tag{3.1}$$

where $\psi_0(t_j)$ gives the baseline reversed hazard rate at time t_j. For the given frailty W_j=w_j, the conditional distribution function of T is given as

$$\begin{aligned} F(t_j|w_j) &= e^{-\int_{t_j}^{\infty}\psi(t_j|w_j)dt} \\ &= e^{-w_j\Psi_0(t_j)e^{\underline{K}'_j\underline{\beta}_0}} \end{aligned} \tag{3.2}$$

where $\Psi_0(t_j)$ is the cumulative baseline reversed hazard at time t_j. Now, we integrate $f_W(w_j)$ over the range of frailty variable W_j and we find the marginal distribution as follows

$$\begin{aligned} F(t_j) &= \int_0^{\infty} F(t_j|w_j)f_{W_j}(w_j)dw_j \\ &= \int_0^{\infty} e^{-w_j\Psi_0(t_j)\rho_j}f_{W_j}(w_j)dw_j \\ &= L_{W_j}(\Psi_0(t_j)\rho_j), \end{aligned} \tag{3.3}$$

where $L_{W_j}(.)$ is a Laplace transformation of distribution of W_j and $\rho_j = e^{\underline{K}'_j\underline{\beta}_0}$.

The shared frailty model is relevant to event time of related individuals, similar organs and repeated measurements. For example, the failure time of paired organs like lungs, ears, eyes and kidney is the example of bivariate survival data. In this model, common covariates are shared by individuals from a group. The survival times are conditionally independent in the shared frailty models. The survival times are connected through some unobservable covariates or frailty.

Here, for the study we observed n individuals. Let first and second life times of jth individual be represented by bivariate random vector (T_{1j}, T_{2j}). Let observed covariates be denoted by $\underline{K_0}$, $\underline{K_1}$ and $\underline{K_2}$, where $\underline{K_0}$ is vector of common covariates for both survival times T_{1j} and T_{2j} and the $\underline{K_1}$ and $\underline{K_2}$ are the vector of covariates corresponding to survival times T_{1j} and T_{2j}. Let a vector $\underline{K_{lj}} = (K_{1lj}, \ldots, K_{k_l lj})$, $(l = 0, 1, 2)$ for j^{th} individual where K_{alj} $(a = 1, 2, 3, \ldots, k_l)$ represents the value of ath observed covariate for j^{th} individual. Let W_j be the shared frailty for the jth individuals. Multiplicative effect can be seen in baseline reversed hazard of frailty. For given frailty, the survival times of individuals are conditionally independent. For given frailty W_j=w_j, the conditional reversed hazard rate for j^{th} individuals at i^{th} survival time $t_{ij} > 0$ is given as

$$\psi(t_{ij} \mid w_j, \underline{K}_j) = w_j \psi_0(t_{ij}) e^{\underline{K_{0j}} \underline{\beta_0} + \underline{K_{ij}} \underline{\beta_i}} \qquad ; i = 1, 2. \tag{3.4}$$

where $\psi_0(t_{ij})$ is baseline reversed hazard at time $t_{ij} > 0$ and $\underline{\beta}$ is a vector of order k, of regression coefficients. The conditional cumulative reversed hazard rate for j^{th} individual at i^{th} lifetime $t_{ij} > 0$ for a given frailty $W_j = w_j$ is

$$\Psi(t_{ij} \mid w_j, \underline{K}_j) = w_j \Psi_0(t_{ij}) \rho_{0j} \rho_{ij} \tag{3.5}$$

where $\rho_{0j} = e^{K_{0j} \underline{\beta_0}}$, $\rho_{ij} = e^{K_{ij} \underline{\beta_i}}$, i= 1,2 and $\Psi_0(t_{ij})$ is cumulative baseline reversed hazard rate at time $t_{ij} > 0$. For given frailty W_j=w_j, the conditional distribution function for j^{th} individuals at i^{th} survival time $t_{ij} > 0$ is given as

$$\begin{aligned} F(t_{ij} \mid w_j, \underline{K}_j) &= e^{-\Psi(t_{ij} \mid w_j, \underline{K}_j)} \\ &= e^{-w_j \Psi_0(t_{ij}) \rho_{0j} \rho_{ij}} \end{aligned} \tag{3.6}$$

Under the assumption of independence, bivariate conditional distribution function at time $t_{1j} > 0$ and $t_{2j} > 0$ for given frailty W_j=w_j is

$$\begin{aligned} F(t_{1j}, t_{2j} \mid w_j, \underline{K}_j) &= F(t_{1j} \mid w_j, \underline{K}_j) F(t_{2j} \mid w_j, \underline{K}_j) \\ &= e^{-w_j (\Psi_{01}(t_{1j}) \rho_{1j} + \Psi_{02}(t_{2j}) \rho_{2j}) \rho_{0j}} \end{aligned} \tag{3.7}$$

By integrating probability function $f_W(w_j)$ over frailty variable W_j, we obtain the unconditional bivariate distribution function as follows

$$\begin{aligned} F(t_{1j}, t_{2j} \mid \underline{K}_j) &= \int_{W_j} F(t_{1j}, t_{2j} \mid w_j) f(w_j) dw_j \\ &= \int_{W_j} e^{-w_j (\Psi_{01}(t_{1j}) \rho_{1j} + \Psi_{02}(t_{2j}) \rho_{2j}) \rho_{0j}} f(w_j) dw_j \\ &= L_{W_j}[(\Psi_{01}(t_{1j}) \rho_{1j} + \Psi_{02}(t_{2j}) \rho_{2j}) \rho_{0j}] \end{aligned} \tag{3.8}$$

where $L_{W_j}(.)$ is Laplace transform of frailty variable of W_j for j^{th} individual.

4 Generalized Lindley Frailty Model

Lindley (1958) proposed a distribution with one parameter. Because of having only one parameter, the Lindley distribution does not provide enough flexibility for modeling purposes. It will be useful to consider further alternatives of this distribution. Elbatal et al. (2013) proposed generalized Lindley distribution which generalizes Lindley distribution and includes exponential and gamma distributions as special cases. For a frailty distribution, generalized Lindley distribution has been considered in this chapter. This distribution is the mixture of two gamma distributions G(θ,μ) and G(θ,η) with mixing coefficient $\theta/(\theta+1)$. Because of the mixture of two gamma densities, quite bit of suppleness can be seen during analysis of time to event data. That is the reason why GL frailty model is more adaptable in comparison with gamma frailty model. Probability density function of GL distribution has been specified below:

$$f_W(w) = \begin{cases} \frac{1}{(1+\theta)}\left[\frac{\theta^{\mu+1}w^{\mu-1}}{\Gamma\mu} + \frac{\theta^{\eta}w^{\eta-1}}{\Gamma\eta}\right]e^{-\theta w} & ; w \in \mathbb{R}^+, \mu, \eta, \theta \in \mathbb{R}^+ \\ 0 & ; otherwise \end{cases}$$

with mean $E[W] = \frac{1}{1+\theta}\left[\mu + \frac{\eta}{\theta}\right]$. And corresponding variance is

$$V(W) = \frac{1}{(1+\theta)}\left[\left(\mu^2 + \frac{\eta^2}{\theta}\right)\left(\frac{1}{\theta(1+\theta)}\right) + \left(\frac{\mu+\eta}{\theta}\right) - \left(\frac{2\mu\eta}{\theta(1+\theta)}\right)\right]$$

After applying identifiability property, i.e., $E[W] = 1$, we get a relation between parameters $\eta = \theta\,(1+\theta-\mu)$. Consequently, the density function, Laplace transformation and variance for GL reduced to

$$f_W(w) = \begin{cases} \frac{1}{(1+\theta)}\left[\frac{\theta^{\mu+1}w^{\mu-1}}{\Gamma\mu} + \frac{\theta^{\theta(1+\theta-\mu)}w^{\theta(1+\theta-\mu)-1}}{\Gamma\theta(1+\theta-\mu)}\right]e^{-\theta w} & ;\ w, \theta \in \mathbb{R}^+, \mu \in (0, 1+\theta) \\ 0 & ;\ otherwise. \end{cases}$$

$$L_W(s) = \frac{1}{(1+\theta)}\left[\frac{\theta^{\mu+1}}{(s+\theta)^{\mu}} + \frac{\theta^{\theta(1+\theta-\mu)}}{(s+\theta)^{\theta(1+\theta-\mu)}}\right] \tag{4.1}$$

$$V(W) = \frac{\theta^4 - \theta^3\mu + 3\theta^2(1+\theta) - 4\theta^2\mu + 3\theta\mu(\mu-1) + \mu^2}{\theta(1+\theta)^2} \tag{4.2}$$

Let n be the number of observations under study. Let (T_{1j}, T_{2j}) be the first and second survival times of pairs of components of jth $(1, 2, ..., n)$ objects. The unconditional bivariate distribution function at time $t_{1j} \in \mathbb{R}^+$ and $t_{2j} \in \mathbb{R}^+$ using Equations (3.8) and (4.1) can be written as

$$
\begin{aligned}
F(t_{1j}, t_{2j}) &= L_{W_j}[(\Psi_{01}(t_{1j})\rho_{1j} + \Psi_{02}(t_{2j})\rho_{2j})\rho_{0j}] \\
&= \frac{1}{(1+\theta)}\left[\frac{\theta^{\mu+1}}{([(\Psi_{01}(t_{1j})\rho_{1j} + \Psi_{02}(t_{2j})\rho_{2j})\rho_{0j}] + \theta)^{\mu}} + \right. \\
&\left. \frac{\theta^{\theta(\theta+1-\mu)}}{([(\Psi_{01}(t_{1j})\rho_{1j} + \Psi_{02}(t_{2j})\rho_{2j})\rho_{0j}] + \theta)^{\theta(\theta+1-\mu)}}\right]
\end{aligned}
\tag{4.3}
$$

where $\Psi_{01}(t_{1j})$ and $\Psi_{02}(t_{2j})$ are the cumulative baseline reversed hazard rate functions of the lifetime T_{1j} and T_{2j}, respectively. In the absence of frailty effect, models will be

$$
F(t_{1j}, t_{2j}) = e^{-(\rho_{0j}(\Psi_{01}(t_{1j})\rho_{1j} + \Psi_{02}(t_{2j})\rho_{2j}))} \tag{4.4}
$$

One can have different baseline distributions for T_1 and T_2. After substituting different cumulative hazard functions in (4.3), we get different generalized Lindley frailty distributions.

5 Dependence Measure

Sometimes due to complex form of frailty models, it is difficult to compare the degree of dependence between different frailty models. Kendall's τ can be used to quantify dependence because it is independent of transformations on the timescale and the frailty model used. It is a rank-based dependence measure.

$$
\tau = \int_{s \in \mathbb{R}^+} 4s L''_W(s) L_W(s) ds - 1 \tag{5.1}
$$

After using Equations (4.2) and (5.1), we get

$$
\tau = \int_{s \in \mathbb{R}^+} R(s \mid \theta, \mu) ds - 1 \tag{5.2}
$$

where $R(s \mid \theta, \mu) = \frac{4\theta s\left(\theta^{\mu+1} A^{-\mu} + \theta^{\theta B} A^{\theta(\mu-\theta-1)}\right)\left(\mu(\mu+1)\theta^{\mu} A^{-\mu} + \theta^{\theta B} B\left(-\mu\theta + \theta^2 + \theta + 1\right) A^{\theta(\mu-\theta-1)}\right)}{(\theta+1)^2 A^2}$.
$A = (\theta + s)$, $B = (1 + \theta - \mu)$.
Kendall's τ cannot be found in closed form for GL frailty. Some numerical approaches can be utilized to obtain Kendall's τ dependence measure.

6 Baseline Distributions

6.1 Modified Inverse Weibull Distribution

Modified inverse Weibull distribution has been chosen as baseline distribution due to more utilitarian for computational point of view for left censored data (Kumar and Singh (2011)). A continuous random variable T follows modified inverse Weibull distribution with corresponding distribution function, reversed hazard and cumulative reversed hazard functions given below

$$F(t) = \begin{cases} e^{-\zeta t^{-\delta} e^{-\xi t}} & ; t \in \mathbb{R}^+, \delta, \zeta, \xi \in \mathbb{R}^+ \\ 0 & ; otherwise \end{cases} \tag{6.1}$$

$$\psi_0(t) = \begin{cases} \zeta e^{-\xi t} t^{-1-\delta} (\delta + \xi t) & ; t \in \mathbb{R}^+, \delta, \zeta, \xi \in \mathbb{R}^+ \\ 0 & ; otherwise \end{cases} \tag{6.2}$$

$$\Psi_0(t) = \begin{cases} \zeta t^{-\delta} e^{-\xi t} & ; t \in \mathbb{R}^+, \delta, \zeta, \xi \in \mathbb{R}^+ \\ 0 & ; otherwise \end{cases} \tag{6.3}$$

6.2 Generalized Rayleigh Distribution

Generalized Rayleigh distribution has been chosen as second baseline distribution. If shape parameter is less than 0.5, PDF of generalized Rayleigh distribution is a decreasing function, and if shape parameter is greater than 0.5, it is a right-skewed unimodal function (see Raqab & Kundu, 2006). In modeling strength data as well as general lifetime data, it can be utterly dominant (Surles and Padgett (2001)). A continuous random variable T follows generalized Rayleigh distribution with corresponding distribution function, reversed hazard and cumulative reversed hazard functions given below

$$F(t) = \begin{cases} \left(1 - e^{-(\delta t)^2}\right)^{\zeta} & ; t \in \mathbb{R}^+, \delta, \zeta, \in \mathbb{R}^+ \\ 0 & ; otherwise \end{cases} \tag{6.4}$$

$$\psi_0(t) = \begin{cases} \frac{2\zeta\delta^2 t e^{-(\delta t)^2}}{\left(1-e^{-(\delta t)^2}\right)} & ; t \in \mathbb{R}^+, \delta, \zeta, \in \mathbb{R}^+ \\ 0 & ; otherwise \end{cases} \tag{6.5}$$

$$\Psi_0(t) = \begin{cases} -\log\left(1 - e^{-(\delta t)^2}\right)^{\zeta} & ; t \in \mathbb{R}^+, \delta, \zeta, \in \mathbb{R}^+ \\ 0 & ; otherwise \end{cases} \tag{6.6}$$

7 Proposed Models

After substituting cumulative reversed hazard function for modified inverse Weibull and generalized Rayleigh baseline distributions in Equations (4.3) and (4.4), we get

$$F(t_{1j}, t_{2j}) = \frac{1}{(1+\theta)}\left[\frac{\theta^{\mu+1}}{([(\zeta_1 t_{1j}^{-\delta_1} e^{-\xi_1 t_{1j}} \rho_{1j} + \zeta_2 t_{2j}^{-\delta_2} e^{-\xi_2 t_{2j}} \rho_{2j})\rho_{0j}] + \theta)^{\mu}} + \frac{\theta^{\theta(\theta+1-\mu)}}{([(\zeta_1 t_{1j}^{-\delta_1} e^{-\xi_1 t_{1j}} \rho_{1j} + \zeta_2 t_{2j}^{-\delta_2} e^{-\xi_2 t_{2j}} \rho_{2j})\rho_{0j}] + \theta)^{\theta(\theta+1-\mu)}}\right] \tag{7.1}$$

$$F(t_{1j}, t_{2j}) = e^{-(\rho_{0j}(\zeta_1 t_{1j}^{-\delta_1} e^{-\xi_1 t_{1j}} \rho_{1j} + \zeta_2 t_{2j}^{-\delta_2} e^{-\xi_2 t_{2j}} \rho_{2j}))} \tag{7.2}$$

$$F(t_{1j}, t_{2j}) = \frac{1}{(1+\theta)}\left[\frac{\theta^{\mu+1}}{([(-\log\left(1 - e^{-(\delta_1 t_{1j})^2}\right)^{\zeta_1} \rho_{1j} + -\log\left(1 - e^{-(\delta_2 t_{2j})^2}\right)^{\zeta_2} \rho_{2j})\rho_{0j}] + \theta)^{\mu}} + \frac{\theta^{\theta(\theta+1-\mu)}}{([(-\log\left(1 - e^{-(\delta_1 t_{1j})^2}\right)^{\zeta_1} \rho_{1j} + -\log\left(1 - e^{-(\delta_2 t_{2j})^2}\right)^{\zeta_2} \rho_{2j})\rho_{0j}] + \theta)^{\theta(\theta+1-\mu)}}\right] \tag{7.3}$$

$$F(t_{1j}, t_{2j}) = e^{-(\rho_{0j}(-\log\left(1-e^{-(\delta_1 t_{1j})^2}\right)^{\zeta_1} \rho_{1j} + -\log\left(1-e^{-(\delta_2 t_{2j})^2}\right)^{\zeta_2} \rho_{2j}))} \tag{7.4}$$

Here, Equations (7.1) and (7.2) can be called as Model-I and Model-III, respectively, which are modified inverse Weibull baseline distribution with and without frailty and Equations (7.3) and (7.4) can be called as Model-II and Model-IV, respectively, which are generalized Rayleigh baseline distribution with and without frailty. Figures 1 and 2 show the probability density functions of Model-I and Model-II, respectively, for three different sets of the values of the parameters.

Fig. 1 PDF Model-I

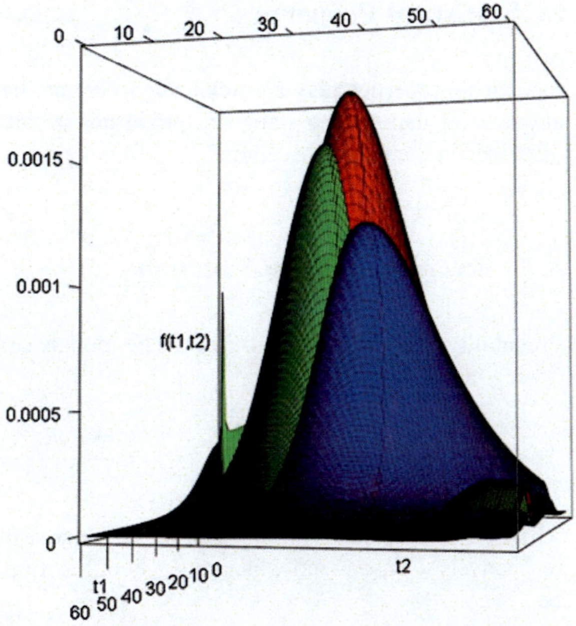

Fig. 2 PDF for Model-II

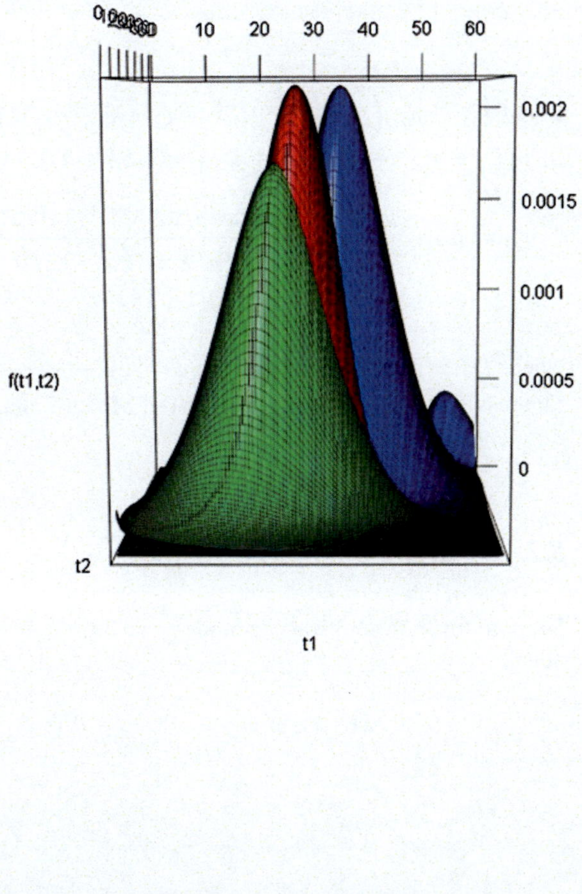

8 Statistical Properties

Statistical properties play a crucial role to define the applicability of any model on any kind of data. Some statistical properties of Model-I and Model-II have been discussed in the next subsections.

8.1 Bivariate Density Function

Probability density function for bivariate models can be obtained by using the following mechanism

$$f(t_{1j}, t_{2j}) = \frac{\partial^2 F(t_{1j}, t_{2j})}{\partial t_{1j} \partial t_{2j}}$$

where $F(t_{1j}, t_{2j})$ stands for the bivariate cumulative distribution function of (t_{1j}, t_{2j}), respectively. Consequently, hazard gradient function for Model-I and Model-II will be

$$f_1(t_{1j}, t_{2j}) = \frac{1}{(\theta+1)} \left[\zeta_1 \zeta_2 \rho_{0j}^2 \rho_{1j} \rho_{2j} \theta t_{1j}^{-\delta_{1j}-1} t_{2j}^{-\delta_{2j}-1} (\xi_1 t_{1j} + \delta_{1j})(\xi_2 t_{2j} + \delta_{2j}) P_{11}^{-\mu-B\theta-2} \omega_1 e^{-\xi_1 t_{1j} - \xi_2 t_{2j}} \right] \tag{8.1}$$

where $P_{11} = \rho_{0j} \left(\zeta_1 \rho_{1j} e^{-\xi_1 t_{1j}} t_{1j}^{-\delta_1} + \zeta_2 \rho_{2j} e^{-\xi_2 t_{2j}} t_{2j}^{-\delta_2} \right)$
$\omega_1 = \left(\mu^2 \theta^\mu P_{11}^{B\theta} + \mu \theta^\mu P_{11}^{B\theta} + B P_{11}^{\mu} \theta^{B\theta} (B\theta + 1) \right)$

$$f_2(t_{1j}, t_{2j}) = \frac{4 \zeta_1 \zeta_2 \delta_1^2 \delta_2^2 \rho_{0j}^2 \rho_{1j} \rho_{2j} \theta t_{1j} t_{2j} P_{21}^{-\mu-B\theta-2} \left(\mu^2 \theta^\mu P_{21}^{B\theta} + \mu \theta^\mu P_{21}^{B\theta} + B P_{21}^{\mu} \theta^{B\theta} (B\theta+1) \right)}{(\theta+1) \left(e^{\delta_1^2 t_{1j}^2} - 1 \right) \left(e^{\delta_2^2 t_{2j}^2} - 1 \right)} \tag{8.2}$$

where $P_{21} = \rho_{0j} \left(-\zeta_1 \rho_{1j} \log \left(1 - e^{-\delta_1^2 t_{1j}^2} \right) - \zeta_2 \rho_{2j} \log \left(1 - e^{-\delta_2^2 t_{2j}^2} \right) \right) + \theta.$
Figures 1 and 2 show the shape of pdf of Model-I and Model-II, respectively, for the different values of parameters.

8.2 Bivariate Survival Function

Survival function for bivariate models can be obtained by using the following mechanism

$$S(t_{1j}, t_{2j}) = 1 - F(t_{1j}) - F(t_{2j}) + F(t_{1j}, t_{2j})$$

where $F(t_{1j})$ and $F(t_{1j})$ stand for the marginal cumulative distribution function of t_{1j} and t_{2j}, respectively. Consequently, bivariate survival function for Model-I and Model-II will be

$$S_1(t_{1j}, t_{2j}) = \frac{\theta^{\mu+1}\left(P_{11}^{-\mu} - P_{12}^{-\mu} - P_{13}^{-\mu}\right) + \theta^{B\theta}\left(P_{11}^{-B\theta} - P_{12}^{-B\theta} - P_{13}^{-B\theta}\right) + \theta + 1}{\theta + 1} \tag{8.3}$$

where $P_{12} = \zeta_1 \rho_{0j} \rho_{1j} e^{-\xi_1 t_{1j}} t_{1j}^{-\delta_1} + \theta$; $P_{13} = \zeta_2 \rho_{0j} \rho_{2j} e^{-\xi_2 t_{2j}} t_{2j}^{-\delta_2} + \theta$

$$S_2(t_{1j}, t_{2j}) = \frac{\theta^{\mu+1}\left(P_{21}^{-\mu} - P_{22}^{-\mu} - P_{23}^{-\mu}\right) + \theta^{B\theta}\left(P_{21}^{-B\theta} - P_{22}^{-B\theta} - P_{23}^{-B\theta}\right) + \theta + 1}{\theta + 1} \tag{8.4}$$

where $P_{22} = \theta - \zeta_1 \rho_{0j} \rho_{1j} \log\left(1 - e^{-\delta_1^2 t_{1j}^2}\right)$; $P_{23} = \theta - \zeta_2 \rho_{0j} \rho_{2j} \log\left(1 - e^{-\delta_2^2 t_{2j}^2}\right)$

8.3 Hazard Gradient Function

Hazard gradient function for bivariate models can be obtained by using the following mechanism

$$\eta(t_1, t_2) = -\frac{\partial \log S(t_1, t_2)}{\partial t_2}$$

where $S(t_1, t_2)$ stands for the bivariate survival function. Consequently, hazard gradient function for Model-I and Model-II will be

$$\eta_1(t_1, t_2) = \frac{\zeta_2 P_{14} \rho_{0j} \rho_{2j} \theta e^{-\xi_2 t_{2j}} t_{2j}^{-\delta_2 - 1} (\xi_2 t_{2j} + \delta_2) P_{12}^{\mu+B\theta}}{P_{13}(P_{11}+\theta)\left(-\theta^{\mu+1} P_{12}^{B\theta} (P_{11}+\theta)^{\mu+B\theta} P_{13}^{\mu+B\theta} + P_{12}^{\mu} \theta^{B\theta} (P_{11}+\theta)^{\mu+B\theta}\left(-P_{13}^{\mu+B\theta}\right) + P_{15} P_{12}^{\mu+B\theta}\right)} \tag{8.5}$$

where

$$P_{14} = -\mu\theta^{\mu}(P_{11}+\theta)^{B\theta} P_{13}^{B\theta}\left(P_{13}^{\mu+1} - (P_{11}+\theta)^{\mu+1}\right) - B P_{13}^{\mu} \theta^{B\theta} (P_{11}+\theta)^{\mu}\left(P_{13}^{B\theta+1} - (P_{11}+\theta)^{B\theta+1}\right)$$

$$P_{15} = \left(\theta^{B\theta}(P_{11}+\theta)^{\mu} + \theta^{\mu+1}(P_{11}+\theta)^{B\theta} + \theta(P_{11}+\theta)^{\mu+B\theta} + (P_{11}+\theta)^{\mu+B\theta}\right) P_{13}^{\mu+B\theta} - \theta^{\mu+1} P_{13}^{B\theta} (P_{11}+\theta)^{\mu+B\theta} + P_{13}^{\mu}\left(-\theta^{B\theta}\right)(P_{11}+\theta)^{\mu+B\theta}$$

$$\eta_2(t_1, t_2) = \frac{2\zeta_2 \delta_2^2 P_{24} \rho_{0j} \rho_{2j} \theta t_{2j} P_{22}^{\mu+B\theta}}{P_{21} P_{23}\left(e^{\delta_2^2 t_{2j}^2} - 1\right)\left(\theta^{\mu+1} P_{21}^{B\theta} P_{22}^{\mu+B\theta} P_{23}^{\mu+B\theta} + P_{25} P_{21}^{\mu+B\theta} + P_{21}^{\mu} \theta^{B\theta} P_{22}^{\mu+B\theta} P_{23}^{\mu+B\theta}\right)} \tag{8.6}$$

where

$$P_{24} = \mu\theta^{\mu}\left(P_{21}^{\mu+1} - P_{23}^{\mu+1}\right)P_{21}^{B\theta}P_{23}^{B\theta} + BP_{21}^{\mu}P_{23}^{\mu}\theta^{B\theta}\left(P_{21}^{B\theta+1} - P_{23}^{B\theta+1}\right)$$

$$P_{25} = -\theta^{\mu+1}P_{22}^{B\theta}P_{23}^{\mu+B\theta} + P_{22}^{\mu+B\theta}\left(-\theta^{\mu+1}P_{23}^{B\theta} + (\theta+1)P_{23}^{\mu+B\theta} + P_{23}^{\mu}\left(-\theta^{B\theta}\right)\right) + P_{22}^{\mu}\theta^{B\theta}\left(-P_{23}^{\mu+B\theta}\right)$$

8.4 *Conditional Probability Measure*

Conditional probability measure for bivariate models can be obtained by using the following mechanism

$$\chi(t_1, t_2) = \frac{P\left(T_1 > t_1 \mid T_2 > t_2\right)}{P(T_1 > t_1)} = \frac{S(t_1, t_2)}{S(t_1)S(t_2)}$$

where $S(t_1)$, $S(t_2)$ stands for the marginal survival functions of T_1 and T_2, respectively. Consequently, conditional probability measure for Model-I and Model-II will be

$$\chi_1(t_1, t_2) = \frac{(\theta+1)\left(\theta^{\mu+1}\left((P_{11}+\theta)^{-\mu} - P_{12}^{-\mu} - P_{13}^{-\mu}\right) + \theta^{B\theta}\left((P_{11}+\theta)^{-B\theta} - P_{12}^{-B\theta} - P_{13}^{-B\theta}\right) + \theta + 1\right)}{\left(P_{12}^{-\mu}\left(-\theta\mu+1\right) - \theta^{B\theta}P_{12}^{-B\theta} + \theta + 1\right)\left(P_{13}^{-\mu}\left(-\theta\mu+1\right) - \theta^{B\theta}P_{13}^{-B\theta} + \theta + 1\right)} \tag{8.7}$$

$$\chi_2(t_1, t_2) = \frac{(\theta+1)\left(\theta^{\mu+1}\left(P_{21}^{-\mu} - P_{22}^{-\mu} - P_{23}^{-\mu}\right) + \theta^{B\theta}\left(P_{21}^{-B\theta} - P_{22}^{-B\theta} - P_{23}^{-B\theta}\right) + \theta + 1\right)}{\left(P_{22}^{-\mu}\left(-\theta\mu+1\right) - \theta^{B\theta}P_{22}^{-B\theta} + \theta + 1\right)\left(P_{23}^{-\mu}\left(-\theta\mu+1\right) - \theta^{B\theta}P_{23}^{-B\theta} + \theta + 1\right)} \tag{8.8}$$

8.5 *Cross-ratio Function*

Cross-ratio function [see Clayton 1978 and Oakes 1989] can be interpreted as the relative risk for an individual if the other one has experienced the event rather than being event free at a given time

$$\kappa(t_1, t_2) = \frac{\frac{\partial^2 S(t_1,t_2)}{\partial t_1 \partial t_2} S(t_1, t_2)}{\frac{\partial S(t_1,t_2)}{\partial t_1} \frac{\partial S(t_1,t_2)}{\partial t_2}} \tag{8.9}$$

Consequently, cross-ratio function for Model-I and Model-II will be

$$\kappa_1(t_1, t_2) = \frac{P_{12} P_{13} Z_{12} \left(B^2 \theta^{B\theta+1} (P_{11}+\theta)^{\mu} + B\theta^{B\theta} (P_{11}+\theta)^{\mu} + \mu(\mu+1)\theta^{\mu} (P_{11}+\theta)^{B\theta}\right)}{\theta Z_{13} \left(\mu\theta^{\mu} (P_{11}+\theta)^{B\theta} P_{12}^{B\theta} \left(P_{12}^{\mu+1} - (P_{11}+\theta)^{\mu+1}\right) + B P_{12}^{\mu} \theta^{B\theta} (P_{11}+\theta)^{\mu} \left(P_{12}^{B\theta+1} - (P_{11}+\theta)^{B\theta+1}\right)\right)} \tag{8.10}$$

where

$$Z_{11} = \left(\theta^{B\theta}(P_{11}+\theta)^{\mu} + \theta^{\mu+1}(P_{11}+\theta)^{B\theta} + \theta(P_{11}+\theta)^{\mu+B\theta} + (P_{11}+\theta)^{\mu+B\theta}\right) P_{13}^{\mu+B\theta}$$
$$- \theta^{\mu+1} P_{13}^{B\theta} (P_{11}+\theta)^{\mu+B\theta} + P_{13}^{\mu} \left(-\theta^{B\theta}\right) (P_{11}+\theta)^{\mu+B\theta}$$

$$Z_{12} = -\theta^{\mu+1} P_{12}^{B\theta} (P_{11}+\theta)^{\mu+B\theta} P_{13}^{\mu+B\theta} + P_{12}^{\mu} \theta^{B\theta} (P_{11}+\theta)^{\mu+B\theta} \left(-P_{13}^{\mu+B\theta}\right) + Z_{11} P_{12}^{\mu+B\theta}$$

$$Z_{13} = \mu\theta^{\mu} (P_{11}+\theta)^{B\theta} P_{13}^{B\theta} \left(P_{13}^{\mu+1} - (P_{11}+\theta)^{\mu+1}\right) + B P_{13}^{\mu} \theta^{B\theta} (P_{11}+\theta)^{\mu} \left(P_{13}^{B\theta+1} - (P_{11}+\theta)^{B\theta+1}\right)$$

$$\kappa_2(t_1, t_2) = \frac{P_{22} P_{23} Z_{22} \left(\mu^2 \theta^{\mu} P_{21}^{B\theta} + \mu\theta^{\mu} P_{21}^{B\theta} + B P_{21}^{\mu} \theta^{B\theta} (B\theta+1)\right)}{\theta Z_{23} \left(\mu\theta^{\mu} \left(P_{21}^{\mu+1} - P_{22}^{\mu+1}\right) P_{21}^{B\theta} P_{22}^{B\theta} + B P_{21}^{\mu} P_{22}^{\mu} \theta^{B\theta} \left(P_{21}^{B\theta+1} - P_{22}^{B\theta+1}\right)\right)} \tag{8.11}$$

where

$$Z_{21} = -\theta^{\mu+1} P_{22}^{B\theta} P_{23}^{\mu+B\theta} + P_{22}^{\mu+B\theta} \left(-\theta^{\mu+1} P_{23}^{B\theta} + (\theta+1) P_{23}^{\mu+B\theta} + P_{23}^{\mu}\left(-\theta^{B\theta}\right)\right)$$
$$+ P_{22}^{\mu} \theta^{B\theta} \left(-P_{23}^{\mu+B\theta}\right)$$

$$Z_{22} = \theta^{\mu+1} P_{21}^{B\theta} P_{22}^{\mu+B\theta} P_{23}^{\mu+B\theta} + Z_{21} P_{21}^{\mu+B\theta} + P_{21}^{\mu} \theta^{B\theta} P_{22}^{\mu+B\theta} P_{23}^{\mu+B\theta}$$

$$Z_{23} = \mu\theta^{\mu} \left(P_{21}^{\mu+1} - P_{23}^{\mu+1}\right) P_{21}^{B\theta} P_{23}^{B\theta} + B P_{21}^{\mu} P_{23}^{\mu} \theta^{B\theta} \left(P_{21}^{B\theta+1} - P_{23}^{B\theta+1}\right).$$

Figures 3 and 4 show the graph of cross-ratio function which show that there is positive association between time T_{1j} and T_{2j} for both Model-I and Model-II with three different sets of the values of parameters.

9 Likelihood Design and Bayesian Paradigm

For the study, n individuals have been considered. Observed failure times have been indicated by (t_{1j}, t_{2j}). We are using the left censoring scheme. Censoring time supposed to be indicated by c_{1j} and c_{2j} for j^{th} individual $(j = 1, 2, 3, ..., n)$. Independence between censoring schemes and lifetimes of individuals has been presumed. Likelihood function can be described for bivariate lifetime random variable of the jth individual as

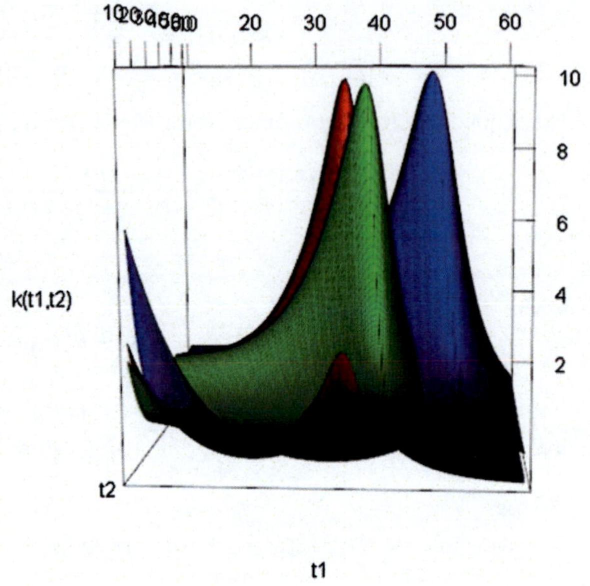

Fig. 3 Cross-ratio function for Model-I

$$L_j(t_{1j}, t_{2j}) = \begin{cases} f_1(t_{1j}, t_{2j}), & t_{1j} > c_{1j}, t_{2j} > c_{2j} \\ f_2(t_{1j}, c_{2j}), & t_{1j} > c_{1j}, t_{2j} < c_{2j} \\ f_3(c_{1j}, t_{2j}), & t_{1j} < c_{1j}, t_{2j} > c_{2j} \\ f_4(c_{1j}, c_{2j}), & t_{1j} < c_{1j}, t_{2j} < c_{2j} \end{cases} \tag{9.1}$$

and the likelihood function is given by

$$L(\underline{\Theta}, \underline{\beta}, \theta, \mu) = \prod_{j=1}^{n_1} f_1(t_{1j}, t_{2j}) \prod_{j=1}^{n_2} f_2(t_{1j}, c_{2j}) \prod_{j=1}^{n_3} f_3(c_{1j}, t_{2j}) \prod_{j=1}^{n_4} f_4(c_{1j}, c_{2j}) \tag{9.2}$$

where $\underline{\Theta}$, $\underline{\beta}$, θ and μ are vector of baseline parameters and the vector of regression coefficients and frailty parameters, respectively. Likelihood function for without frailty model is

$$L(\underline{\Theta}, \underline{\beta}) = \prod_{j=1}^{n_1} f_1(t_{1j}, t_{2j}) \prod_{j=1}^{n_2} f_2(t_{1j}, c_{2j}) \prod_{j=1}^{n_3} f_3(c_{1j}, t_{2j}) \prod_{j=1}^{n_4} f_4(c_{1j}, c_{2j}) \tag{9.3}$$

Let n_1, n_2, n_3 and n_4 be the number of pairs for which first and second failure times (t_{1j}, t_{2j}) lie in the ranges $t_{1j} > c_{1j}$, $t_{2j} > c_{2j}$; $t_{1j} > c_{1j}$,$t_{2j} < c_{2j}$; $t_{1j} < c_{1j}$,$t_{2j} > c_{2j}$; and $t_{1j} < c_{1j}$,$t_{2j} < c_{2j}$, respectively, and let

Fig. 4 Cross-ratio function for Model-II

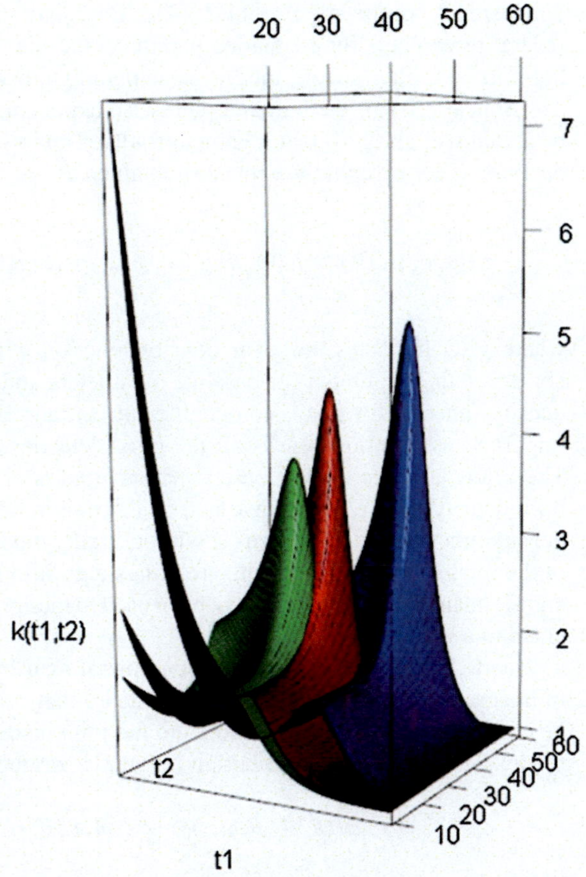

$$
\begin{aligned}
f_1(t_{1j}, t_{2j}) &= \frac{\partial^2 F(t_{1j}, t_{2j})}{\partial t_{1j} \partial t_{2j}} \\
f_2(t_{1j}, c_{2j}) &= \frac{\partial F(t_{1j}, c_{2j})}{\partial t_{1j}} \\
f_3(c_{1j}, t_{2j}) &= \frac{\partial F(c_{1j}, t_{2j})}{\partial t_{2j}} \\
f_4(c_{1j}, c_{2j}) &= F(c_{1j}, c_{2j})
\end{aligned} \tag{9.4}
$$

Substituting cumulative reversed hazard rates $\Psi_{01}(t_{1j})$ and $\Psi_{02}(t_{2j})$ and distribution function $F(t_{1j}, t_{2j})$ in Equation (9.4) for Model-I and Model-II and by differentiating, we get the likelihood function. The maximum likelihood method has a crucial importance in computing efficient estimators. Inappropriately, due to a convergence problem, maximum likelihood failed to estimate the parameters, because Model-I and Model-II have nine-dimensional and Model-III and Model-IV have seven-dimensional optimization problem. The Bayesian scenario has been discussed by several researchers for estimating parameters of the frailty models. For gamma and log-normal frailty models, the Bayesian paradigm has been contemplated by Santos and Achcar (2010). Weibull and piecewise exponential model have been discussed by Ibrahim et al. (2001) with gamma frailty. The joint posterior density function of parameters for given failure times is obtained as

$$\pi(\Theta, \theta, \mu, \underline{\beta_0}) \propto L(\Theta, \theta, \mu, \underline{\beta}_0) g_1(\zeta) g_2(\delta) g_3(\theta) g_4(\mu) \prod_{i=1}^{3} p_i(\beta_{0i \times 1})$$

where $g_i(.)$ indicates the prior density function with known hyperparameters of corresponding argument for baseline parameters and frailty variance; $p_i(.)$ is prior density function for regression coefficient β_{0i}, and likelihood function is $L(.)$. An important assumption here is all the prior densities are independently distributed. The prior densities for $g_i(.)$ are taken as gamma or uniform distribution, and the prior density for $p_i(.)$ is taken as normal distribution. In a similar way, joint posterior density function can be written for without frailty models. To estimate the parameters of the models, hyper-Metropolis–Hastings algorithms have been used. Geweke test and Gelman–Rubin statistics have been used to monitor the convergence of a Markov chain to a stationary distribution.

Due to the high dimensions of conditional distributions, it is not unproblematic to integrate out. Thus, it has been considered that full conditional distributions can be obtained as they are proportional to the joint distribution of the parameter of the model. The conditional distribution for single parameter δ with frailty is

$$\varphi_1(\delta \mid \zeta, \theta, \mu, \underline{\beta_0}) \propto L(\delta, \zeta, \theta, \mu, \underline{\beta_0}) \cdot g_1(\delta)$$

and the conditional distribution for single parameter δ without frailty is

$$\varphi_1(\delta \mid \zeta, \underline{\beta_0}) \propto L(\delta, \zeta, \underline{\beta_0}) \cdot g_1(\delta)$$

Similarly, full conditional distributions for all the parameters can be obtained.

10 Simulation Study

A simulation study has been executed to appraise the Bayesian estimation paradigm for Model-I and Model-II. Single covariate K_1 has been considered as follows: normal distribution. The frailty variable W is assumed to follow generalized Lindley distribution. Independence between lifetimes of individuals has been considered. Samples are generated using the subsequent mechanism

1. From the binomial distribution with probability 0.6, 25 values for K_1 have been generated.

2. For known covariate, compute $\rho = e^{K_0\beta_0}$.
3. Lifetimes reckoned to follow modified inverse Weibull and generalized Rayleigh baseline distributions for given frailty W_j. Twenty-five values of lifetimes have been generated.
   Conditional distribution function for lifetime t_j $(j = 1, 2, ..., n)$ for given frailty $W_j = w_j$ and covariate K_1 is

$$F(t_j|w, K_0) = e^{-w\Psi_0(t_j)\rho}$$

   Equating $F(t_j|w, K_0)$ to random number, say $v_j(0 < v_j < 1)$ spawned from $U(0, 1)$ over $t_j > 0$ we get
   for Model-I,

$$t_j = \frac{-1}{\delta}\log\left(1 - v^{\frac{1}{w\zeta\rho}}\right) \tag{10.1}$$

   for Model-II,

$$t_j = \left[\frac{-\delta}{\log\left(1-\left(1-v^{\frac{1}{w\rho}}\right)^{\frac{1}{\zeta}}\right)}\right]^{\frac{1}{2}} \tag{10.2}$$

4. Censoring time c_j has been generated from $G(0.9, 0.01)$ for Model-I.
5. Observe the jth survival time $t_j^* = min(t_j, c_j)$ and the censoring indicator Υ_j for the jth individual $(j = 1, 2, ..., 25)$ where

$$\Upsilon_j = \begin{cases} 1, \; ; \; t_j < c_j \\ 0, \; ; \; t_j > c_j \end{cases}$$

   Thus, we have data consisting of 25 pairs of survival times t_j^* and the censoring indicator Υ_j.

We consider gamma (0.0001,0.0001) or uniform (0,100) prior for the distribution of the unknown parameters of all the frailty distribution. We also consider normal (0,1000) prior for the distribution of the unknown regression parameters. Concurrently, with different priors and starting points, two chains based on two priors (one is based on gamma prior, and another is based on uniform prior) have been operated. Both chains recapitulated 100,000 times. Gelman–Rubin test (Gelman and Rubin 1992) values are very close to one. Due to small values of Geweke test statistic (see Geweke 1992) and corresponding p-values, the chains reach stationary distribution for both prior sets. In view, estimates of parameters were about the same, and no impact of prior distributions has been found on posterior summaries. Here, the analysis for one chain has been exhibited because both the chains have shown similar

Table 1 Posterior summary of generalized Lindley frailty with baseline modified inverse Weibull (simulation study: Model-I)

Parameter	Estimate	s.e.	L.C.L.	U.C.L.	Geweke test	p-value	Gelman–Rubin test
$\zeta_1(70)$	69.29948	5.85707	60.41352	79.00819	−0.01475	0.49412	1.00176
$\delta_1(0.024)$	0.02381	0.00297	0.01929	0.02878	−0.00190	0.49924	1.00274
$\xi_1(0.15)$	0.12442	0.01001	0.10801	0.14573	−0.00745	0.49703	1.00151
$\zeta_2(70)$	67.62318	4.93882	60.33410	77.83040	−0.00431	0.49828	1.00020
$\delta_2(0.024)$	0.02390	0.00269	0.01924	0.02866	0.01723	0.50687	1.00019
$\xi_2(0.15)$	0.12794	0.00985	0.10800	0.14772	−0.00440	0.49824	1.00052
$\theta(11)$	11.33853	1.52686	8.41236	13.86792	−0.00117	0.49953	1.00138
$\mu(2)$	2.09611	0.15760	1.83462	2.42991	0.00944	0.50377	1.00009
$\beta_1(-0.02)$	−0.04951	0.10860	−0.19131	0.21751	0.00613	0.50245	1.00090

Table 2 Posterior summary of generalized Lindley frailty with baseline generalized Rayleigh (simulation study: Model-II)

Parameter	Estimate	s.e.	L.C.L.	U.C.L.	Geweke test	p-value	Gelman–Rubin test
$\zeta_1(350)$	354.26160	28.60295	301.38660	398.17020	0.01640	0.50654	1.00245
$\delta_1(0.14)$	0.13671	0.00343	0.13061	0.14378	0.00390	0.50156	1.00325
$\zeta_2(350)$	354.11490	27.04054	306.67280	398.12180	0.02299	0.50917	1.00010
$\delta_2(0.14)$	0.13992	0.00379	0.13259	0.14672	0.00684	0.50273	1.00107
$\theta(10.2)$	9.60497	1.32983	7.44799	12.59520	−0.00515	0.49794	1.00054
$\mu(2.8)$	2.98979	0.42008	2.20360	3.73705	−0.01113	0.49556	1.00006
$\beta_1(0.02)$	0.03066	0.05487	−0.06309	0.12161	−0.00981	0.49609	1.00051

results. Tables 1 and 2 present the estimates and the credible intervals of the parameters for the Model-I and Model-II based on the simulation study. The Gelman–Rubin convergence statistic values are nearly equal to one; also, the Geweke test values are quite small and the corresponding p-values are large enough to say that the chain attains stationary distribution.

11 Analysis of Australian Twin Data

We run test for this model to Australian twin data/appendectomy patient data (Duffy et al. (1990)). The zygote category 4 has been considered out of six zygote categories. 9 and 11 observations have been censored, respectively, in twins 1 and 2 out of 350 pairs of twins. Left censored observations are considered to those individuals who are

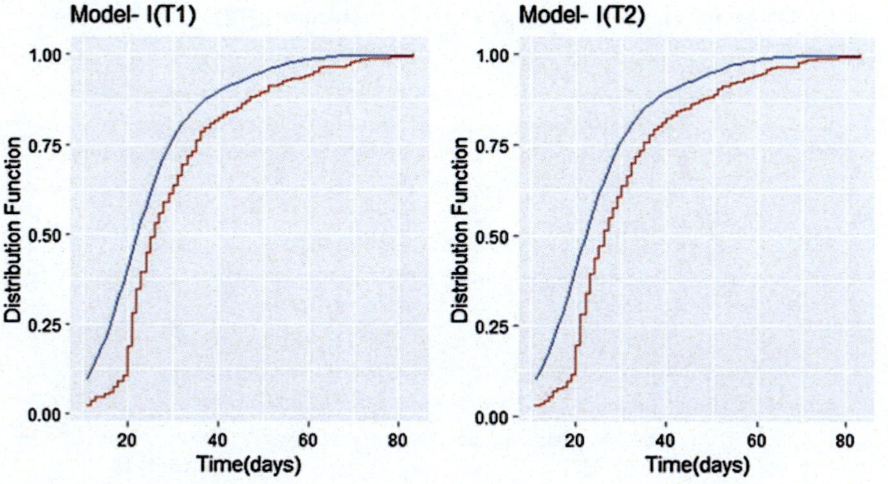

Fig. 5 K-S plot for Model-I

Table 3 p-value of K-S statistics for goodness of fit test for Australian twin data set

Model	T_1 p-value	T_2 p-value
Model-I	0.27600	0.19670
Model-II	0.38790	0.28490

having age at onset less than 11. Information on the age at onset of the appendectomy of twins is given in the data which are lifetimes of twins (T_1 and T_2). In the risk of appendectomy, we take hereditary and environmental factors as the frailty variable. Here, there is a common covariate age for both T_1 and T_2 and one covariate each for T_1 and T_2, i.e., presence or absence of appendectomy.

The Australian twin data is left censored data which is suitable to analyze bivariate survival data with reversed hazard rate. The use of the reversed hazard rate is more relevant in the left censored data. To check goodness of fit of Australian twin data set, we consider Kolmogorov–Smirnov (K-S) test for two baseline distributions. Table 3 gives the p-values of goodness of fit test for Model-I and Model-II. Thus from p-values of K-S test, we can say that there is no statistical evidence to reject the hypothesis that data are from the Model-I and Model-II in the marginal case and we assume that they also fit for bivariate case. Figures 5 and 6 show the parametric plot with semi-parametric plot for Model-I and Model-II taken separately for T_1 and T_2, and both lines are close to each other.

For frailty parameters, gamma distribution with very small shape and scale parameters (say, 0.0001) has been used. Additionally, it can be considered regression coefficients are normally distributed with mean zero and high variance (say 1000). A similar type of prior was used in Ibrahim et al. (2001) and Santos and Achcar (2010). That is why for frailty parameters θ, μ and regression coefficients β_{0i}, $i = 1, 2, 3,$

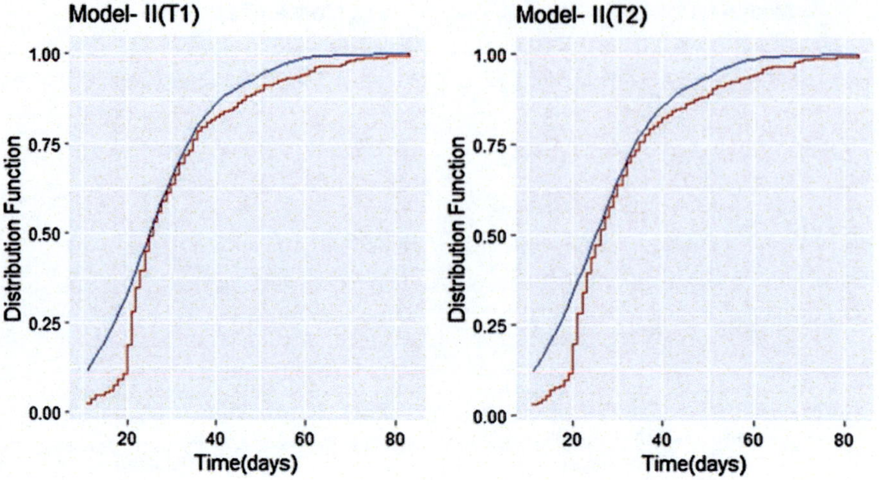

Fig. 6 K-S plot for Model-II

Table 4 Posterior summary of generalized Lindley frailty with baseline modified inverse Weibull for Australian twin data (Model-I)

Parameter	Estimate	s.e.	L.C.L.	U.C.L.	Geweke test	p-value	Gelman–Rubin test
ζ_1	212.04090	20.25532	160.35860	239.22250	−0.00004	0.49998	1.00192
δ_1	0.02497	0.00530	0.01571	0.03400	0.00365	0.50146	1.00273
ξ_1	0.15884	0.00432	0.14970	0.16756	−0.00305	0.49879	1.00305
ζ_2	200.15810	23.62432	152.47470	237.29860	0.00221	0.50088	1.00504
δ_2	0.02520	0.00532	0.01570	0.03432	−0.00143	0.49943	1.00049
ξ_2	0.15666	0.00479	0.14705	0.16598	−0.00282	0.49887	1.00014
θ	14.41710	1.03652	12.27471	16.17366	−0.00060	0.49976	1.00342
μ	1.85515	0.19233	1.50561	2.26569	−0.00461	0.49816	1.00112
β_1	0.02003	0.00366	0.01243	0.02660	−0.00088	0.49965	1.00740
β_2	−0.04577	0.04730	−0.13603	0.03760	0.00704	0.49965	1.00081
β_3	0.01171	0.04752	−0.07242	0.10314	−0.00188	0.49925	0.99996

vague priors have been used. We consider the prior distribution of the baseline parameters as flat because we do not have any information on these baseline parameters. We considered two different vague prior distributions for the distribution of parameters of all frailty models; one is gamma distribution with shape and scale hyperparameters as (0.0001,0.0001), and another is uniform distribution with interval (0,100). We consider normal (0,1000) prior for the distribution of regression coefficients. Under the Bayesian paradigm, for both models, two parallel chains have been run. Also, two sets of prior distributions have been used with different starting points using the

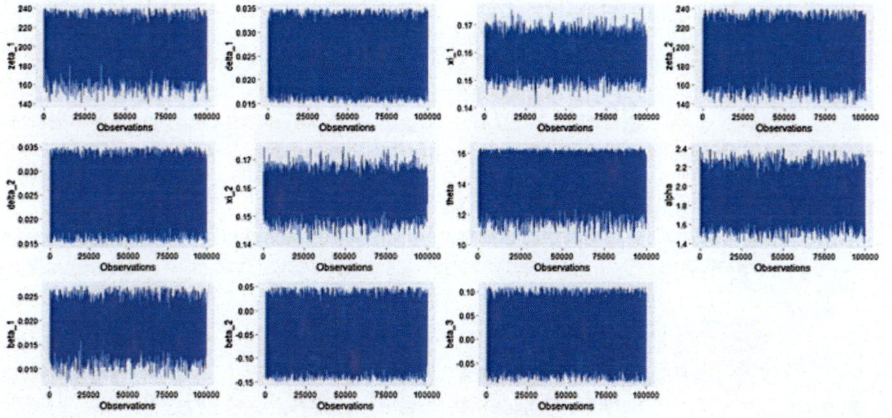

Fig. 7 Trace plots for Model-I

hyper-Metropolis–Hastings algorithm based on normal transition kernels. It can be said that estimates are independent of the different prior distributions because, for both sets of priors, estimates of parameters are approximately similar. We got almost similar convergence rate of Gibbs sampler for both sets of priors. Here, the analysis for one chain has been exhibited because both the chains have shown similar results.

Markov chain has attained the stationary state because the zigzag pattern of the trace plots for all the parameters indicates that the parameters move and mix more freely (see Fig. 7). Coupling from the past plot has been applied to fix up the burn-in period (see Fig. 8). A sequence of draws may have serial correlation after the burn-in period. Randomness may not be shown in successive draws. But almost independence can be seen in values at the extensive split (see Fig. 9). After using the values from the single run of the Markov chain, a vague sample can be obtained from the posterior distribution. Because of the burn-in period, it has been found at extensive spaced time points. After a certain lag, the serial correlation of the parameters turns out to almost negligible for all the parameters. Observations are shown independently after thinning the serial correlation function plot. Autocorrelation function (ACF) plots can be utilized to examine the appropriate blend of our chains. ACF plot for each parameter is converging to the posterior mean of the parameter and, thus, represents a good mixing of the chain. Thus, our diagnostic plots suggest that the MCMC chains are mixing very well. Posterior density plots are drawn for the estimates of the parameters in the Model-I to visualize (see Fig. 10). From Fig. 10, it is observed that some of posterior densities of the parameters have unimodal and bimodal shapes which are quite possible in the mixture models. Figure 11 shows conditional predictive ordinate (CPO) plot for Model-I against Model-III, and Fig. 12 shows CPO plot for Model-II against Model-IV. From Figs. 11 and 12, we can observe that more than 50% of the points fall on the positive side (upper side) of the plot which supports Model-I and Model-II against Model-III and Model-IV, respectively.

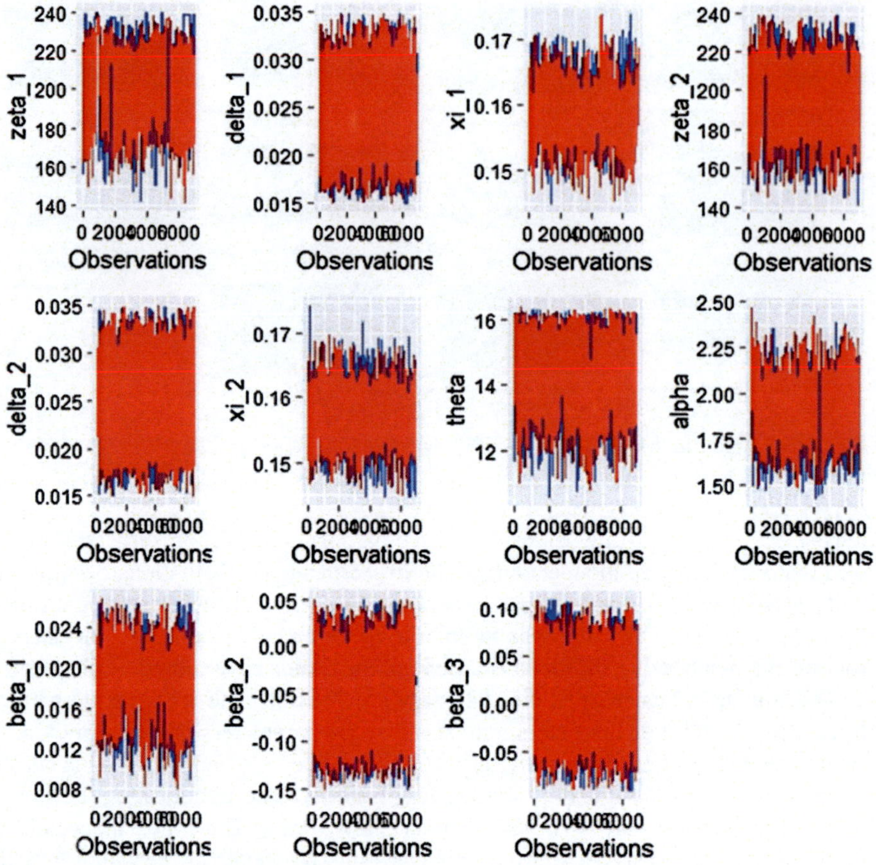

Fig. 8 Coupling from the past plots for Model-I

The Gelman–Rubin convergence statistic values are closely equal to one. The Geweke test statistic values are somewhat small, and the corresponding p-values are large enough to say that the chains reach stationary distribution. Tables 4 and 5 give the values of posterior mean and the standard error with 95% credible intervals, the Gelman–Rubin statistics values and the Geweke test with p-values for Model-I, Model-II, Model-III and Model-IV. Table 6 shows Kendall's τ values for Model-I and Model-II which show that there is a dependence between T_1 and T_2. Table 7 presents the values of AIC, BIC and DIC for both models.

Values of AIC, BIC, and DIC, given in Table 7, have been used for the comparison of all models. Model-I holds the lowest possible values of AIC, BIC and DIC. For all models, regression coefficients contained different values. For all the models, the credible intervals of β_1 (age of twins) do not contain zero. It indicates that the covariate age of twins has significant effect on the lifetimes (age at onset).

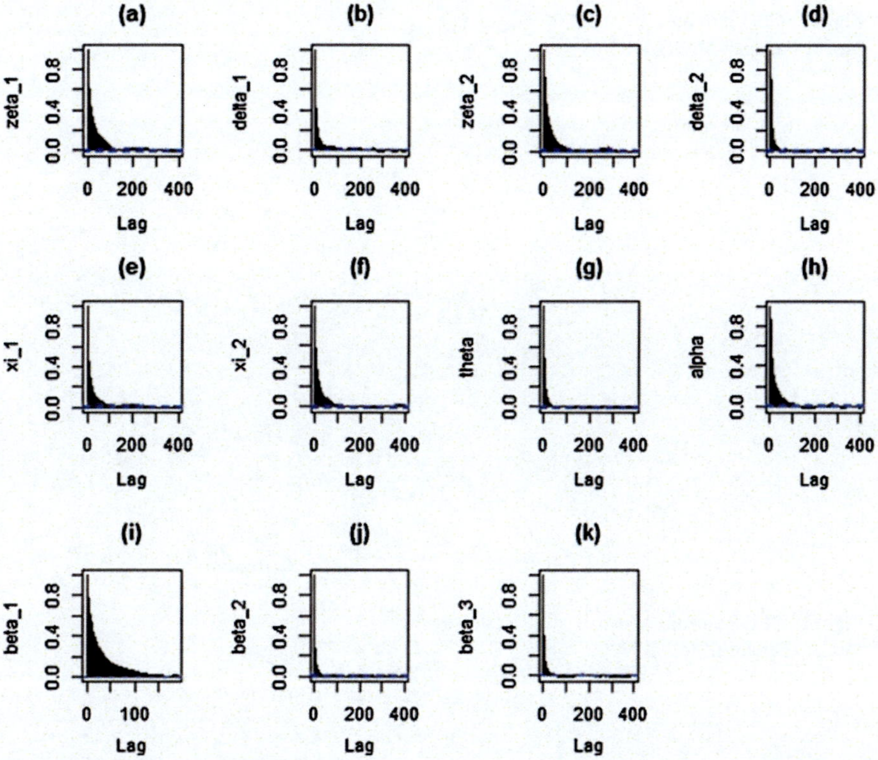

Fig. 9 ACF plots for Model-I

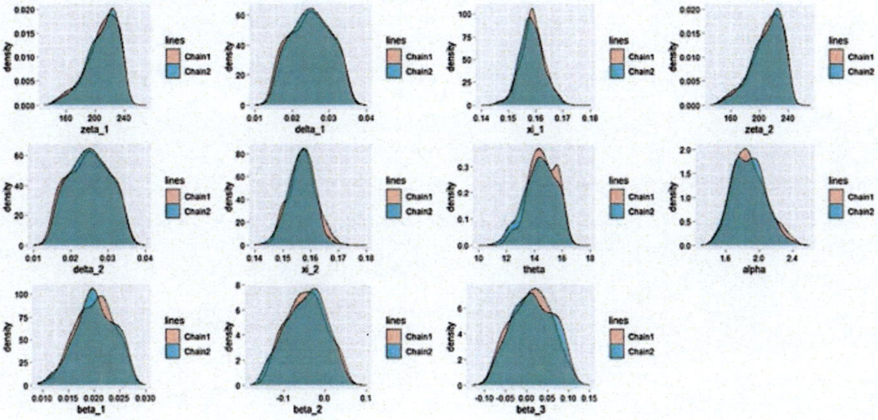

Fig. 10 Posterior density plots for Model-I

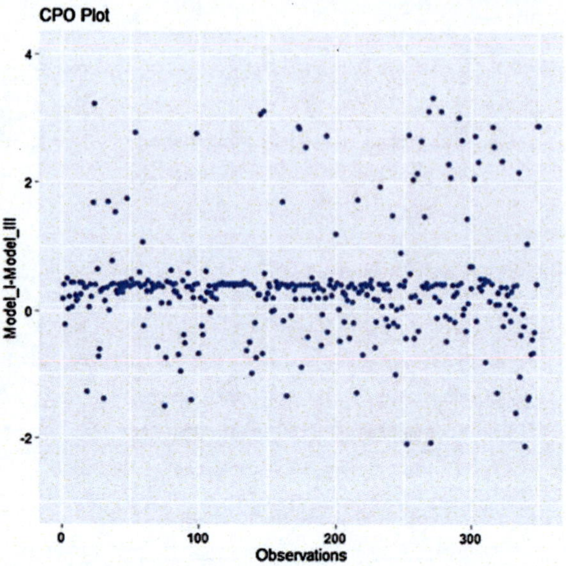

Fig. 11 CPO plot for Model-I against Model-III

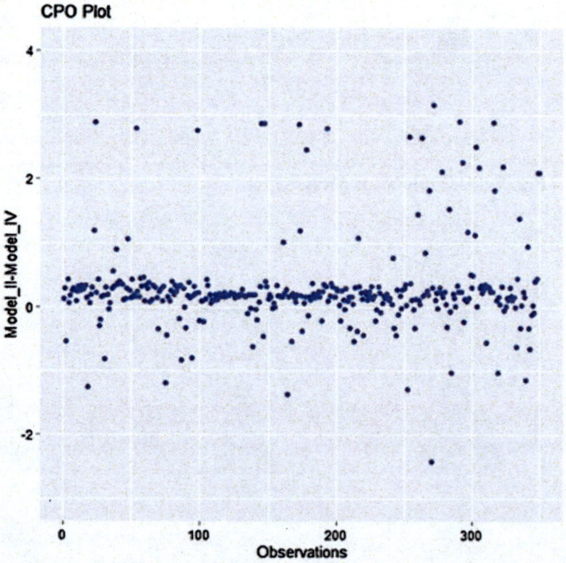

Fig. 12 CPO plot for Model-II against Model-IV

Table 5 Posterior summary of generalized Lindley frailty with baseline generalized Rayleigh for Australian twin data (Model-II)

Parameter	Estimate	s.e.	L.C.L.	U.C.L.	Geweke	p-value test	Gelman Rubin test
ζ_1	14.15306	1.08294	12.21941	15.94604	−0.00579	0.49769	1.04435
δ_1	0.04428	0.00082	0.04260	0.04618	−0.00236	0.49906	1.00108
ζ_2	13.66598	1.05881	12.08083	15.74751	−0.00028	0.49989	1.02802
δ_2	0.04397	0.00091	0.04220	0.04588	0.00394	0.50157	1.00051
θ	15.97924	0.47291	14.75272	16.55289	0.00311	0.50124	1.00133
μ	2.43047	0.20181	2.04625	2.81012	0.00286	0.50114	1.01476
β_1	0.01029	0.00335	0.00367	0.01659	0.00355	0.50142	1.00021
β_2	−0.07414	0.13910	−0.36440	0.16239	0.01134	0.50142	1.00135
β_3	0.04978	0.05017	−0.03710	0.14436	0.00560	0.50223	1.00362

Table 6 Kendall's τ measure of dependence

Model	Kendall's τ value
Model-I	0.30248
Model-II	0.25746

Table 7 AIC, BIC and DIC comparison for all four models

Model	AIC	BIC	DIC
Model-I	5125.697	5176.856	5070.713
Model-II	5282.064	5322.080	5274.653
Model-III	5385.744	5425.27	5377.353
Model-IV	5476.058	5507.883	5471.697

12 Conclusions

To authenticate the influence of frailty, generalized Lindley frailty model under modified inverse Weibull and generalized Rayleigh baseline distributions have been proposed. To fit the proposed models, M-H algorithm has been applied. Analysis has been done in R statistical software with self-written programs. The value of both frailty parameters for Model-I ($\theta = 14.41710$, $\mu = 1.85515$) and Model-II ($\theta = 15.97924$, $\mu = 2.43047$) is very high, and corresponding variances are 13.36152 and 14.32177 by using Equation (2.2). In Table 6, we calculate Kendall's τ measure of dependence by using Equation (5.2). This exhibits that there is a strong indication of heterogeneity among the patient in the population for the data set. For appendectomy patient data, age of twins has been found statistically significant factor for Model-I, Model-II and Model-III. To take the decision about all models, different tools have been utilized. With the lowest value of AIC, BIC and DIC, carried by Table 6 and CPO plots (see Figs. 11 and 12), it can be said that Model-I and

Model-II are more beneficial to use in comparison with Mode-III and Model-IV for appendectomy patient data. The generalized Lindley frailty models with modified inverse Weibull and generalized Rayleigh baseline distributions perform better than two models discussed by Hanagal and Pandey (2017b) and Hanagal and Bhambure (2017) when we compare the model selection criteria (AIC, BIC and DIC values). Between the Model-I and Model-II, Model-I (generalized inverse Weibull baseline with generalized Lindley frailty) is better as compared to Model-II (generalized Rayleigh baseline with generalized Lindley frailty).

Acknowledgements We thank both the referees for the valuable suggestions and comments.

References

Block, H. W., Savits, T. H., & Singh, H. (1998). On the reversed hazard rate function. *Probability in the Engineering and Informational Sciences, 12*, 69–90.

Cox, D. R. (1972). Regression Models and Life Tables (with Discussion). *Journal of Royal Statistical Society, Series B, 34*, 187–220.

Duffy, D. L., Martin, N. G., & Mathews, J. D. (1990). Appendectomy in Australian twins. *Australian Journal of Human Genetics, 47*(3), 590–92.

Elbatal, I., Merovei, F., & Elgarhy, M. (2013). A new generalized Lindley distribution. *Mathematical Theory and Modeling, 3*(13), 30–47.

Gelman, A. and Rubin, D. B. (1992). A single series from the Gibbs sampler provides a false sense of security. In Bayesian Statistics 4 (J. M. Bernardo, J. 0.Berger, A. P. Dawid and A. F. M. Smith, eds.). Oxford: Oxford University Press. pp. 625-632.

Geweke, J. (1992). Evaluating the Accuracy of Sampling-Based Approaches to the Calculation of Posterior Moments. In J. M. Bernardo, J. Berger, A. P. Dawid, & A. F. M. Smith (Eds.), *Bayesian Statistics 4* (pp. 169–193). Oxford: Oxford University Press.

Hanagal, D. D. (2008). Frailty regression models in mixture distributions. *Journal of Statistical Planning and Inference, 138*(8), 2462–68.

Hanagal, D. D. (2011). *Modeling Survival Data Using Frailty Models*. New York: Chapman & Hall/CRC.

Hanagal, D.D. (2017). Frailty Models in Public Health. *Handbook of Statistics, 37(B)*, 209-247. Elsevier Publishers; Amsterdam.

Hanagal, D. D. (2019). *Modeling Survival Data Using Frailty Models* (2nd ed.). Singapore: Springer.

Hanagal, D. D., & Bhambure, S. M. (2014). Analysis of kidney infection data using shared positive stable frailty models. *Advances in Reliability, 1*, 21–39.

Hanagal, D. D., & Bhambure, S. M. (2015). Comparison of shared gamma frailty models using Bayesian approach. *Model Assisted Statistics & Applications, 10*, 25–41.

Hanagal, D. D., & Bhambure, S. M. (2016). Modeling bivariate survival data using shared inverse Gaussian frailty model. *Communications in Statistics, Theory & Methods, 45*(17), 4969–4987.

Hanagal, D. D., & Bhambure, S. M. (2017). Modeling Australian twin data using shared positive stable frailty models based on reversed hazard rate. *Communications in Statistics-Theory and Methods, 46*(8), 3754–3771.

Hanagal, D. D., & Dabade, A. D. (2013). Modeling of inverse Gaussian frailty model for bivariate survival data. *Communications in Statistics-Theory and Methods, 42*(20), 3744–3769.

Hanagal, D. D., & Dabade, A. D. (2015). Comparison of shared frailty models for kidney infection data under exponential power baseline distribution. *Communications in Statistics-Theory and Methods, 44*(23), 5091–5108.

Hanagal, D. D., & Kamble, A. T. (2015). Bayesian estimation in shared compound poisson frailty models. *Journal of Reliability and Statistical Studies, 8*(1), 159–180.

Hanagal, D. D., & Pandey, A. (2014a). Inverse Gaussian shared frailty for modeling kidney infection data. *Advances in Reliability, 1*, 1–14.

Hanagal, D. D., & Pandey, A. (2014b). Gamma shared frailty model based on reversed hazard rate for bivariate survival data. *Statistics & Probability Letters, 88*, 190–196.

Hanagal, D. D., & Pandey, A. (2015a). Gamma frailty models for bivariate survival data. *Journal of Statistical Computation and Simulation, 85*(15), 3172–3189.

Hanagal, D. D., & Pandey, A. (2015b). Inverse Gaussian shared frailty models with generalized exponential and generalized inverted exponential as baseline distributions. *Journal of Data Science, 13*(2), 569–602.

Hanagal, D. D., & Pandey, A. (2016a). Gamma shared frailty model based on reversed hazard rate. *Communications in Statistics-Theory and Methods, 45*(7), 2071–2088.

Hanagal, D. D., & Pandey, A. (2016b). Inverse Gaussian shared frailty models based on reversed hazard rate. *Model Assisted Statistics and Applications, 11*, 137–151.

Hanagal, D. D., & Pandey, A. (2017a). Shared inverse Gaussian frailty models based on additive hazards. *Communications in Statistics-Theory and Methods, 46*(22), 11143–11162.

Hanagal, D. D., & Pandey, A. (2017b). Shared frailty models based on reversed hazard rate for modified inverse Weibull distribution as baseline distribution. *Communications in Statistics-Theory and Methods, 46*(1), 234–246.

Hanagal, D. D., & Sharma, R. (2013). Modeling heterogeneity for bivariate survival data by shared gamma frailty regression model. *Model Assisted Statistics and Applications, 8*, 85–102.

Hanagal, D. D., & Sharma, R. (2015a). Bayesian inference in Marshall-Olkin bivariate exponential shared gamma frailty regression model under random censoring. *Communications in Statistics, Theory and Methods, 44*(1), 24–47.

Hanagal, D. D., & Sharma, R. (2015b). Comparison of frailty models for acute leukaemia data under Gompertz baseline distribution. *Communications in Statistics, Theory & Methods, 44*(7), 1338–1350.

Hanagal, D. D., & Sharma, R. (2015c). Analysis of bivariate survival data using shared inverse Gaussian frailty model. *Communications in Statistics, Theory and Methods, 44*(7), 1351–1380.

Hougaard, P. (2000). *Analysis of Multivariate Survival Data*. New York: Springer.

Ibrahim, J. G., Ming-Hui, C., & Sinha, D. (2001). *Bayesian Survival Analysis*. Springer Verlag.

Kumar, D., & Singh, A. (2011). *General Mathematics Notes, 3*(1), 26–31.

Lindley, D. V. (1958). Fiducial distributions and Bayes's theorem. *Journal of the Royal Statistical Society, B, 20*, 102–107.

Raqab, M. Z., & Kundu, D. (2006). Burr type X distribution. *Journal of Probability and Statistics, 4*(2), 179–193.

Sankaran, P. G., & Gleeja, V. L. (2011). On proportional reversed hazards frailty models. *Metron, 69*(2), 151–173.

Santos, C. A., & Achcar, J. A. (2010). A Bayesian analysis for multivariate survival data in the presence of covariates. *Journal of Statistical Theory and Applications, 9*, 233–253.

Shaked, M., & Shantikumar, J. G. (1994). *Stochastic Orders and Their Applications*. New York: Academic Press.

Surles, J. G., & Padgett, W. J. (2001). Inference for reliability and stress-strength for a scaled Burr Type X distribution. *Lifetime Data Analysis, 7*, 187–200.

Vaupel, J. W., Manton, K. G., & Stallaed, E. (1979). The impact of heterogeneity in individual frailty on the dynamics of mortality. *Demography, 16*, 439–454.

Wienke, A. (2011). *Frailty Models in Survival Analysis*. New York: Chapman & Hall/CRC.

Ultimate Ruin Probability for Benktander Gibrat Risk Model

Kanchan Jain and Harmanpreet Singh Kapoor

Abstract In actuarial science and finance, the derivation of ultimate ruin probability for various loss distributions is of key interest. There are many methods available in literature for evaluating ultimate ruin probability for different distributions. Probability of ultimate ruin is derived for a risk model under Benktander Gibrat (BG) distribution, also known as Benktander Type I distribution. Laplace transform, generalized exponential integrals, MeijerG function and Bromwich Integral are used to find ultimate ruin probability.

Keywords Risk model · Ruin theory · Laplace transform · Exponential integral · Bromwich integral · MeijerG function

1 Introduction

In insurance sector, an insured has to pay premium to the insurer for taking out an insurance policy for providing security to the insured as well as the insurer against future risks. The insurer has to protect herself/himself against the liabilities that can arise at anytime. Hence, it is important for the insurance company to have sufficient funds to avoid a situation of insolvency. The study of probability of ultimate ruin has great significance for suggesting ways of protection to the insurer against insolvency. Lot of research has been carried out for finding the probability of ultimate ruin when the loss random variables are modeled using different Statistical distributions (Asmussen & Albrecher, 2010).

In economics and actuarial science, the positively skewed or heavy tailed distributions are used to model income or claim amount. There are many examples of such distributions in Lambert (1993), Johnson et al. (1995) and Arnold (2015). Most

K. Jain (✉)
Department of Statistics, Panjab University, Chandigarh 160014, India

H. S. Kapoor
Department of Mathematics and Statistics, Central University of Punjab, Bathinda 151001, India

D. D. Hanagal et al. (eds.), *Applied Statistical Methods*, Springer Proceedings in Mathematics & Statistics 380, https://doi.org/10.1007/978-981-16-7932-2_11

commonly used distribution to model catastrophic losses in an insurance portfolio is Pareto distribution (Hogg & Klugman, 1984 and Daykin et al., 1994).
Bohman (1971) discussed the problem of ultimate ruin probabilities. This has been further studied for Gamma distribution by Grandell and Segerdahl (1971) and Willmot (1988), Lognormal distribution by Thorin and Wikstad (1977) and Ramsay and Usábel (1997) and Generalized Gamma distribution by Usábel (2001). For exponential and mixture of two exponential distributions, ultimate ruin probability was derived by Dufresne and Gerber (1988). Ramsay and Usábel (1997) and Ramsay (2003) provided methodology for evaluating the ultimate ruin probability for Pareto distribution. Wei and Hai-liang (2004) gave explicit expressions for the ultimate ruin probabilities of Erlang risk processes with Pareto individual claim distributions. For work related to distribution of sum of Pareto random variables, one can refer to Ramsay (2009). Adekambi and Essiomle (2020) studied ultimate ruin probability by assuming a phase-type distribution for the time of default of payment in banks and provided the Cramer Lundberg type bounds.

In many situations, heavy-tailed distributions provide an appropriate fit to actual claim data. Due to difficulty in the evaluation of exact expressions of ultimate ruin probability, asymptotic properties of ultimate ruin probability for various distributions have been studied extensively in the literature (Embrechets and Veraverbeke (1982), Grübel (1987), Baltrünas (1999) and Barbe and McCormick (2009)).

In literature, there are two commonly used approaches for evaluation of ultimate ruin probability.

- The first approach uses Laplace Transformation and has been discussed in Sect. 3. One can also refer to Seal (1969), Bohman (1975), Thorin and Wikstad (1977), Usábel (1999, 2001) and references therein for more information.
- Goovaerts and De Vijylder (1984), Panjer (1986), Dickson (1989), Dickson and Waters (1991), Panjer and Wang (1993), Dickson et al. (1995) and Ramsay and Usábel (1997) have used the second approach wherein limits of integration are discretized and recursive procedure applied to get desired values of ultimate ruin probability.

In this chapter, the ultimate ruin probability is evaluated by implementing the Laplace Transformation method, for Benktander Gibrat (BG) distribution which represents a Benktander distribution of type I with parameters a and b. This is one of two distributions introduced by Benktander (1970) to model heavy-tailed losses commonly found in non-life/casualty actuarial science, using various forms of mean excess functions (Benktander & Segerdahl, 1960). The distribution of first type is "close" to the log-normal distribution (Klieber & Kotz, 2003). Benktander Type I and Type II distributions were further studied by Beirlant et al. (1996) and Embrechts et al. (1997). Benktander Type II distribution is also known as Benktander Weibull (BW) distribution. BG and BW Distributions are considered to be intermediate between Pareto and Exponential Distributions.

For a random variable X, following Benktander Gibrat (BG) distribution with parameters $a, b > 0$, the probability density function (pdf), cumulative distribution function (cdf) and expectation are given as

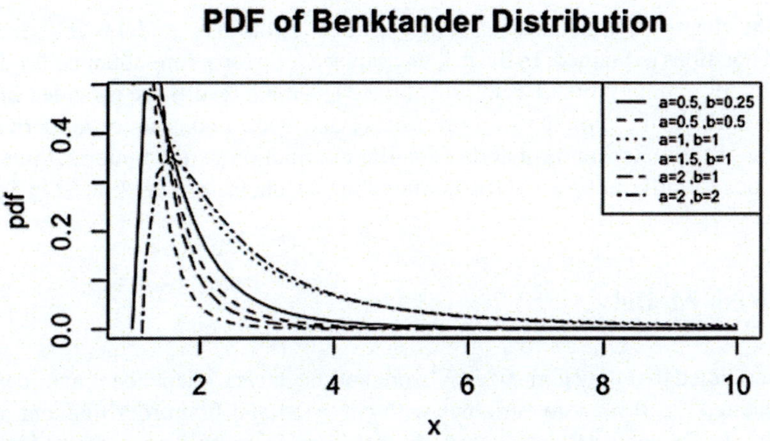

Fig. 1 PDF of Benktander Gibrat Distribution

$$f(x) = e^{-blog(x)^2} x^{-2-a}\Big(- \frac{2b}{a} + (1 + a + 2blog(x))(1 + \frac{2blog(x)}{a})\Big), x \geq 1 \tag{1.1}$$

$$F(x) = 1 - e^{-blog(x)^2} x^{-1-a}\Big(1 + \frac{2blog(x)}{a}\Big), \tag{1.2}$$

$$E(X) = p_1 = 1 + \frac{1}{a}. \tag{1.3}$$

The plots of the probability distribution function of BG distribution for different combinations of parameters a and b are shown in Fig. 1.

In Fig. 1, it can be seen that BG distribution has heavy tail, and hence is useful for modelling the insurance data. So ultimate ruin probability has been obtained assuming that claims follow the BG distribution.

Pareto Distribution (Arnold, 2015) is a limiting case of Benktander Gibrat (BG) Distribution, in the sense that probability density function (pdf) of Benktander Gibrat Distribution [a,b] tends to that of Pareto Distribution [1, a+1] when b approaches 0. This fact implies that pdfs of both Benktander Gibrat Distribution [a,b] and Benktander Weibull Distribution [a,b] tend to the same limiting function as b tends to 0. Furthermore, the stationary renewal distribution associated with a Benktander Gibrat distribution is same on its domain as the pdf of Benini Distribution [a,b,1] (Kleiber & Kotz, 2003). The asymptotic behaviour of BG distribution lies between light tailed exponential distribution and heavy tailed Pareto distribution. Claim amounts of vehicles under general insurance in urban areas follow BG distribution. Hence, it is of practical importance to study the probability of ultimate ruin when claim distribution is BG.

The chapter is organized as follows. In Sect. 2, the risk model is introduced and used notations explained. In Sect. 3, the Laplace transforms are obtained for deriving results for finding ultimate ruin probability. The main results are provided in Sect. 4 for evaluating the ultimate ruin probability and Sect. 5 consists of numerical evaluation of ultimate ruin probability for BG distribution using numerical integration methods for different parametric combinations. Conclusions are placed in Sect. 6.

2 Risk Model

It is assumed that the claim sizes $X_i's$ are non-negative, independent and identically distributed (i.i.d) random variables with cdf $F(x)$ and first order moment as $p_1 = E[X_i]$. In the actuarial risk theory, the risk surplus process denoted by U(t) under classical risk model is defined as

$$U(t) = u + ct - \sum_{i=1}^{N(t)} X_i$$

where $u \geq 0$ is the initial reserve (risk surplus) at $t = 0$;

$N(t)$, representing claims frequency in $[0, t)$, is a time homogeneous Poisson process with intensity λ;
$c = \lambda(1+\theta)p_1$ denotes premium rate with safety loading factor $\theta > 0$ which is percentage of expected loss that helps the insurer to cover administrative expenses. θ can be positive, negative or zero but in this study, it is assumed to be positive.

The surplus process of an insurance portfolio is inflow from the premium payments minus the outflow in form of claims and other administrative expenses. If this process becomes negative, it is said that ruin has occurred. For an initial reserve (surplus) of $u \geq 0$, the probability of ultimate ruin is written as

$$\psi(u) = Pr[U(t) < 0 \text{ for some } t > 0 \mid U(0) = u]. \text{ (Asmussen and Albrecher, 2010)}$$

The probability of ultimate ruin is the probability that the reserve ever drops below zero whereas the probability of ruin is the probability that reserve becomes negative in a finite period of time.

It is known from Seal (1969), Gerber (1979), Bowers et al. (1997), Hazewinkel (2001) that $\psi(u)$ satisfies the Volterra integral equation written as

$$\begin{aligned}\psi(u) &= \frac{1}{(1+\theta)p_1}\left(\int_0^u (1-F(x))\psi(u-x)\,dx + \int_u^\infty (1-F(x))\,dx\right)\\ &= \frac{1}{(1+\theta)}\left(\int_0^u H(x)\psi(u-x)\,dx + K(u)\right). \qquad (2.1)\end{aligned}$$

where

$$H(x) = \frac{1 - F(x)}{p_1} \tag{2.2}$$

and

$$K(u) = \int_u^\infty H(x)\,dx \tag{2.3}$$

Explicit expressions for ultimate ruin probability have been derived for limited number of claim size distributions such as exponential and mixed exponential (Klugman et al., 1998). Asymptotic results are available for Pareto and Lognormal distributions (Von Bahr, 1975, Embrechets & Veraverbeke, 1982 and Embrechets et al., 1997). In the next section, the expressions for Laplace transforms of $H(x)$, $K(x)$ and $\psi(u)$ have been derived.

3 Laplace Transformation

Let C represent the finite complex plane and $z \in C$. For $Re(z) > 0$, Laplace transform of $H(u)$ is given by

$$\begin{aligned} H^*(z) &= \int_0^\infty e^{-zu} H(u)\,du \\ &= \frac{1}{p_1}\int_0^\infty e^{-zu}(1 - F(u))\,du, \text{ using (2.2).} \end{aligned} \tag{3.1}$$

Laplace transform of $K(u)$ is written as

$$\begin{aligned} K^*(z) &= \int_0^\infty e^{-zu} K(u)\,du = \int_0^\infty e^{-zu} \int_u^\infty H(v)\,dv\,du = \int_0^\infty H(v)\left(\int_0^v e^{-zu}\,du\right) dv \\ &= \frac{1}{z}\int_0^\infty H(v)(1 - e^{-zv})\,dv = \frac{1}{z}\left(\int_0^\infty H(v)\,dv - \int_0^\infty e^{-zv} H(v)\,dv\right) \\ &= \frac{1}{z}[1 - H^*(z)], \text{ since } \int_0^\infty H(v)\,dv = \int_0^\infty \frac{1 - F(v)}{p_1}\,dv = 1. \end{aligned}$$

Hence using (2.1), Laplace transform of $\psi(u)$ is given by

$$\begin{aligned} \psi^*(z) &= \int_0^\infty e^{-zu}\psi(u)\,du \\ &= \frac{1}{(1+\theta)}\left[\int_0^\infty e^{-zu} K(u)\,du + \int_0^\infty e^{-zu}\left(\int_0^u \psi(u - x) H(x)\,dx\right) du\right]. \end{aligned}$$

Interchanging the order of integration, we have $x < u < \infty$ and $0 < x < \infty$ and hence

$$\begin{aligned}\psi^*(z) &= \frac{1}{(1+\theta)}\left[K^*(z) + \int_0^\infty \left(\int_x^\infty e^{-zu}\psi(u-x)\,du\right) H(x)\,dx\right]\\ &= \frac{1}{(1+\theta)}\left[K^*(z) + \int_0^\infty \left(\int_x^\infty e^{-z(u-x)}\psi(u-x)\,du\right) e^{-zx} H(x)\,dx\right]\\ &= \frac{1}{(1+\theta)}\left[K^*(z) + \int_0^\infty \left(\int_0^\infty e^{-zy}\psi(y)\,dy\right) e^{-zx} H(x)\,dx\right],\ \text{obtained by substituting } u-x=y\\ &= \frac{1}{(1+\theta)}\left(K^*(z) + \int_0^\infty \psi^*(z)e^{-zx} H(x)\,dx\right)\\ &= \frac{1}{(1+\theta)}\left[K^*(z) + \psi^*(z)H^*(z)\right].\end{aligned}$$

This gives $K^*(z) = [(1+\theta) - H^*(z)]\psi^*(z)$ which implies that

$$\psi^*(z) = \frac{K^*(z)}{(1+\theta - H^*(z))}. \tag{3.2}$$

4 Ultimate Ruin Probability for BG Distribution

It is assumed that each X_i, the ith claim size, follows Benktander Gibrat (BG) distribution with parameters a and b.

Using (1.2), (1.3) and (3.1), Laplace transform of $H(u)$ can be written as

$$H^*(z) = \frac{a}{a+1}\int_1^\infty e^{-zx-blog(x)^2}x^{-1-a}(1+\frac{2blog(x)}{a})\,dx.$$

Using Mathematica Software, we get

$$\begin{aligned}H^*(z) = &\frac{a}{a+1}\{E_{1+a+2b}(z)\\ &+\frac{2b}{a}\text{MeijerG}[\{1+a+2b, 1+a+2b\}, \{0, a+2b, a+2b\}, z]\end{aligned}$$

where for $Re(z) > 0$,

$$E_n(z) = \int_1^\infty \frac{e^{-zt}}{t^n}\,dt, n = 1, 2, \ldots \tag{4.1}$$

is Generalized exponential Integral (Abramowitz & Stegun, 1964). MeijerG function (Bateman & Erdély, 1953) is defined as

$$\begin{aligned}&\text{MeijerG}[\{\{a_1,\ldots,a_n\},\{a_{n+1},\ldots,a_p\}\},\{\{b_1,\ldots,b_m\},\{b_{m+1},\ldots,b_q\}\},z]\\ &= G^{m\ n}_{p\ q}(z \mid {}^{a_1,\ldots,a_p}_{b_1,\ldots,b_q}) = \frac{1}{2\pi i}\int_L \frac{\prod_{j=1}^m \Gamma(b_j - s)\prod_{k=1}^n \Gamma(1-a_k+s)}{\prod_{j=m+1}^q \Gamma(1-b_j+s)\prod_{k=n+1}^p \Gamma(a_k - s)} z^s\,ds\end{aligned} \tag{4.2}$$

where L is a loop (as defined in the following expression) beginning and ending at $+\infty$, encircling all poles of $\Gamma(b_j - s)$, $j = 1, 2, \ldots, m$ exactly once in the negative direction but not encircling any pole of $\Gamma(1 - a_k + s)$, $k = 1, 2, \ldots, n; 0 \leq m \leq q$; $0 \leq n \leq p$; m, n, p and q are integers, $z \neq 0$ and Γ denotes the gamma function. The integral converges for all z if $q > p \geq 0$.
Using (4.1) and (4.2), we get

$$H^*(z) = \frac{a}{a+1}\left[\int_1^\infty \frac{e^{-zt}}{t^{(1+a+2b)}}\,dt + \frac{2b}{a}\frac{1}{2\pi i}\int_L \frac{\prod_{j=1}^m \Gamma(s+b_j)}{\prod_{j=n+1}^p \Gamma(s+a_j)} z^s\,ds\right].$$

This gives

$$\begin{aligned} K^*(z) &= \frac{1 - H^*(z)}{z} \\ &= \frac{a + 1 - aE_{1+a+2b}(z) - 2bG_{2\,3}^{3\,0}\left(z \mid {}^{1+a+2b,1+a+2b}_{0,\ a+2b,\ a+2b}\right)}{z(a+1)}. \end{aligned} \tag{4.3}$$

Using (3.2) and (4.3), we get

$$\begin{aligned} \psi^*(z) &= \frac{a + 1 - aE_{1+a+2b}(z) - 2bG_{2\,3}^{3\,0}\left(z \mid {}^{1+a+2b,1+a+2b}_{0,\ a+2b,\ a+2b}\right)}{z\left((1+\theta)(a+1) - aE_{1+a+2b}(z) - 2bG_{2\,3}^{3\,0}\left(z \mid {}^{1+a+2b,1+a+2b}_{0,\ a+2b,\ a+2b}\right)\right)} \\ &= \frac{1}{z} - \frac{(a+1)\theta}{z\left((1+\theta)(a+1) - aE_{1+a+2b}(z) - 2bG_{2\,3}^{3\,0}\left(z \mid {}^{1+a+2b,1+a+2b}_{0,\ a+2b,\ a+2b}\right)\right)} \\ &= \frac{1}{z} - A^*(z) \end{aligned} \tag{4.4}$$

where

$$A^*(z) = \frac{(a+1)\theta}{z\left((1+\theta)(a+1) - aE_{1+a+2b}(z) - 2bG_{2\,3}^{3\,0}\left(z \mid {}^{1+a+2b,1+a+2b}_{0,\ a+2b,\ a+2b}\right)\right)}.$$

Two possibilities are considered

(i) $\theta = 0$
(ii) $\theta > 0$.

For $\theta = 0$, $\psi^*(z) = \frac{1}{z}$ which implies that $\psi(u) = 1$. For second possibility, we prove the following Lemma.

Lemma 1 *Let* ***D*** *denote the finite complex plane, with zero and negative real axis removed, that is,* $\boldsymbol{D} = \boldsymbol{C}/(-\infty, 0]$. *Then for* $(1+\theta)(a+1) > 0$, *the equation*

$$(1+\theta)(a+1) - aE_{1+a+2b}(z) - 2bG^{3\,0}_{2\,3}(z \mid {}^{1+a+2b,1+a+2b}_{0,\;a+2b,\;a+2b}) = 0, a+2b = 3, 4, \ldots. \tag{4.5}$$

has no solution in **D***.*

Proof Let $z = x + iy \in \mathbf{D}$ where x and y are real numbers.
$E_{1+a+2b}(z)$ is analytic in **D** and is symmetric for $z \in \mathbf{D}$ (Abramowitz & Stegun, 1964, Chap. 5) which gives that $E_{1+a+2b}(\bar{z}) = \bar{E}_{1+a+2b}(z)$, where $(\bar{z})$ represents the complex conjugate of z.
Taking complex conjugate on both sides of (4.5), we get

$$(1+\theta)(a+1) - a\bar{E}_{1+a+2b}(z) - 2b\bar{G}^{3\,0}_{2\,3}(z \mid {}^{1+a+2b,1+a+2b}_{0,\;a+2b,\;a+2b}) = 0.$$

Since $\bar{E}_{1+a+2b}(z) = E_{1+a+2b}(\bar{z})$ and $\bar{G}^{3\,0}_{2\,3}(z \mid {}^{1+a+2b,1+a+2b}_{0,\;a+2b,\;a+2b}) = -G^{3\,0}_{2\,3}(\bar{z} \mid {}^{1+a+2b,1+a+2b}_{0,\;a+2b,\;a+2b})$, hence

$$(1+\theta)(a+1) - aE_{1+a+2b}(\bar{z}) + 2bG^{3\,0}_{2\,3}(\bar{z} \mid {}^{1+a+2b,1+a+2b}_{0,\;a+2b,\;a+2b}) = 0. \tag{4.6}$$

Subtracting (4.6) from (4.5) , we get

$$aE_{1+a+2b}(\bar{z}) - 2bG^{3\,0}_{2\,3}(\bar{z} \mid {}^{1+a+2b,1+a+2b}_{0,\;a+2b,\;a+2b}) - aE_{1+a+2b}(z) - 2bG^{3\,0}_{2\,3}(z \mid {}^{1+a+2b,1+a+2b}_{0,\;a+2b,\;a+2b}) = 0.$$

This gives

$$a[E_{1+a+2b}(\bar{z}) - E_{1+a+2b}(z)] - 2b[G^{3\,0}_{2\,3}(\bar{z} \mid {}^{1+a+2b,1+a+2b}_{0,\;a+2b,\;a+2b}) + G^{3\,0}_{2\,3}(z \mid {}^{1+a+2b,1+a+2b}_{0,\;a+2b,\;a+2b})] = 0. \tag{4.7}$$

For $y = 0$, the L.H.S of (4.7) will be

$$a[E_{1+a+2b}(x) - E_{1+a+2b}(x)] - 2b[G^{3\,0}_{2\,3}(x \mid {}^{1+a+2b,1+a+2b}_{0,\;a+2b,\;a+2b}) + G^{3\,0}_{2\,3}(x \mid {}^{1+a+2b,1+a+2b}_{0,\;a+2b,\;a+2b})]$$
$$= -4bG^{3\,0}_{2\,3}(x \mid {}^{1+a+2b,1+a+2b}_{0,\;a+2b,\;a+2b}) < 0.$$

So, no solution is possible for $y = 0$.

Now for $y > 0$,

$$a[E_{1+a+2b}(\bar{z}) - E_{1+a+2b}(z)] - 2b[G^{3\,0}_{2\,3}(\bar{z} \mid {}^{1+a+2b,1+a+2b}_{0,\;a+2b,\;a+2b}) + G^{3\,0}_{2\,3}(z \mid {}^{1+a+2b,1+a+2b}_{0,\;a+2b,\;a+2b})] \neq 0$$

because $E_{1+a+2b}(\bar{z}) \neq E_{1+a+2b}(z)$ and

$G^{3\,0}_{2\,3}(\bar{z} \mid {}^{1+a+2b,1+a+2b}_{0,\;a+2b,\;a+2b}) \neq -G^{3\,0}_{2\,3}(z \mid {}^{1+a+2b,1+a+2b}_{0,\;a+2b,\;a+2b})$.

Hence for $y > 0$, L.H.S of (4.7) is not equal to zero and no solution is possible.

$\psi(u)$ can be obtained using Bromwich complex integral (Inverse Laplace Transform) as

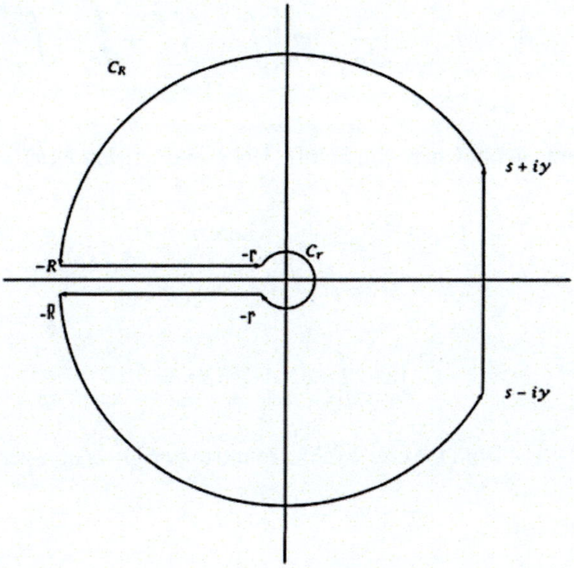

Fig. 2 Closed contour Ω

$$\psi(u) = \frac{1}{2\pi i}\int_{s-i\infty}^{s+i\infty} e^{uz}\psi^*(z)\,dz = 1 - \frac{1}{2\pi i}\int_{s-i\infty}^{s+i\infty} e^{uz}A^*(z)\,dz \tag{4.8}$$

and $s > 0$ is an arbitrary large constant value such that all singularities of $\psi^*(z)$ lie to the left of vertical line $Re(z) = s$ (Davies, 2002). Along the negative real axis, $E_{1+a+2b}(z)$ has a logarithmic branch cut and branch point at $z = 0$. Hence the Bromwich integral can be evaluated in the counter-clockwise direction around the deformed closed contour Ω as shown in Fig. 2.

The integral around Ω can be represented as

$$\frac{1}{2\pi i}\int_{\Omega} e^{uz}A^*(z)\,dz = \frac{1}{2\pi i}\left[\int_{s-iy}^{s+iy} + \int_{C_R} + \int_{-r}^{-R} + \int_{C_r} + \int_{-R}^{-r}\right] e^{uz}A^*(z)\,dz. \tag{4.9}$$

Using Lemma 1, $(1+\theta)(a+1) - aE_{1+a+2b}(z) - 2bG^{3\,0}_{2\,3}\left(z \,\middle|\, {1+a+2b,\,1+a+2b \atop 0,\, a+2b,\, a+2b}\right)=0$

has no root in Ω. Hence Cauchy's residue theorem of complex analysis (Brown & Churchill, 2013) gives that

$$\frac{1}{2\pi i}\int_{\Omega} e^{uz}A^*(z)\,dz = 0. \tag{4.10}$$

Using (4.9) and (4.10), we get

$$\frac{1}{2\pi i}\int_{s-i\infty}^{s+i\infty} e^{uz}A^*(z)\,dz = -\frac{1}{2\pi i}\lim_{\substack{R\to\infty\\ r\to 0}}\left[\int_{C_R}+\int_{-R}^{-r}+\int_{C_r}+\int_{-r}^{-R}\right]e^{uz}A^*(z)\,dz. \tag{4.11}$$

Using Lemma 4.1 of Chap. 4 in Schiff (1999) and writing $F(u) = A^*(z)$, we have

$$A^*(z) \le \frac{1}{z^p},\ \text{where } p > 0.$$

Hence

$$\lim_{R\to\infty}\int_{C_R} e^{uz}A^*(z)\,dz = 0. \tag{4.12}$$

The contribution from the circle (C_r) around the origin is non-zero (Schiff, 1999).

$$\lim_{r\to 0}\int_{C_r} e^{uz}A^*(z)\,dz$$

$$= \lim_{r\to 0}\int_{C_r}\frac{(a+1)\theta e^{uz}}{z((1+\theta)(a+1) - aE_{1+a+2b}(z) - 2bG^{3\,0}_{2\,3}(z \mid {}^{1+a+2b,1+a+2b}_{0,\ a+2b,\ a+2b}))}dz$$

$$= \lim_{r\to 0}\int_{\pi}^{-\pi}\frac{(a+1)\theta i r e^{iw} e^{ure^{iw}}}{re^{iw}((1+\theta)(a+1) - aE_{1+a+2b}(re^{iw}) - 2bG^{3\,0}_{2\,3}(re^{iw} \mid {}^{1+a+2b,1+a+2b}_{0,\ a+2b,\ a+2b}))}dw$$

$$= \int_{\pi}^{-\pi}\frac{\theta i(a+1)(a+2b)}{(1+\theta)(a+1)(a+2b) - a}dw$$

$$= -\frac{2\pi i\theta(a+1)(a+2b)}{(1+\theta)(a+1)(a+2b) - a}. \tag{4.13}$$

Now $z = xe^{i\pi} = -x$ along the contour above the negative real axis. As z approaches $-R$ to $-r$, x approaches R to r and $dz = -dx$. Hence for $R \to \infty$ and $r \to 0$, we have

$$\int_{-\infty}^{0} e^{uz}A^*(z)\,dz = \int_{0}^{\infty} e^{-ux}A^*(xe^{i\pi})\,dx. \tag{4.14}$$

Similarly, $z = xe^{-i\pi} = -x$ along the contour just below the negative real axis, and as z approaches $-R$ from $-r$, x approaches R from r and $dz = -dx$. So as $R \to \infty$ and $r \to 0$, we get

$$\int_0^{-\infty} e^{uz} A^*(z)\, dz = -\int_0^{\infty} e^{-ux} A^*(xe^{-i\pi})\, dx. \tag{4.15}$$

Using (4.12)–(4.15), we get

$$\begin{aligned}
&\frac{1}{2\pi i}\int_{s-i\infty}^{s+i\infty} e^{uz} A^*(z)\, dz \\
&= -\frac{1}{2\pi i}\int_{C_R} \frac{(a+1)\theta e^{uz}}{z\left((1+\theta)(a+1) - aE_{1+a+2b}(z) - 2bG_{2\,3}^{3\,0}\left(z \mid {}^{1+a+2b,1+a+2b}_{0,\, a+2b,\, a+2b}\right)\right)} dz \\
&- \frac{1}{2\pi i}\left(\int_0^{\infty} e^{-ux} A^*(xe^{i\pi})\, dx - \int_0^{\infty} e^{-ux} A^*(xe^{-i\pi})\, dx\right) \\
&- \frac{1}{2\pi i}\left(\int_{C_r} \frac{(a+1)\theta e^{uz}}{z\left((1+\theta)(a+1) - aE_{1+a+2b}(z) - 2bG_{2\,3}^{3\,0}\left(z \mid {}^{1+a+2b,1+a+2b}_{0,\, a+2b,\, a+2b}\right)\right)} dz\right)
\end{aligned}$$

$$\begin{aligned}
&= -\frac{1}{2\pi i}\left(\int_0^{\infty} e^{-ux} A^*(xe^{i\pi})\, dx - \int_0^{\infty} e^{-ux} A^*(xe^{-i\pi})\, dx\right) \\
&+ \frac{\theta(a+1)(a+2b)}{(1+\theta)(a+1)(a+2b) - a} \\
&= \frac{\theta(a+1)(a+2b)}{(1+\theta)(a+1)(a+2b) - a} - \frac{1}{2\pi i}\int_0^{\infty} e^{-ux}\left[A^*(xe^{i\pi}) - A^*(xe^{-\pi i})\right] dx.
\end{aligned} \tag{4.16}$$

Using (4.16) in (4.8), we have

$$\begin{aligned}
\psi(u) &= 1 - \frac{\theta(a+1)(a+2b)}{(1+\theta)(a+1)(a+2b) - a} \\
&\quad + \frac{1}{2\pi i}\int_0^{\infty} e^{-ux}\left[A^*(xe^{i\pi}) - A^*(xe^{-i\pi})\right] dx
\end{aligned}$$

$$\begin{aligned}
&= 1 - \frac{\theta(a+1)(a+2b)}{(1+\theta)(a+1)(a+2b) - a} \\
&+ \frac{1}{2\pi i}\int_0^{\infty} \frac{(a+1)e^{-ux}\theta}{xe^{i\pi}\left((\theta+1)(a+1) - aE_{1+a+2b}(xe^{i\pi}) - 2bG_{2\,3}^{3\,0}\left(xe^{i\pi} \mid {}^{1+a+2b,1+a+2b}_{0,\, a+2b,\, a+2b}\right)\right)} \\
&- \frac{(a+1)e^{-ux}\theta}{xe^{-i\pi}\left((\theta+1)(a+1) - aE_{1+a+2b}(xe^{-i\pi}) - 2bG_{2\,3}^{3\,0}\left(xe^{-i\pi} \mid {}^{1+a+2b,1+a+2b}_{0,\, a+2b,\, a+2b}\right)\right)} dx.
\end{aligned} \tag{4.17}$$

Exponential Integral is not analytic along the negative real axis.

For $x > 0$, $E_{1+a+2b}(-x)$ takes on different values along the negative real axis depending on whether $(-x)$ lies above the branch cut or below the cut.
Using series expansion of Exponential Integral (Abramowitz & Stegun, (1964, Chap. 5)), we get

$$E_{1+a+2b}(z) = \frac{z^{a+2b}}{(a+2b)!}[log(z) + \gamma - \sum_{r=1}^{a+2b} \frac{1}{r}] + \sum_{\substack{l=0 \\ l \neq a+2b}}^{\infty} \frac{z^l}{(l-a-2b)l!}.$$

Putting $z = xe^{\pm i\pi}$ and solving, we get

$$E_{1+a+2b}(xe^{\pm i\pi}) = -Ei_{1+a+2b}(x) \mp i\pi \frac{x^{a+2b}}{(a+2b)!}$$

where $Ei_{1+a+2b}(x) = \frac{x^{a+2b}}{(a+2b)!}[log(x) + \gamma - \sum_{r=1}^{a+2b} \frac{1}{r}] + \sum_{\substack{l=0 \\ l \neq a+2b}}^{\infty} \frac{x^l}{(l-a-2b)l!}$

for $a + 2b = 3, 4, \ldots,$ is

$(1 + a + 2b)^{th}$ order generalization of the exponential integral $Ei(x)$.
Thus, from (4.17), we have

$$\psi(u) = \frac{(a+1)(a+2b)-a}{(1+\theta)(a+1)(a+2b)-a} + \frac{\theta(a+1)}{2\pi i}$$
$$\int_0^\infty \frac{e^{-ux}[-x((\theta+1)(a+1) + aEi_{1+a+2b}(x) - ia\pi \frac{x^{a+2b}}{(a+2b)!} - 2bG_{2\,3}^{3\,0}(xe^{i\pi} \mid {}^{1+a+2b,1+a+2b}_{0,\,a+2b,\,a+2b})) + x((\theta+1)(a+1) + aEi_{1+a+2b}(x) + ia\pi \frac{x^{a+2b}}{(a+2b)!} - 2bG_{2\,3}^{3\,0}(xe^{-i\pi} \mid {}^{1+a+2b,1+a+2b}_{0,\,a+2b,\,a+2b}))]}{x^2\left[\left((\theta+1)(a+1) + aEi_{1+a+2b}(x) - 2bG_{2\,3}^{3\,0}(-x \mid {}^{1+a+2b,1+a+2b}_{0,\,a+2b,\,a+2b})\right)^2 + \left(\frac{a\pi x^{2a+4b}}{(a+2b)!}\right)^2\right]} dx$$
$$= \frac{(a+1)(a+2b)-a}{(1+\theta)(a+1)(a+2b)-a} + \theta$$
$$\int_0^\infty \frac{e^{-ux}\left[\frac{a(a+1)x^{a+2b-1}}{(a+2b)!}\right]}{\left[\left((\theta+1)(a+1) + aEi_{1+a+2b}(x) - 2bG_{2\,3}^{3\,0}(-x \mid {}^{1+a+2b,1+a+2b}_{0,\,a+2b,\,a+2b})\right)^2 + \left(\frac{a\pi x^{2a+4b}}{(a+2b)!}\right)^2\right]} dx. \quad (4.18)$$

This is the exact form of the ultimate ruin probability $\psi(u)$, but it is difficult to find its solution analytically. So, we use numerical approximation method to solve the integral in (4.18). Numerical values of probability of ultimate ruin are calculated in the next section for some combinations of values of a and b.

5 Calculation of Ultimate Ruin Probability

In this section, we evaluate ultimate ruin probability for BG distribution by solving (4.18) using numerical integration in Mathematica for some parametric combinations

Table 1 Ultimate ruin Probability for BG risk model

	$\theta = 0.10$	$\theta = 0.25$	$\theta = 0.50$	$\theta = 0.75$
		$a = 1$ and $b = 2$		
u=10	0.9000+4.66E-8	0.7826+1.10E-7	0.6428+1.08E-7	0.5454+1.17E-7
u=20	0.9000+3.73E-9	0.7826+2.11E-8	0.6428+3.74E-9	0.5454+1.60E-8
u=40	0.9000+5.06E-10	0.7826+6.37E-9	0.6428+3.93E-10	0.5454+1.19E-10
		$a = 3$ and $b = 2$		
	$\theta = 0.10$	$\theta = 0.25$	$\theta = 0.50$	$\theta = 0.75$
u=10	0.8990+8.87E-11	0.7812+1.74E-10	0.6410+2.59E-10	0.5435+3.16E-10
u=20	0.8990+2.58E-11	0.7812+9.82E-12	0.6410+3.22E-12	0.5435+3.11E-11
u=40	0.8990+2.36E-12	0.7812+1.56E-13	0.6410+1.06E-13	0.5435+2.09E-12
		$a = 3$ and $b = 5$		
	$\theta = 0.10$	$\theta = 0.25$	$\theta = 0.50$	$\theta = 0.75$
u=10	0.9041+7.21E-17	0.7903+1.61E-16	0.6533+2.71E-16	0.5568+3.47E-16
u=20	0.9041+2.83E-20	0.7903+5.75E-20	0.6533+8.68E-20	0.5568+1.01E-19
u=40	0.9041+4.57E-23	0.7903+3.43E-23	0.6533+3.67E-23	0.5568+3.01E-23

and different values of u and θ. The values of $\psi(u)$ for these combinations are shown in Table 1.

The values of ultimate ruin probability are calculated by using Mathematica software and during evaluation, it was observed that some values of probability are complex but the coefficient of i is 10^{-12} which is very very small. So, we ignore the effect of i and consider only the real part to show the ultimate ruin probabilities. From Table 1, it is concluded that

- as loading factor increases, the probability of ultimate ruin decreases for fixed u;
- an increase in initial reserves leads to no significant decrease in ultimate ruin probability implying that initial reserves do not have a significant influence on ultimate ruin probability;
- probability of ultimate ruin decreases as value of a increases for fixed value of b;
- probability of ultimate ruin increases as value of b increases for fixed a;
- when values of both the parameters a and b increase, the probability of ultimate ruin is slightly higher.

Hence, in case the claim distribution is Benktander Gibrat, the companies should pay more attention to the loading factor than initial reserves in order to remain solvent.

6 Conclusions

Benktander Gibrat risk model is not very well known in literature but due to characteristics of BG distribution lying between light tailed exponential distribution and heavy tailed Pareto distribution, its importance cannot be ignored. We evaluate the probability of ultimate ruin by assuming claim distribution to be BG and conclude that companies have to concentrate more on the loading factor than initial reserves to be solvent.

Acknowledgements The authors are grateful to the Editor and refrees for their valuable comments and suggestions which have helped in improving the contents of this chapter.

References

Abramowitz, M., & Stegun, I. A. (1964). *Handbook of mathematical functions*. New York: Dover Publications.

Adekambi, F., & Essiomle, K. (2020). Ruin probability for stochastic flows of financial contract under phase-type distribution. *Risks, 8*(53), 1–21.

Arnold, B. C. (2015). *Pareto distributions* (2nd ed.). Florida: CRC Press.

Asmussen, S., & Albrecher, H. (2010). *Ruin probabilities* (2nd ed., Vol. 14). Advanced Series on Statistical Science and Applied Probability. Singapore: World Scientific.

Baltrünas, A. (1999). Second order behaviour of ruin probabilities. *Scandinavian Actuarial Journal, 2*, 120–133.

Barbe, P., & McCormick, W. P. (2009). *Asymptotic expansions for infinite weighted convolutions of heavy tail distributions and applications* (Vol. 197). USA: American Mathematical Society.

Bateman, H., & Erdély, A. (1953). *Higher transcendental functions* (Vol. I). New York: Mc-Graw Hill.

Beirlant, J., Teugels, J., & Vynckier, P. (1996). *Practical analysis of extreme values*. Leuven: Leuven University Press.

Benktander, G. (1970): Schadenverteilungen nach Grösse in der Nicht-Lebensversicherung [Loss Distributions by Size in Non-life Insurance], *Bulletin of the Swiss Association of Actuaries* (in German), 263–283.

Benktander, G., & Segerdahl, C. O. (1960): On the analytical representation of claim distributions with special reference to excess of loss reinsurance. In: *Transaction of the 16th International Congress of Actuaries*, vol. 1, pp. 626–648.

Bohman, H. (1971). Ruin probabilities. *Scandinavian Actuarial Journal, 54*, 159–163.

Bohman, H. (1975). Numerical inversion of characteristic functions. *Scandinavian Actuarial Journal, 2*, 121–124.

Bowers, N. L., Gerber, H. U., Hickman, J. C., Jones, D. A., & Nesbitt, C. J. (1997). *Actuarial mathematics*. USA: The Society of Actuaries.

Brown, J. W., & Churchill, R. V. (2013). *Complex variables and applications* (9th ed.). New York: McGraw-Hill.

Davies, B. J. (2002), *Integral transforms and their applications* (3rd ed.). Berlin, New York: Springer-Verlag.

Daykin, C. D., Pentikäinen, T., & Pesonen, M. (1994). *Practical risk theory for actuaries*. London: Chapman and Hall.

Dickson, D. C. M. (1989): Recursive calculation of the probability and severity of ruin. *Insurance: Mathematics and Economics, 8*, 145–148.

Dickson, D. C. M., & Waters, H. R. (1991). Recursive calculation of survival probabilities. *ASTIN Bulletin, 21*, 199–221.

Dickson, D. C. M., Egidio dos Reis, A. D., & Waters, H. R. (1995). Some stable algorithms in ruin theory and their application. *ASTIN Bulletin, 25*, 153–175.

Dufresne, F., & Gerber, H. U. (1988). The probability and severity of ruin for combination of exponential claim amount distributions and their translations. *Insurance: Mathematics and Economics 7*, 75-80.

Embrechts, P., Klüppelberg, C., & Mikosch, T. (1997). *Modelling extremal events for insurance and finance*. Springer-Verlag.

Embrechets, P., & Veraverbeke, N. (1982). Estimates for the probability of ruin with special emphasis on the probability of large claims. *Insurance: Mathematics and Economics, 1*(1), 55–72.

Gerber, H. U. (1979). *An introduction to mathematical risk theory*. Huebner Foundation Monographs. Philadelphia, S.S: University of Pennsylvania.

Goovaerts, M. J., & De Vijylder, F. (1984). A stable recursive algorithm for evaluation of ultimate ruin probabilities. *ASTIN Bulletin, 14*, 53–59.

Grandell, J., & Segerdahl, C. O. (1971). A comparison of some applications of ruin probabilities. *Scandinavian Actuarial Journal, 3–4*, 143–158.

Grübel, R. (1987). On subordinated distributions and generalized renewal measures. *The Annals of Probability, 15*(1), 394–415.

Hazewinkel, M. (ed.) (2001). Volterra equation,*Encyclopedia of Mathematics*, Springer Science+Business Media B.V./Kluwer Academic Publishers.

Hogg, R. V., & Klugman, S. A. (1984). *Loss distributions*. New York: John Wiley & Sons.

Johnson, N. L., Kotz, S., & Balakrishnan, N. (1995). *Distribution in statistics: continuous univariate distributions* (Vol. 2). John Wiley, New York: Second Edition.

Lambert, P. J. (1993). *The distribution and redistribution of income* (2nd ed.). Manchester: Manchester University Press.

Kleiber, C., & Kotz, S. (2003). *Statistical size distributions in economics and actuarial sciences*. New Jersey: Hoboken.

Klugman, S. A., Panjer, H. H., & Willmot, G. E. (1998). *Loss models*. New York: Wiley.

Panjer, H. H. (1986). Direct calculation of ruin probabilities. *Journal of Risk and Insurance, 53*, 521–529.

Panjer, H. H., & Wang, S. (1993). On the stability of recursive formulas. *ASTIN Bulletin, 23*, 227–258.

Ramsay, C. M., & Usábel, M. (1997). Calculating ruin probabilities via product integration. *ASTIN Bulletin, 27*(2), 263–271.

Ramsay, C. M. (2003). A solution to the ruin problem for pareto distributions. *Insurance: Mathematics and Economics, 33*(1), 109–116.

Ramsay, C. M. (2009). The distribution of compound sums of pareto distributed losses. *Scandinavian Actuarial Journal, 1*, 27–37.

Schiff, J. L. (1999). *The laplace transform*. New York: Springer.

Seal, H. L. (1969). *Stochastic theory of a risk business*. New York: Wiley.

Seal, H. L. (1977). Numerical inversion of characteristic functions. *Scandinavian Actuarial Journal, 1*, 48–53.

Thorin, O., & Wikstad, N. (1977). Calculation of ruin probabilities when the claim distribution is lognormal. *ASTIN Bulletin, 9*, 231–246.

Usábel, M. (1999). Calculating multivariate ruin probabilities via gaver-stehfest inversion technique. *Insurance: Mathematics and Economics, 25*, 133–142.

Usábel, M. (2001). Ultimate ruin probabilities for generalized gamma-convolution claim sizes. *ASTIN Bulletin, 31*(1), 59–79.

Von Bahr, B. (1975). Asymptotic ruin probabilities when exponential moments do not exist. *Scandinavian Actuarial Journal*, 6–10.

Wei, L., & Hai-liang, Y. (2004). Explicit expressions for the ruin probabilities of Erlang risk processes with Pareto individual claim distributions. *Acta Mathematicae Applicatae Sinica, 20*, 495–506.

Willmot, G. E. (1988): Further use of shiu's approach to the evaluation of ultimate ruin probabilities. *Insurance: Mathematics and Economics, 7*, 275–281.

Test of Homogeneity of Scale Parameters Based on Function of Sample Quasi Ranges

Kalpana K. Mahajan, Sangeeta Arora, and Anil Gaur

Abstract A multi-sample test for homogeneity of scale parameters against simple ordered alternative based on function of sample quasi ranges is proposed for censored data, as well as for data contaminated with outliers. These problems find a number of applications in different fields, such as agriculture, competitive markets, engineering and quality control. The critical points have been computed through simulation for the given test in case of standard exponential, standard logistic and standard uniform distributions. Nonetheless, the proposed test also finds applications in distributions such as Laplace, Pareto, Weibull, etc. Herein construction is proposed for simultaneous one-sided confidence intervals (SOCIs) along with an illustration. Besides computing the power of the proposed test, some power comparisons have also been undertaken.

Keywords Ordered alternative · Power comparisons · Quasi ranges · Simulation

1 Introduction

More often than not, in real-life situations the interest is in weeding out the process or mechanism leading to high variations. For example, a soft-drink filling unit will like to discard the machines giving high variations in the context, or in animal husbandry, the interest may be on scaling variations while attempting with different breeding methods, for successive improvements. In general, such problems are quite common in almost all walks of life, be it hailing from agriculture, quality control, engineering or the like, wherein the interest lies in scale parameter of distributions. Let $\pi_1, \pi_2, \ldots, \pi_k$ be k (k $\geq$ 3) independent populations wherein an observation from population π_i follows a distribution with cumulative distribution function (cdf) $F_i(x) = F[(x - \mu_i)/\theta_i]$, such that μ_i $(-\infty < \mu_i < \infty)$ is location parameter, θ_i $(\theta_i > 0)$ is scale parameter and $F(.)$ is any absolutely continuous cdf. Thus, $F(.)$ is a member of location-scale family, $i = 1, 2, \ldots, k$. Herein the interest lies in

K. K. Mahajan (✉) · S. Arora · A. Gaur
Department of Statistics, Panjab University, Chandigarh 160014, India
e-mail: mahajan_kr@pu.ac.in

D. D. Hanagal et al. (eds.), *Applied Statistical Methods*, Springer Proceedings in Mathematics & Statistics 380, https://doi.org/10.1007/978-981-16-7932-2_12

to "ordering" of the scale parameters of k populations—more precisely to see if the scale parameter of one distribution is less/more than the other. The null hypothesis of interest is to test $H_0 : \theta_1 = \theta_2 = \cdots = \theta_k$ against the simple ordered alternative $H_A : \theta_1 \leq \theta_2 \leq \cdots \leq \theta_k$ with at least one strict inequality.

For this problem, many tests have been proposed in the literature including those of Kochar and Gupta (1985), Kusum and Bagai (1988), Hayter (1990), Gill and Dhawan (1999), Shetty and Bhatt (2003), Singh and Gill (2004), Bansal et al. (2011), Gaur et al., (2012, 2013), Mahajan et al. (2012) and others. For more details, one may refer to Barlow et al. (1972), Robertson et al. (1988) and Hochberg and Tamhane (1987). Gill and Dhawan (1999) proposed a test procedure based on maximum likelihood estimators (MLE) of the scale parameters of two-parameter exponential distribution. Computations of critical points, simultaneous one-sided confidence intervals (SOCIs) construction and power computations were carried out using simulation technique for two-parameter exponential distribution only. Singh and Gill (2004) proposed a test procedure based on sample quasi ranges and computed the critical points in case of exponential distribution and uniform distribution using Hayter (1990) technique, and power computation and comparisons are carried out for equal sample sizes.

The test proposed by Singh and Gill (2004) is more suitable when the practitioner or experimenter has smaller samples or samples containing outliers to draw inferences. In such cases, the use of sample quasi ranges as a measure of dispersion has been recommended (see David & Nagaraja, 2003). Further, because of their computational and distributional conveniences, sample quasi ranges are considered good measures of dispersion in comparison to measures based on maximum likelihood estimator (MLE) or best linear unbiased estimator (BLUE).

David and Nagaraja (2003) and Budescu (1980) mentioned that "weighted function of quasi ranges could be considered a good measure of dispersion". David and Nagaraja (2003) showed that "weighted function of quasi ranges is more efficient than quasi ranges alone". With this backdrop, a multi-sample test for testing homogeneity of scale parameters against simple ordered alternative based on weighted function of sample quasi ranges is proposed for the situations: (i) when it is not possible to measure the observations below and above some limits, i.e. experimenter has censored data for statistical inference; or (ii) there is suspicion that data is contaminated with outliers or (iii) when the practitioner or experimenter has smaller samples to draw statistical inferences. Herein the critical points have been computed for three distributions, viz. two-parameter standard exponential, standard logistic and standard uniform. These critical points could also be applied in case of some other distributions such as Gamma, Laplace, Pareto, Weibull, etc., while using certain transformations.

The scheme of the chapter is as follows: Sect. 1 accounts for the brief introduction to the problem, and in Sect. 2, a test is proposed for the multi-sample scale problem under order alternative using weighted function of quasi ranges. The computations of critical points are shown in detail in Sect. 3, and construction of SOCIs and a simulated example are taken up in Sect. 4. In Sect. 5, the power of the proposed test is carried out, and finally conclusion is given in Sect. 6.

2 Proposed Test Procedure

Sample quasi ranges R_{ir} are defined as $R_{ir} = X_{i[n-r]} - X_{i[r+1]}, r = 0, 1, \ldots, [n/2] - 1$, where $[x]$ is the greatest integer less than or equal to x and $X_{i[1]} \leq X_{i[2]} \leq \cdots \leq X_{i[n]}$ be the corresponding order statistics, $i = 1, 2, \ldots, k$ based on a random sample $X_{i1}, X_{i2}, \ldots, X_{in}$ of size n from the ith population π_i where $i = 1, 2, \ldots, k$.

The weighted function of quasi ranges is thus defined as

$$Q_i = \frac{\sum_{r=1}^{m} (n - 2r + 1) R_{ir}}{n(n-1)/2}, \tag{2.1}$$

where $m = [n/2]$, the integral part of $[n/2]$. The weighted function of quasi ranges is in fact a measure of dispersion and this is used in the test statistics proposed below.

Consider null hypothesis of homogeneity of scale parameters

$$H_0 : \theta_1 = \theta_2 = \cdots = \theta_k,$$

against the simple ordered alternative

$$H_A : \theta_1 \leq \theta_2 \leq \cdots \leq \theta_k,$$

with at least one strict inequality.

The proposed test statistic is given as

$$T_Q = \max_{1 \leq i < j \leq k} (Q_j / Q_i). \tag{2.2}$$

and the test is to reject H_0 at level α *iff*

$$T_Q \geq q_{k,\alpha,n}, \tag{2.3}$$

where $q_{k,\alpha,n}$ are the critical points for k samples each of size n obtained by

$$P_0[T_Q \geq q_{k,\alpha,n}] = \alpha, \tag{2.4}$$

and $P_0(A)$ indicates that the probability of event A is computed under H_0 at level of significance α.

The proposed test captures the concept that if null hypothesis is true, i.e. all the scale parameters are equal, the ratio in the test statistics will be equal to 1 otherwise under alternative hypothesis, these ratios will be greater than 1 and maximum of these ratios should lead to the rejection of the hypothesis.

The proposed statistics is based on functions of sample quasi ranges, and distribution of sample quasi ranges does not depend upon parent distribution (see David and Nagaraja (2003)) and hence is distribution free.

3 Calculation of Critical Points for Some Specific Distributions: Simulation Method

The exact distribution of weighted function of quasi ranges is not available in literature. Of course, some results regarding distribution of weighted function of quasi ranges are given in David and Nagaraja (2003), but not the exact distribution. To compute the exact critical points for the proposed test statistics (stated in Eq. 2.1), the approach suggested by Hayter (1990) is not adopted here as the exact distribution of the test statistic which uses ratio of two Q_i really gets too complicated without having any precise form.

Therefore, we have adopted the other technique, i.e. the technique of statistical simulation. The critical points of the proposed test for the above said problem are obtained through statistical simulation taking 2×10^5 repetitions for different configuration of n_i's (sample sizes).

The critical points are computed for different distributions, viz. two-parameter standard exponential, standard logistic and standard uniform distributions. It is interesting to see that these critical points can also be used for some other well-known distributions using certain transformations. These critical points are provided for three and more populations.

3.1 *Critical Points for Standard Exponential, Standard Logistic and Standard Uniform Distributions*

The critical points are provided for the proposed test statistics as given in Eq. 2.2, for three distributions: standard exponential, standard logistic and standard uniform.

We have generated 2×10^5 values of test statistic $T_Q = \max\limits_{1 \leq i < j \leq k} (Q_j/Q_i)$ taking equal samples of size $n = 6, 7, \ldots, 25, 30, 35$ and 40 and using k as 3, 4, …, 0.10 from the above three distributions. The critical points are calculated for 0.05 and 0.01, two different values of level of significance α.

The simulated critical points are reported in Tables 1, 2 and 3 for all the three distributions for level of significance $\alpha = 0.05$ and for different samples, i.e. $k = 3(1)10$.

Remark 3.1 For the proposed test, critical constants are also computed for $\alpha = 0.01$ and $\alpha = 0.10$, for standard exponential distribution, standard logistic distribution and standard uniform distribution for some different samples, i.e. $k = 3(1)10$. The tables are available at https://kkmofpu.blogspot.com/p/tables_22.html?m=1.

Remark 3.2 For each of the three distributions, the size of the test is computed to check the validity of the critical points $q_{k,\alpha,n}$. It was observed that the actual sizes were less than the nominal level α. The tables are available at https://kkmofpu.blogspot.com/p/tables_22.html?m=1.

Table 1 Simulated critical points for exponential distribution when $\alpha = 0.05$

n/k	2	3	4	5	6	7	8	9
6	3.2834	4.5324	5.5526	6.2838	6.8237	7.5313	8.0394	8.6553
7	3.0190	3.9617	4.7565	5.3166	5.9110	6.3161	6.6577	7.0068
8	2.7161	3.5861	4.1614	4.7763	5.0677	5.4726	5.7534	6.0740
9	2.5593	3.3097	3.8309	4.2217	4.6049	4.8394	5.1155	5.3508
10	2.4532	3.1014	3.5531	3.8657	4.2235	4.4570	4.6269	4.8737
11	2.2954	2.9361	3.3529	3.6469	3.9062	4.0962	4.2746	4.4897
12	2.2305	2.7531	3.1440	3.4410	3.6490	3.8599	4.0265	4.2241
13	2.1694	2.6919	2.9861	3.2678	3.4379	3.6304	3.7813	3.9422
14	2.0858	2.5622	2.8589	3.0984	3.2858	3.4844	3.5943	3.7422
15	2.0447	2.4412	2.7438	2.9667	3.1634	3.3154	3.4321	3.5967
16	1.9813	2.391328	2.6509	2.8720	3.0487	3.1918	3.2888	3.3983
17	1.9400	2.3430	2.5821	2.7910	2.9165	3.0421	3.1846	3.2908
18	1.9149	2.2646	2.5281	2.6941	2.8292	2.9855	3.0647	3.1513
19	1.8803	2.2275	2.4491	2.6087	2.7632	2.8787	2.9823	3.0508
20	1.8447	2.1764	2.3817	2.5513	2.6802	2.7921	2.8768	2.9798
21	1.8284	2.1521	2.3530	2.4745	2.6262	2.7254	2.7927	2.8949
22	1.7897	2.1038	2.2915	2.4314	2.5519	2.6480	2.7424	2.8151
23	1.7516	2.0477	2.2586	2.3832	2.5079	2.6013	2.6937	2.7359
24	1.7458	2.0343	2.2272	2.3359	2.4570	2.5421	2.6293	2.6512
25	1.7308	2.0077	2.1862	2.3151	2.4186	2.4946	2.5575	2.6172
30	1.6387	1.8775	2.0109	2.1309	2.2125	2.2970	2.3587	2.4247
35	1.5970	1.7843	1.9258	2.0258	2.0841	2.1414	2.2108	2.2565
40	1.5316	1.7288	1.8345	1.9154	1.9886	2.0466	2.1034	2.1400

Remark 3.3 Using some suitable transformations, the critical points obtained above can also be used for some other distributions, as suggested by Bansal et al. (2011).

4 Simultaneous One-Sided Confidence Intervals (SOCI's) of the Proposed Test

One may also construct the SOCI*s* for the ordered pairwise ratios of scale parameters (see Singh and Gill (2004), Bansal et al. (2011)). The statistical analysis may not be further necessary if the null hypothesis of homogeneity H_0 is not significant. However, if the null hypothesis is significant, then one may wish to determine which θ_i's differ and by how much. The test statistic $T_Q = \max_{1 \leq i < j \leq k} (Q_j/Q_i)$ and using Eq. (2.4) allows us to construct $1 - \alpha$ level SOCIs for the ordered pairwise ratios

Table 2 Simulated critical points for logistic distribution when $\alpha = 0.05$

n/k	2	3	4	5	6	7	8	9
6	2.477	3.1796	3.7078	4.1404	4.4026	4.6964	4.9289	5.2753
7	2.2678	2.8581	3.2454	3.5589	3.8617	4.0915	4.2449	4.3974
8	2.1293	2.6312	2.9547	3.2302	3.4096	3.6469	3.7867	3.9362
9	2.0247	2.4732	2.7666	2.9563	3.1617	3.3083	3.4276	3.5275
10	1.9329	2.3011	2.551	2.7695	2.937	3.0624	3.1857	3.2959
11	1.8799	2.2142	2.4412	2.6358	2.7809	2.8688	2.9956	3.1113
12	1.8205	2.1513	2.3599	2.4946	2.6453	2.7319	2.8292	2.9204
13	1.7839	2.0707	2.2667	2.4088	2.5012	2.6343	2.7298	2.7853
14	1.7249	2.0006	2.1871	2.3062	2.4297	2.5302	2.5877	2.6877
15	1.696	1.973	2.1328	2.2362	2.3498	2.4191	2.5007	2.5723
16	1.6602	1.9166	2.0603	2.1837	2.2813	2.3434	2.4224	2.503
17	1.63	1.8923	2.0127	2.1397	2.2155	2.2852	2.3597	2.4178
18	1.6113	1.84	1.9802	2.0808	2.1663	2.2448	2.3023	2.3375
19	1.5839	1.7981	1.9407	2.0418	2.1199	2.1829	2.225	2.2843
20	1.5747	1.7839	1.9011	1.9907	2.0783	2.1402	2.1913	2.2503
21	1.5481	1.7446	1.858	1.9545	2.0272	2.0956	2.144	2.1905
22	1.5345	1.715	1.8317	1.9275	1.9958	2.0573	2.1065	2.1585
23	1.5279	1.7018	1.8062	1.8988	1.9704	2.0173	2.0737	2.1112
24	1.5086	1.6843	1.7953	1.8818	1.9426	1.9821	2.0255	2.0671
25	1.4845	1.6576	1.7689	1.837	1.9098	1.9648	2.0033	2.0472
30	1.4359	1.5868	1.6747	1.7482	1.7978	1.8398	1.8769	1.913
35	1.4016	1.5248	1.6135	1.6697	1.719	1.7555	1.7862	1.82
40	1.3677	1.4902	1.564	1.6229	1.6558	1.6961	1.7229	1.7471

θ_j/θ_i, $1 \le i < j \le k$, as follows (Hayter, 1990; Miller, 1981):

$$\begin{aligned} 1 - \alpha &= P_0(T_Q \le q_{k,\alpha,n}) \\ &= P_0\left[\max_{1 \le i < j \le k}\left(\frac{Q_j}{Q_i}\right) \le q_{k,\alpha,n}\right] \\ &= P_0\left[\max_{1 \le i < j \le k}\left(\frac{Q_j/\theta_j}{Q_i/\theta_i}\right) \le q_{k,\alpha,n}\right] \end{aligned}$$

as under $H_0: \theta_1 = \theta_2 = \cdots = \theta_k$, $\forall i = 1, 2, \ldots, k$ or $\frac{\theta_j}{\theta_i} = 1$, $1 \le i < j \le k$, where $P_0(A)$ indicates that the probability of event A is computed under H_0 at level of significance α and Q_i, for $i = 1, 2, \ldots, k$ is defined in Eq. (2.1).

Therefore,

Table 3 Simulated critical points for uniform distribution when $\alpha = 0.05$

n/k	2	3	4	5	6	7	8	9
6	1.8896	2.3224	2.6073	2.8556	3.0872	3.2381	3.3503	3.4915
7	1.7553	2.0864	2.3102	2.4764	2.613	2.7133	2.8499	2.9384
8	1.6421	1.9108	2.0967	2.2482	2.3407	2.4182	2.5312	2.6083
9	1.5756	1.8059	1.9425	2.0769	2.1795	2.2523	2.3198	2.3689
10	1.5292	1.7099	1.847	1.9518	2.0403	2.1037	2.1578	2.2133
11	1.4843	1.6755	1.7739	1.8731	1.9268	2.0008	2.039	2.0814
12	1.4371	1.6077	1.7022	1.8021	1.8561	1.9186	1.9465	1.9912
13	1.4184	1.5707	1.6604	1.735	1.7922	1.8363	1.8817	1.9159
14	1.4015	1.5271	1.6174	1.6901	1.7427	1.7729	1.8165	1.8408
15	1.3692	1.499	1.5838	1.6501	1.6852	1.7312	1.7638	1.7872
16	1.356	1.4752	1.5492	1.609	1.6557	1.6896	1.7214	1.7518
17	1.338	1.4528	1.5221	1.5716	1.6166	1.6524	1.678	1.7069
18	1.3217	1.4295	1.503	1.5539	1.5801	1.6173	1.6464	1.6745
19	1.3015	1.4208	1.4815	1.5268	1.5703	1.5903	1.6231	1.6517
20	1.2895	1.4014	1.4579	1.5073	1.5415	1.5738	1.5934	1.616
21	1.2902	1.3874	1.437	1.4894	1.5185	1.5485	1.5727	1.59
22	1.2791	1.3783	1.4343	1.4668	1.4996	1.5241	1.5528	1.5654
23	1.2744	1.3631	1.4199	1.4578	1.4859	1.5114	1.5363	1.5526
24	1.2692	1.3533	1.4022	1.4451	1.4704	1.4925	1.513	1.5353
25	1.2564	1.3365	1.3943	1.4267	1.4573	1.4779	1.4986	1.5162
30	1.2211	1.3033	1.3406	1.3769	1.3998	1.4215	1.437	1.4508
35	1.2091	1.2698	1.3115	1.3383	1.3559	1.3749	1.3907	1.4026
40	1.1899	1.2479	1.2888	1.3063	1.3304	1.3431	1.3593	1.3689

$$
\begin{aligned}
1-\alpha &= P\left[\left(\frac{Q_j/\theta_j}{Q_i/\theta_i}\right) \le q_{k,\alpha,n},\ 1 \le \mathrm{i} < \mathrm{j} \le \mathrm{k}\right] \\
&= P\left[\frac{\theta_j}{\theta_i} \ge \frac{Q_j}{Q_i}\frac{1}{q_{k,\alpha,n}},\ 1 \le \mathrm{i} < \mathrm{j} \le \mathrm{k}\right] \qquad (4.1)
\end{aligned}
$$

It may be noted that the validity of the simultaneous confidence intervals given in (4.1) does not depend on the assumption whether heterogeneity among θ_i's follow the simple ordering and are valid only if this ordering is specified independently without any examination of data. Although if the knowledge of the likely ordering $\theta_1 \le \theta_2 \le \cdots \le \theta_k$ is known in advance, then this information may be used to improve the confidence intervals given by (4.1). One can see that the lower end of the confidence intervals given by (4.1) with a value less than one, in this case, be non-informative, and may be truncated at one. Close values of the lower end points of the SOCIs given by (4.1) indicate closeness among the scale parameters. These

SOCIs are of interest to the experimenter in agriculture, engineering, quality control, etc., to see which of the treatments are heterogeneous.

4.1 Simulated Example to Compute Test Statistic and SOCIs

Computation of test statistic and construction of SOCIs for the ordered pairwise ratios of scale parameters with the help of the following data generated from four logistic distributions is shown

$L(0, 1)$: −0.25538, 1.41174, −0.16191, 0.35481, −0.92192, 0.86421, 2.73712, 0.25074, −0.01846, 0.30758, −1.18180, 0.67576, 1.17131, −0.89851, −2.82519

$L(1, 3)$: 6.00785, 7.19795, 1.03843, −4.38304, 9.59191, −10.17868, −0.54060, −5.86195, 0.38682, −6.48398, −4.69253, 6.91611, 1.62835, 5.15529, 11.25271

$L(3, 5)$: −5.41441, −4.59933, 3.02097, −9.81064, −7.67931, 5.00972, 12.87510, 9.53253, 5.46022, −3.08865, 0.49221, 3.20816, −1.30717, 5.73115, 4.90288

$L(2, 8)$: 27.21326, 13.76259, 12.61635, −10.64983, −9.71614, −7.04021, 9.32855, 18.79243, −22.04662, −12.76820, 10.85363, 9.91979, −8.58689, −10.87091, 11.32912

where $L(a, b)$ denotes the logistic distribution with location (scale) parameter $a(b)$. Here $Q_1 = 1.45081$, $Q_2 = 7.65743$, $Q_3 = 7.57082$ and $Q_4 = 16.55985$. Using these values, we can get $T_Q = 11.41421$ which is significant at 5% level of significance, as the critical value $q_{k,\alpha,n} = q_{4,0.05,15}$, from Table 3, is 2.1328. The set of 95% SOCIs, using (4.1) with lower end of the confidence interval truncated at one if it is less than one, for Q_2/Q_1, Q_3/Q_1, Q_4/Q_1, Q_3/Q_2, Q_4/Q_2 and Q_4/Q_3 are computed as:

$$[[2.4747, \infty), [2.4467, \infty), [5.3517, \infty), [1, \infty), [1.0139, \infty), [1.0255, \infty)].$$

5 Power of the Proposed Test

Earlier, in Sect. 2, the test was proposed, and in this section, statistical simulation is carried out for power computations. For the exponential, logistic and uniform distributions, power is computed under some particular configuration of scale parameters for $k = 3$ & 5 at different levels of α and n_i's. These tables are available at https://kkmofpu.blogspot.com/p/tables_22.html?m=1. The tables suggest that the power of the test is quite high and increases with the increasing sample size.

For testing the homogeneity of scale parameters against the simple ordered alternative, based on sample quasi ranges, Singh and Gill (2004) proposed the test $W_r = \max_{1 \le i < j \le k} (R_{jr}/R_{ir})$, where R_{ir} is as defined in Sect. 2.

A class of tests (T_k) was proposed by Kusum and Bagai (1988) for testing the homogeneity of scale parameters against simple ordered alternative based on linear combination of two-sample U-statistics which was proposed by Deshpande and Kusum (1984). Kusum and Bagai (1988) in turn proposed test when the k ($k \ge 3$) populations differing in scale parameters only. Later, Shetty and Bhatt (2003) proposed a class of tests ($L_{c,d}$) for testing homogeneity of scale parameters against simple ordered alternative based on linear combination of two-sample U-statistics which was proposed by Shetty and Bhatt (1993) using sub-sample median and extremes.

Herein power comparisons are undertaken for the proposed test with respect to the Singh–Gill test (2004), the Kusum–Bagai test (1988) and Shetty–Bhatt test (2003) when the samples have outliers using Monte Carlo simulation technique. Simulation was undertaken for 2×10^5 repetitions, where in each repetition, a fresh random sample of size n contaminated with two outliers from ith distribution $F(\theta)$, $i = 1, 2, \ldots, k$ was generated as given below:

Let δ ($\delta > 1$) be a given constant.

1. $n - 2$ observations generated from distribution, $F(\theta)$.
2. A contaminated observation was generated from distribution $F(\theta\delta)$, a distribution with scale parameter greater than θ.
3. Another contaminated observation was generated from distribution $F(\theta/\delta)$, a distribution with scale parameter smaller than θ.

Thus, random samples of sizes n from each population were contaminated with two outliers. One outlier was observation from the distribution with small-scale parameter, and another is observation from distribution with large-scale parameter. We take four samples ($k = 4$) from exponential, logistic and uniform distributions with different parameters as:

$$\mathrm{I} \sim E(0, 1);\ \mathrm{II} \sim E(0, 2.1);\ \mathrm{III} \sim E(0, 3.2);\ \text{and IV} \sim E(0.4.3)$$
$$\mathrm{I} \sim L(0, 1);\ \mathrm{II} \sim L(0, 2.2);\ \mathrm{III} \sim L(0, 3.2);\ \text{and IV} \sim L(0.4.1)$$
$$\mathrm{I} \sim U(0, 1);\ \mathrm{II} \sim U(0, 1.2);\ \mathrm{III} \sim U(0, 1.5);\ \text{and IV} \sim U(0.1.7)$$

where $E(a, b)$, $L(a, b)$ and $U(a, b)$ are exponential, logistic and uniform distributions, respectively.

The power comparisons of the proposed test (T_Q) with that of the Singh–Gill test (W_r, $r = 0, 1, 2$), the Kusum–Bagai test (T_k) and the Shetty–Bhatt test ($L_{3,3}$) are done in Table 4 when $\alpha = 0.05$, $\delta = 1.0, 1.5$, $k = 4$ (for other table, when $\alpha = 0.05$, $\delta = 2.0, 2.5$, $k = 4$, see https://kkmofpu.blogspot.com/p/tables_22.html?m=1). From these tables, it can be noticed that the power of the proposed test is considerably higher than the other tests for all the distributions, except for uniform distribution when power of the proposed test is on the lower side compared to test W_0 when

Table 4 Power comparison of the proposed test when $\alpha = 0.05$, $\delta = 1.0, 1.5$, $k = 4$

Distribution	N	δ	T_Q	W_O	W_I	W_2	T_k	$L_{3,3}$
Exponential	6	1.0	0.5352	0.3931	0.2876	0.1083	0.1174	0.0523
	8	1.0	0.6552	0.4981	0.4397	0.2745	0.1598	0.0456
	10	1.0	0.7655	0.5756	0.5596	0.4688	0.1806	0.0442
	12	1.0	0.8507	0.6182	0.6631	0.5906	0.2056	0.0375
Logistic	6	1.0	0.7467	0.6565	0.3932	0.1094	0.4726	0.2292
	8	1.0	0.9034	0.8054	0.6838	0.3778	0.6572	0.3445
	10	1.0	0.9916	0.8662	0.8298	0.6662	0.7882	0.4483
	12	1.0	1.0000	0.9206	0.9144	0.8329	0.8672	0.5680
Uniform	6	1.0	0.3486	0.2573	0.1177	0.0706	0.2656	0.1155
	8	1.0	0.5011	0.4673	0.2128	0.1280	0.3504	0.1226
	10	1.0	0.6752	0.7441	0.3494	0.1935	0.4308	0.1512
	12	1.0	0.8070	0.9238	0.5046	0.2873	0.4658	0.1869
Exponential	6	1.5	0.5612	0.4172	0.2912	0.1051	0.1271	0.0498
	8	1.5	0.6553	0.5152	0.4486	0.2815	0.1568	0.0434
	10	1.5	0.7612	0.5849	0.5651	0.4712	0.1758	0.0417
	12	1.5	0.8613	0.6311	0.6640	0.6017	0.2014	0.0387
Logistic	6	1.5	0.7537	0.6682	0.4042	0.1030	0.4362	0.2580
	8	1.5	0.9109	0.7847	0.6744	0.3642	0.6516	0.3540
	10	1.5	0.9858	0.8691	0.8377	0.6614	0.7802	0.4481
	12	1.5	1.000	0.9115	0.9181	0.8391	0.8640	0.5354
Uniform	6	1.5	0.4219	0.3843	0.1135	0.0679	0.1698	0.1054
	8	1.5	0.5964	0.5792	0.2081	0.1157	0.2452	0.1427
	10	1.5	0.7113	0.7598	0.3424	0.1908	0.3102	0.1626
	12	1.5	0.7951	0.8573	0.5216	0.2967	0.3612	0.1688

sample size is higher but power is on higher side compared to W_0 when sample size is small.

Remark 5.1 The power and size of the test T_Q can also be computed when sample size (n) varies and also when location parameter varies.

Remark 5.2 In David and Nagaraja (2003) and Budescu (1980), we can notice that Q_i as given in (2.1) is nothing but weighted function of spacing, which is given as

$$Q_i = \frac{\pi^{1/2} \sum_{r=1}^{n-1} r(n-r)(X_{i[r+1]} - X_{i[r]})}{n(n-1)}.$$

The weighted function of spacing is considered as a robust estimator of scale parameter like weighted function of quasi ranges.

6 Conclusion

The proposed test based on weighted function of sample quasi ranges is more useful for a situation wherein, due to limitations of measurement mechanism, it is difficult to measure the observations above or below some reference point, or may be when data is subject to double censoring. The proposed test is also more robust in comparison to other tests based on complete samples and even in the case of small samples, as also when the outliers are suspected. As it is, the proposed test is easy to understand and easy to use by practitioners.

Acknowledgements The authors acknowledge with thanks the Editor and Reviewer for their valuable suggestions.

References

Bansal, P., Arora, S., & Mahajan, K. K. (2011). Testing homogeneity of Gini indices against simple-ordered alternative. *Communications in Statistics - Simulation and Computation, 40*(2), 185–198.

Barlow, R. E., Bartholomew, D. J., Bremner, J. M., & Brunk, H. D. (1972). *Statistical inference under order restrictions*. Wiley.

Budescu, D. V. (1980). Approximate confidence intervals for a robust scale parameter. *Psychometrika, 45*(3), 397–402.

David, H. A., & Nagaraja, H. N. (2003). *Order statistics* (3rd ed.). Wiley.

Deshpande, J. V., & Kusum, K. (1984). A test for the nonparametric two-sample scale problem. *Australian Journal of Statistics, 26*(1), 16–24.

Gaur, A., Mahajan, K. K., & Arora, S. (2012). Nonparametric test for homogeneity of scale parameters against ordered alternatives based on subsample medians. *Probstat Forum, 5*, 133–141.

Gaur, A., Mahajan, K. K., & Arora, S. (2013). A nonparametric test for a multi-sample scale problem using ranked-set data. *Statistical Methodology, 10*(1), 85–92.

Gill, A. N., & Dhawan, A. K. (1999). A one-sided test for testing homogeneity of scale parameters against ordered alternatives. *Communications in Statistics - Theory and Methods, 28*(10), 2417–2439.

Hayter, A. J. (1990). A one-sided studentized range test for testing against a simple ordered alternative. *Journal of the American Statistical Association, 85*, 778–785.

Hochberg, Y., & Tamhane, A. C. (1987). *Multiple comparison procedures*. Wiley.

Kochar, S. C., & Gupta, R. P. (1985). A class of distribution-free tests for testing homogeneity of variances against ordered alternatives. In R. Dykstra, et al. (Eds.) *Proceedings of the Conference on Advances in Ordered Restricted Statistical Inference at Iowa City* (pp. 169–183). Springer.

Kusum, K., & Bagai, I. (1988). A new class of distribution-free procedures for testing homogeneity of scale parameters against ordered alternatives. *Communications in Statistics - Theory and Methods,* 17(5): 1365–1376.

Mahajan, K. K., Arora, A., & Gaur, A. (2012). A class of nonparametric tests for testing homogeneity of variances against ordered alternatives based on subsample extrema. *Metron: International Journal of Statistics, 70*(2–3), 193–206.

Miller, R. G. (1981). *Simultaneous statistical inference* (2nd ed.). Springer.

Robertson, T., Wright, F. T., & Dykstra, R. L. (1988). *Order restricted statistical inference*. Wiley.

Shetty, I. D., & Bhatt, S. V. (1993). A class of distribution-free tests for the two-sample scale problem. In *Proceedings of the Second International Symposium on Optimization and Statistics*, Nov 2–4 (pp. 91–96).

Shetty, I. D., & Bhatt, S. V. (2003). A class of outlier resistant distribution-free tests for testing homogeneity against ordered scale alternatives. *Communications in Statistics - Theory and Methods, 32*(2), 347–379.

Singh, P., & Gill, A. N. (2004). A one-sided test based on sample quasi ranges. *Communications in Statistics - Theory and Methods, 33*(4), 835–849.

A Bayesian Response-Adaptive, Covariate-Balanced and Q-Learning-Decision-Consistent Randomization Method for SMART Designs

Tianjiao Dai and Sanjay Shete

Abstract In clinical trials, various randomization strategies have been developed to allocate subjects to different interventions or treatment arms. The preferred strategy is one that balances the distribution of covariates across the treatment arms and assigns more subjects to the treatments associated with better outcomes. Sequential multiple assignment randomized trial (SMART) designs involve an initial stage in which participants are randomized to a set of intervention options, followed by subsequent stages in which some or all of the individuals are re-randomized to the intervention options available at that stage. For SMART designs, the intervention for each subject can be optimized individually using a Q-learning-based optimization algorithm. However, such response-adaptive or Q-learning-based randomization strategies lead to covariate imbalance that can result in biased inference when the imbalanced covariates are associated with the outcome of interest, particularly with small to moderate sample sizes. To combine the advantages of Q-learning-decision-consistent strategies and response-adaptive designs while controlling for covariate balance, we propose a Bayesian response-adaptive, covariate-balanced and Q-learning-decision-consistent randomization method (RCQ) for SMART designs. Simulation studies to illustrate the performance of the proposed method show that the RCQ randomization method assigned the lowest percentage of subjects to the inferior intervention arms and the highest percentage of subjects to optimal Q-learning-decision-consistent strategies that maximize the long-term outcome for the individual while exhibiting well-controlled covariate balance. The alternative randomization strategies showed pronounced covariate imbalance or assigned higher percentages of subjects to inferior interventions or were not consistent with the Q-learning-based optimal decision strategy.

T. Dai · S. Shete (✉)
Department of Biostatistics, The University of Texas MD Anderson Cancer Center, Houston, TX 77030, USA
e-mail: sshete@mdanderson.org

S. Shete
Department of Epidemiology, The University of Texas MD Anderson Cancer Center, Houston, TX 77030, USA

D. D. Hanagal et al. (eds.), *Applied Statistical Methods*, Springer Proceedings in Mathematics & Statistics 380, https://doi.org/10.1007/978-981-16-7932-2_13

Keywords Covariate-balanced · Q-learning · Response-adaptive · SMART designs

1 Introduction

In the clinical trial literature, various randomization strategies have been developed to allocate subjects to different interventions. In addition to standard equal randomization, in which subjects are randomized to different interventions with equal probability, subjects can be assigned to interventions according to their baseline characteristics, intermediate covariates or intermediate outcomes, or historical information. The goal of such allocation schemes is to have higher probability of assigning subjects to the superior interventions.

Response-adaptive randomization, in which the allocation probabilities are adjusted based on the previous patients' responses in the study, allows more patients to be assigned to the superior treatment as the trial progresses. For example, Efron (1973) proposed a biased coin design to balance the numbers of individuals in the experimental treatment and control arms while avoiding various experimental biases; Berry and Eick (1995) compared a balanced randomization strategy to adaptive randomization in clinical research; Rosenberger et al. (2001) developed an optimal allocation between two treatments in a clinical trial; Thall et al. (2002) proposed an adaptive Bayesian design for patients with hematologic malignancies; Zhang and Rosenberger (2006) evaluated the performance of different response-adaptive randomization procedures in clinical trials with continuous outcomes; and Sverdlov et al. (2014) proposed a multiple-objective response-adaptive design.

However, response-adaptive designs also lead to covariate imbalance, which results in bias when comparing treatment efficacy. Imbalance of prognostic factors, covariates that affect study outcomes, can lead to observed differences in the intervention groups that may reflect patient heterogeneity across the intervention groups rather than the treatment effects. Several methods have been proposed to balance covariate distributions across intervention arms during randomization. For example, Signorini et al. (1993) proposed a randomization method for balancing treatment allocations both within strata and across the trial. This approach was further improved by Heritier et al. (2005) to maintain a marginal balance over important strata. Thall and Wathen (2005) proposed a Bayesian design for a multi-center, randomized clinical trial using covariate-adjusted adaptive randomization. Shao and Yu (2013) established asymptotic results for covariate-adaptive biased coin randomization under generalized linear models. Recent research efforts have combined these two approaches. For example, for trials with binary outcomes, Ning and Huang (2010) developed a patient allocation scheme that adjusts the covariate imbalance during response-adaptive randomization. In particular, Yuan et al. (2011) proposed a randomization procedure and incorporated this method into a group sequential response-adaptive randomization design with a goal of achieving the benefits of the response-adaptive design while balancing the covariates. These methods have the advantage of assigning fewer patients to inferior treatments or controlling the imbalance of covariates across treatments when the sample size is moderate or small. Scott et al. (2002) and Green et al. (2001) provided comprehensive reviews on the allocation method of minimization for balancing treatment groups.

However, these randomization strategies are not straightforward to apply for sequential multiple assignment randomized trial (SMART) designs because participants are re-randomized in multiple stages, and the design involves embedded interventions. A method that is applicable to studies based on the SMART design is Q-learning (Moodie et al., 2012; Nahum-Shani et al. 2012), which uses backward steps to construct a sequence of decision rules that link the patient-specific variables and treatment responses with the most efficient intervention that can maximize the long-term primary outcome for each individual. Importantly, Q-learning reduces the potential bias resulting from unmeasured causes associated with both the tailoring variables and primary outcomes.

In this article, our goal is to develop a randomization strategy for SMART designs that capitalizes on (1) a Q-learning-based approach for constructing high-quality decision rules that maximize the long-term outcome for each individual; (2) adapting to treatment responses in the trial, with the goal of assigning more patients to the superior treatment(s) as the trial progresses; and (3) balancing covariate distributions across intervention arms during randomization so that we can compare treatment efficacy.

Furthermore, in the proposed randomization strategy, instead of randomizing the subjects all at one time, we use the more practical approach of assigning subjects to treatments in sequential groups based on the information obtained from previously enrolled groups. We perform simulation studies to compare the proposed allocation strategy to alternate randomization strategies.

2 Methods

2.1 Overview of the SMART Design

In the SMART design, each individual enrolled in the trial goes through multiple stages with a critical decision point corresponding to each stage (Almirall et al., 2012, 2014; Lei et al., 2012; Kelleher et al., 2017). Specifically, it involves an initial stage in which participants are randomized to all the available intervention options, such as different types of medical treatments or behavioral interventions, followed by subsequent stages in which some or all of the individuals are re-randomized to the intervention options available at that stage. Re-randomization and the intervention options at each subsequent stage depend on the information obtained from previous stages, such as patient adherence or response status. Because the complexity of the SMART design increases with the number of stages, a two-stage SMART design is most commonly used for its simplicity and ability to study various clinical problems.

We motivate our methodology with one of the standard two-stage SMART designs (depicted in Fig. 1) in which subjects in the first stage are randomized to one of two interventions: medication ($A_1 = +1$) or behavioral intervention ($A_1 = -1$). At the second stage, only non-responders from the first stage are re-randomized to either an increased dose of the first-stage intervention ($A_2 = +1$) or augmenting the first-stage intervention with the alternative intervention ($A_2 = -1$). For example, non-responders for those receiving the medication in first-stage (M) are re-randomized to increased dose of medication ($M+$) or added behavioral intervention to the medication ($M + B$).

Let O_1 be the baseline covariates assessed before the first-stage intervention (e.g., level of depression, sex, age) and O_2 be the intermediate covariates assessed prior to the second-stage intervention (e.g., adherence to the first-stage intervention). Let Y be the final outcome value at the end of the trial.

We utilize a strategy to enroll participants in the SMART in a group sequential manner such that the resulting design skews the allocation probability toward the better interventions to optimize the final outcomes and lowers the covariate imbalance based on the information from previous participants. Specifically, subjects enroll in a SMART in sequential groups of sizes $\{N_k\}$, $k = 1, \ldots, K$, where N_k is the sample size of the kth group. The subjects enter the trial sequentially, and the allocation probabilities are updated using the observed data from participants previously enrolled in the study. If little information regarding the superiority of the interventions is known before conducting the trial, subjects in the 1st ($k = 1$) group are allocated randomly

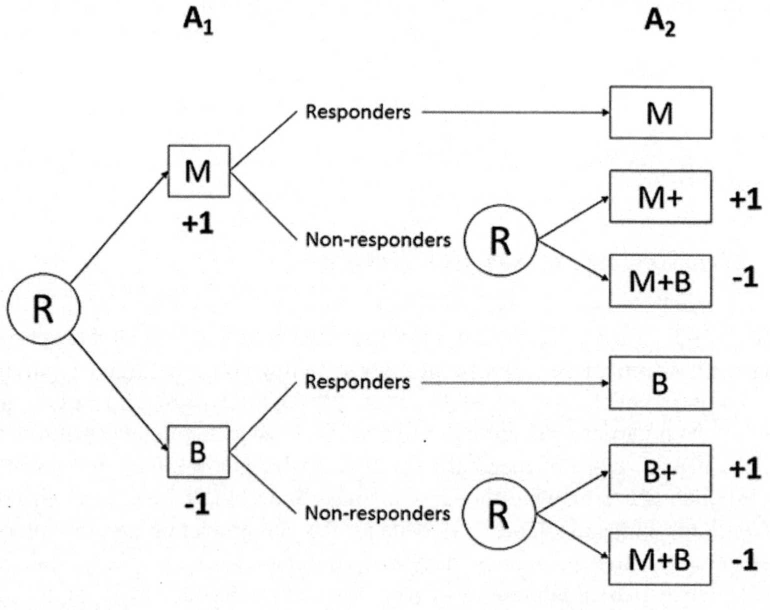

Fig. 1 SMART design, in which re-randomization to different second-stage intervention options *depends* on an intermediate outcome (only non-responders are re-randomized in the second stage)

to the interventions at both stages of the SMART, i.e., the allocation probability is 0.5. Other randomization probabilities can be applied to the first group if historical information regarding the embedded interventions is known before the trial, or based on the baseline characteristics, intermediate covariates or intermediate outcomes of the subjects. For subsequent groups ($k = 2, \ldots, K$), calculations of the allocation probability are based on our proposed approach, described in Sect. 2.5.

2.2 *Randomization Probability Using Q-Learning-Based Optimal Decisions*

First, we briefly introduce the Q-learning approach that can be used to develop adaptive interventions (Moodie et al., 2012; Nahum-Shani et al., 2012). In Q-learning, optimal decisions are derived by maximizing the Q-functions if a higher value of outcome is desired (minimizing the Q-functions if a lower value of the final outcome is desired). In this approach, optimal decisions are constructed using backward induction by first finding the optimal decision rule at the second stage, ${d_2}^*$, using linear regression, as follows

$$d_2^*(O_1, A_1, O_2) = \max Q_2(O_1, A_1, O_2, A_2),$$

where $Q_2(O_1, A_1, O_2, A_2) = E(Y|O_1, A_1, O_2, A_2)$ and maximization is over the values of A_2. Based on the linear regression model, the Q_2-function for the second stage is $Q_2(O_1, A_1, O_2, A_2, \gamma_2, \alpha_2) = \gamma_{20} + \gamma_{21} O_1 + \gamma_{22} O_2 + \gamma_{23} A_1 + \gamma_{24} A_1 \cdot O_1 + (\alpha_{20} + \alpha_{21} A_1 + \alpha_{22} O_2) \cdot A_2$, where $O_1 = (O_{11}, O_{12}, \ldots)$ and $O_2 = (O_{21}, O_{22}, \ldots)$ are the respective vectors of the baseline and intermediate covariates. The parameters $\alpha_2 = (\alpha_{20}, \alpha_{21}, \alpha_{22})$ reflect how the second-stage intervention (A_2) varies as a function of the candidate tailoring variables (here, A_1 and O_2). Based on this equation, the second-stage intervention option (A_2) that maximizes Q_2 is the one that maximizes the term $(\alpha_{20} + \alpha_{21}A_1 + \alpha_{22}O_2)A_2$. If $(\alpha_{20} + \alpha_{21}A_{1i} + \alpha_{22}O_{2i}) > 0$, the term $(\alpha_{20} + \alpha_{21}A_{1i} + \alpha_{22}O_{2i})A_{2i}$ attains its maximal value for $A_{2i} = 1$; and if $(\alpha_{20} + \alpha_{21}A_{1i} + \alpha_{22}O_{2i}) < 0$, the term $(\alpha_{20} + \alpha_{21}A_{1i} + \alpha_{22}O_{2i})A_{2i}$ attains its maximal value for $A_{2i} = -1$. Thus, the optimal second-stage decision for subject i is $d_{2i}^* = sign(\alpha_{20} + \alpha_{21} A_{1i} + \alpha_{22} O_{2i})$. Therefore, the decision rule probability according to Q-learning optimization for assigning subject i to the second-stage intervention $A_2 = +1$ is

$$p_{ki}^{Q(2)} = \Pr(\alpha_{20} + \alpha_{21} A_{1i} + \alpha_{22} O_{2i} > 0|D_k). \tag{2.1}$$

Next, we move backward in time to construct the first-stage decision rule by similarly maximizing the first-stage Q-function: $Q_1(O_1, A_1; \gamma_1, \alpha_1) = \gamma_{10} + \gamma_{11} O_1 + (\alpha_{10} + \alpha_{11} O_1) \cdot A_1$, which leads to the optimal first-stage decision $d_{1i}^* = sign(\alpha_{10} + \alpha_{11} O_{1i})$. Therefore, the first-stage decision rule probability for subject i to $A_1 = +1$ is

$$p_{ki}^{Q(1)} = \Pr(\alpha_{10} + \alpha_{11} O_{1i} > 0 | D_k). \tag{2.2}$$

Excellent papers by Moodie et al. (2012) and Nahum-Shani et al. (2012) provide more details about constructing these decision rules. In our implementation of this approach, we estimate the model parameters using a Bayesian linear regression model with the data from the previous $k - 1$ groups, i.e., D_k. In this Bayesian approach, we assume a vague normally distributed prior, $N(0, 10^6)$, for all model parameters, which is a type of prior that is commonly used when little information is known about the parameters. We call the randomization strategy in Eqs. (2.1) and (2.2) for two stages of the SMART design the Q-learning-decision-consistent randomization method (Q).

2.3 *Covariate-Balanced Randomization Probability According to the Prognostic Score for SMART Designs*

As stated in the introduction, response-adaptive designs lead to covariate imbalance, which results in bias when comparing treatment efficacy. An imbalance of prognostic factors, which are covariates that affect the study outcome, can lead to biased inference, particularly when the sample sizes are small or moderate (Ning & Huang, 2010; Yuan et al., 2011). Next, we extend the prognostic score approach developed by Yuan et al. (2011) to the SMART design.

For each subject i $(i = N \cdot (k-1)+1, \ldots, N \cdot k)$ in the kth $(k = 1, \ldots, K)$ group, the outcome is defined as $E(Y) = \beta_{20}+\beta_{21} O_1 + \beta_{22} O_2+\beta_{23} A_1+\beta_{24} A_2+\beta_{25} A_1 A_2$, where the coefficients $\beta_{21} = (\beta_{21}^1, \beta_{21}^2, \ldots)$ and $\beta_{22} = (\beta_{22}^1, \beta_{22}^2, \ldots)$ reflect the importance of the baseline covariates $O_1 = (O_{11}, O_{12}, \ldots)$ and intermediate covariates $O_2 = (O_{21}, O_{22}, \ldots)$ in predicting the final outcomes. Following the approaches of Stuart et al. (2013) and Yuan et al. (2011), the prognostic score for subjects in group k is defined as $= \beta_{21} O_1 + \beta_{22} O_2$. This definition allows us to balance the covariates through a single variable w while accommodating both categorical and continuous covariates in O_1 and O_2.

All the coefficients in this model are assumed to have normal distributions with vague prior $N(0, 10^6)$, and their posterior means are estimated as $\tilde{\beta_2}= \left(\tilde{\beta}_{20}, \tilde{\beta}_{21}, \tilde{\beta}_{22}, \tilde{\beta}_{23}, \tilde{\beta}_{24}, \tilde{\beta}_{25}\right)$ using Markov chain Monte Carlo (MCMC) regression with $data_k$. Let $\tilde{w}_{A1,A2}$ be the vector of the current prognostic score for the non-responders who received interventions (A_1, A_2), which is calculated using the estimated posterior means $\tilde{\beta_2}$. Then $KS_{A_{1i},A_2,i}$, $A_1 = +1$ or -1 are the two Kolmogorov–Smirnov (KS) statistics (Gail & Green, 1976; Grover, 1977) for subject i who received intervention A_{1i} at the first stage and was then assigned to the second-stage intervention $A_2 (A_2 = +1$ or $-1)$ based on $\tilde{w}_{A1i,A2=+1}$ or $\tilde{w}_{A1i,A2=-1}$. Because higher values of the KS statistics indicate more severe imbalance, the covariate-balanced probability of assigning subject i to intervention $A_2 = +1$ at the second stage is defined as

$$p_{ki}^{C(2)} = \frac{KS_{A_{1i},A_2=-1,i}}{KS_{A_{1i},A_2=-1,i} + KS_{A_{1i},A_2=+1,i}}. \tag{2.3}$$

To avoid extreme values (i.e., values close to 1 or 0) in certain circumstances, a root transformation such as that suggested by Yuan et al. (2011) and Lin et al. (2015) can be applied to stabilize the probability in (2.3) as $p_{ki,\text{stablized}}^{C(2)} = \frac{\sqrt{p_{ki}^{C(2)}}}{\sqrt{p_{ki}^{C(2)}} + \sqrt{1 - p_{ki}^{C(2)}}}$. For simplicity, we use $p_{ki}^{C(2)}$ to denote $p_{ki,\text{stablized}}^{C(2)}$ for simplicity.

Similarly, for the first-stage assignment, the probability of randomizing subject i to the intervention $A_1 = +1$ is

$$p_{ki}^{C(1)} = \frac{KS_{A_{1=-1},i}}{KS_{A_{1=-1},i} + KS_{A_{1=+1},i}}, \tag{2.4}$$

where $KS_{A_{1=+1},i}$ and $KS_{A_{1=-1},i}$ are the two KS statistics calculated for subject i when assigned to the first-stage interventions $A_1 = +1$ and $A_1 = -1$, respectively. The probability in (2.4) may also be stabilized by the aforementioned root transformation. We refer to the randomization strategy in Eqs. (2.3) and (2.4) for the two stages of the SMART design as the covariate-balanced randomization method (C).

2.4 *Response-Adaptive Randomization Probability Based on Outcomes of Previous Groups*

Next, we extend the standard response-adaptive randomization method to the SMART design. As done previously, the outcomes (Y) of the intervention (A_1, A_2) at the second stage are assumed to follow a normal distribution with mean μ_{A_1,A_2}, which are further assumed to be normally distributed with vague priors $N(0, 10^6)$. The subjects in the kth group posterior distribution of μ_{A_1,A_2} follow a normal distribution, $N(\tilde{\mu}_{A_1,A_2}^{(k)}, \tilde{V}_{A_1,A_2}^{(k)})$. Therefore, the posterior distribution of the difference $(\mu_{a_1,A_2=+1} - \mu_{a_1,A_2=-1})$ is normally distributed with mean $(\tilde{\mu}_{a_1,A_2=+1}^{(k)} - \tilde{\mu}_{a_1,A_2=-1}^{(k)})$ and variance $\left(\tilde{v}_{a_1,A_2=+1}^{(k)}\right)^2 + \left(\tilde{v}_{a_1,A_2=-1}^{(k)}\right)^2$. If higher values of the final outcomes are desired, the response-adaptive randomization probability for intervention $= A_2 +1$ at the second stage is calculated as

$$p_k^{R(2)} = \Pr(\mu_{a_1,A_2=+1} - \mu_{a_1,A_2=-1} > 0 | D_k) \tag{2.5}$$

for a subject who received $A_1 = a_1$ as the first-stage intervention. This probability is common for all the subjects in the kth group and may also be stabilized with a root transformation:

$$p_{k,\text{stablized}}^{R(2)}=\frac{\sqrt{p_k^{R(2)}}}{\sqrt{p_k^{R(2)}}+\sqrt{1-p_k^{R(2)}}}.$$

Similarly, the probability of randomizing subjects to interventions $A_1 = +1$ at the first stage is calculated by MCMC regression as

$$p_k^{R(1)} = \Pr\left(\mu_{A_1=+1} - \mu_{A_1=-1} > 0|D_k\right) \tag{2.6}$$

where $\mu_{A_1=+1}$ and $\mu_{A_1=-1}$ are the means of the final outcomes for subjects who received first-stage interventions $A_1 = +1$ and $A_1 = -1$, respectively. We call the randomization strategy in Eqs. (2.5) and (2.6) for two stages of the SMART design the response-adaptive randomization method (R).

2.5 *Response-Adaptive, Covariate-Balanced and Q-Learning-Decision-Consistent (RCQ) Randomization Method*

Our proposed approach is to combine (i) Q-learning-decision-consistent randomization (Q) described in Sect. 2.2; (ii) the extended covariate-balanced randomization for SMART designs (C), described in Sect. 2.3; and (iii) the extended response-adaptive randomization for SMART designs (R), described in Sect. 2.4.

The response-adaptive, covariate-balanced and Q-learning-decision-consistent (RCQ) probability of assigning each individual i in the kth group to intervention $A_1 = +1$ at the first stage is

$$p_{ki}^{(1)} = \frac{p_k^{R(1)} \cdot p_{ki}^{C(1)} \cdot p_{ki}^{Q(1)}}{\sum_k p_k^{R(1)} \cdot p_{ki}^{C(1)} \cdot p_{ki}^{Q(1)}} \tag{2.7}$$

where $p_k^{R(1)}$ is the response-adaptive probability, $p_{ki}^{C(1)}$ is the covariate-balanced probability and $p_{ki}^{Q(1)}$ is the Q-learning-decision-consistent probability. Similarly, the allocation probability of assigning each individual i in the kth group to intervention $A_2 = +1$ at the second stage is

$$p_{ki}^{(2)} = \frac{p_k^{R(2)} \cdot p_{ki}^{C(2)} \cdot p_{ki}^{Q(2)}}{\sum_k p_k^{R(2)} \cdot p_{ki}^{C(2)} \cdot p_{ki}^{Q(2)}} \tag{2.8}$$

Data from the first k groups, i.e., $data_k$, are used for calculating the allocation probability for the subjects in the $(k + 1)$th group. Therefore, the allocation probabilities are updated with data from the ongoing trial. Similar to the approach by Yuan et al. (2011), the allocation of the randomization probability in the proposed strategy is a function of the multiplication of the randomization probabilities from (i) to (iii) above; however, other functional forms of allocation probabilities can also be considered (see Discussion).

3 Simulations Models and Assessment Measures

3.1 Simulation Models

We evaluated the proposed randomization method using the SMART design depicted in Fig. 1. We used simulation studies to evaluate the proposed randomization strategy RCQ and compared it with alternative randomization strategies Q only, as defined in Sect. 2.2; with equal randomization (E), where subject assignment at each stage uses an equal probability of 0.5; with C only, as defined in Sect. 2.3; and with R only, as defined in Sect. 2.4.

We generated data for subject i using the following model

$$Y_i = \gamma_{20} + \gamma_{21} O_{1i} + \gamma_{22} O_{2i} + \gamma_{23} A_{1i} + \alpha_{20} A_{2i} + \alpha_{21} A_{1i} A_{2i} + \gamma_{24} A_{1i} O_{1i} + \alpha_{22} A_{2i} O_{1i} + \varepsilon_i$$

where A_1 and A_2 are intervention indicators of the two stages of SMART, O_1 is the baseline covariate vector, O_2 is the intermediate covariate vector, and errors follow the standard normal distribution. The results reported in Tables 1, 2, 3 and 4 are based on 1000 replicates. For each comparison, we carried out simulations with equal randomization (probability equal to 0.5) and unequal randomization (i.e., the probability of randomizing a non-responder in the initial first group depends on the value of his/her intermediate covariates, O_2). We used the MCMCpack (Martin et al., 2011) package in R to estimate the model parameters.

Table 1 Simulation results based on 1000 replicates for SMART from Fig. 1 with parameter values $\gamma_{20} = 1$, $\gamma_{21} = 1$, $\gamma_{22} = -1$, $\gamma_{23} = -0.3$, $\gamma_{24} = 0.1$, $\alpha_{20} = 0.1$, $\alpha_{21} = -0.2$ and $\alpha_{22} = -0.2$; sample size is 200 with group sizes of 40; non-responders in the first group are assigned to the second-stage intervention based on their intermediate outcome O_2 (i.e., unequal randomization)

Method	ITN%	Percentage of significant covariate imbalance		ODQ%
		$A_1 = 1$	$A_1 = -1$	
RCQ (%)	22.73	2.0	2.0	67.21
E (%)	34.94	5.0	6.5	50.13
Q (%)	26.12	7.5	6.0	62.60
R (%)	27.46	8.5	9.0	60.51
C (%)	35.20	1.5	1.5	50.24

3.2 Assessment Measures

We considered three summary measures to compare these randomization strategies. (i) We used the percentage of the inferior treatment number (ITN%), which is the percentage of subjects who were assigned to the inferior intervention arms (i.e., interventions for which the expected final outcomes are lower than those expected for the competing interventions). (ii) We calculated the KS statistics for the prognostic score. We reported the percentage of the occurrence of significant covariate imbalance for each of the two first-stage interventions, $A_1 = 1$ and $A_1 = -1$, by calculating the percentage of the p-values of the KS statistics that are less than 0.05. (iii) We reported the percentage of subjects assigned to their optimal adaptive strategies according to the Q-learning algorithm (ODQ%).

3.3 Simulation Results

Tables 1, 2, 3 and 4 provide the simulation results based on the SMART design shown in Fig. 1. Data for the simulations were generated using the simulation model with the following parameter choices: $\gamma_{20} = 1$, $\gamma_{21} = 1$, $\gamma_{22} = -1$, $\gamma_{23} = -0.3$, $\gamma_{24} = 0.1$, $\alpha_{20} = 0.1$, $\alpha_{21} = -0.2$ and $\alpha_{22} = -0.2$. The values of O_1 and O_2, which are univariate covariates, were generated from the normal distributions $N(3, 1^2)$ and $N(0.1, 0.3^2)$, respectively.

In Table 1, the total study sample size is 200 individuals, and the group size is 40, i.e., the allocation probability is updated for each sequential group of 40 subjects. The non-responders in the first group were assigned to the second-stage intervention based on their intermediate covariate O_2 (i.e., unequal randomization). Specifically, a non-responder was assigned to the intervention $A_2 = 1$ with probability equal to the cumulative normal distribution at his/her observed O_2 value. As shown in Table 1, the percentage of subjects who were assigned to the inferior intervention

Table 2 Simulation results based on 1000 replicates for SMART from Fig. 1 with parameter values $\gamma_{20} = 1$, $\gamma_{21} = 1$, $\gamma_{22} = -1$, $\gamma_{23} = -0.3$, $\gamma_{24} = 0.1$, $\alpha_{20} = 0.1$, $\alpha_{21} = -0.2$ and $\alpha_{22} = -0.2$; sample size is 200 with group sizes of 40; non-responders in the first group are assigned to the second-stage intervention completely at random (i.e., equal randomization)

Method	ITN%	Percentage of significant covariate imbalance		ODQ%
		$A_1 = 1$	$A_1 = -1$	
RCQ (%)	21.97	2.5	2.5	68.35
E (%)	34.64	6.0	4.5	50.60
Q (%)	25.87	6.5	5.0	63.02
R (%)	27.87	8.0	5.5	60.04
C (%)	34.66	2.5	1.0	50.49

Table 3 Simulation results based on 1000 replicates for SMART in Fig. 1 with parameter values $\gamma_{20} = 1, \gamma_{21} = 1, \gamma_{22} = -1, \gamma_{23} = -0.3, \gamma_{24} = 0.1, \alpha_{20} = 0.1, \alpha_{21} = -0.2$ and $\alpha_{22} = -0.2$; sample size is 500 with group sizes of 100; non-responders in the first group are assigned to the second-stage intervention based on their intermediate outcome O_2 (i.e., unequal randomization)

Method	ITN%	Percentage of significant covariate imbalance		ODQ%
		$A_1 = 1$	$A_1 = -1$	
RCQ (%)	17.49	7.5	9.0	74.41
E (%)	35.03	4.0	7.0	49.74
Q (%)	21.79	10.5	11.0	68.95
R (%)	22.38	4.5	5.5	67.32
C (%)	34.84	3.5	3.0	49.96

Table 4 Simulation results based on 1000 replicates for SMART from Fig. 1 with parameter values $\gamma_{20} = 1, \gamma_{21} = 1, \gamma_{22} = -1, \gamma_{23} = -0.3, \gamma_{24} = 0.1, \alpha_{20} = 0.1, \alpha_{21} = -0.2$ and $\alpha_{22} = -0.2$; sample size is 500 with group sizes of 100; non-responders in the first group are assigned to the second-stage intervention completely at random (i.e., equal randomization)

Method	ITN%	Percentage of significant covariate imbalance		ODQ%
		$A_1 = 1$	$A_1 = -1$	
RCQ (%)	18.44	4.0	4.0	73.39
E (%)	35.02	3.0	5.0	49.97
Q (%)	21.47	9.5	6.0	69.30
R (%)	23.85	4.5	5.5	65.75
C (%)	35.17	2.5	2.5	49.75

arms was 22.73% for the proposed RCQ randomization method, which is lower than the percentages for the other randomization methods: 34.94% for E, 26.12% for Q, 27.46% for R, and 35.20% for C. Our method also had the highest percentage of subjects assigned to the most optimal adaptive strategy: 67.21%, compared to 62.60% for Q and 60.51% for R. The ODQ% was 50.13% and 50.24% for the E and C methods, respectively. Importantly, the proposed RCQ method had an acceptable percentage of significant covariate imbalance (2%, 2%) compared with that of the Q (7.5%, 6%), and R (8.5%, 9%) methods for each of the two first-stage interventions: $A_1 = 1$ and $A_1 = -1$. The levels of imbalance for the E and C methods were also acceptable.

In Table 2, we present the results for the same sample sizes and the same parameter configurations as in Table 1, except instead of unequal randomization, we applied equal randomization to the initial group (the first group of 40 individuals) in the trial (i.e., subjects were randomized to the two interventions with probability of 0.5). Overall, the results are similar to those shown in Table 1. The proposed RCQ method had the lowest ITN% (21.97%), an acceptable level (2.5%) of significant covariate imbalance, and the highest ODQ% (68.35%). The E and C methods showed the highest ITN% (34.64% and 34.66%, respectively) and lowest ODQ% (50.60% and 50.49%) among all the methods. The percentages of significant covariate imbalance for the alternative randomization strategies were C (2.5% and 1%), Q (6.5% and 5%), E (6% and 4.5%) and R (8% and 5.5%).

Table 3 shows the results for the same simulation parameters shown in Table 1, except that the group size is 100, with a total sample size of 500. Similar to the results in Table 1, the proposed RCQ had the lowest ITN% (17.49%) and highest ODQ% (74.41%) of all the methods. The percentage of significant covariate imbalance for the RCQ method was 7.5%, 9%. Although the percentages of significant covariate imbalance were acceptable for the E and C methods, 4%, 7% and 3.5%, 3% respectively, these methods had higher ITN% (35.03% and 34.84%, respectively) and lower ODQ% (49.74% and 49.96%, respectively) when compared to those values for the proposed RCQ.

Table 4 shows the results of the same simulation parameters shown in Table 2, except that the group size is 100 and the total sample size is 500. The proposed RCQ method again showed the best performance with respect to all three metrics, i.e., lowest ITN% (18.44%), highest ODQ% (73.39%) and acceptable percentages of significant covariate imbalance (4%, for both first stages).

4 Discussion

We have proposed a Bayesian response-adaptive, covariate-balanced and optimal Q-learning-decision-consistent randomization strategy for SMART designs (RCQ) that successfully combines the advantages of the response-adaptive randomization strategy and the covariate-balanced randomization strategy while having the highest consistency for the optimal interventions derived using the Q-learning algorithm. In this approach, the assignment probability for a new subject who enters the SMART depends on his/her covariates and the previous subjects' treatment assignments and responses. In this method, more subjects are assigned to the better interventions because the randomization probability uses both the response-adaptive randomization strategy and the individual's optimal decisions under the Q-learning algorithm, which leads to higher ODQ%, maximizes the long-term primary outcome, and reduces bias resulting from unmeasured covariates.

One can also consider an approach such as combining the Q-learning-decision-consistent probability with only the response-adaptive probability. Such a method would have the advantages of the two strategies: response-adaptive and Q-learning-decision-consistent randomization, i.e., low ITN% and high ODQ%; however, we found that such an approach leads to higher covariate imbalance, especially when the first group in the trial was randomized according to the intermediate covariates.

For all the methods, when the group sizes were larger, the observed imbalance was higher, as seen by comparing Tables 3 and 4 to Tables 1 and 2. The parameters can be better estimated using the larger group sizes; however, because each subject was allocated based on only the covariate imbalance of the previous groups, the assignment of the current group of subjects was "over-skewed," which resulted in higher covariate imbalance. This trade-off needs to be taken into account when randomizing subjects into a SMART. We can use simulations to determine the appropriate group size. In many behavioral interventions, relatively smaller group sizes (e.g., 10 or 20, which is likely the most practical approach) are used; therefore, the proposed methods are appropriate in such scenarios.

We illustrated the proposed methods and compared them to other randomization strategies by applying them to the two-stage SMART design depicted in Fig. 1. They can also be applied to multi-stage SMART designs (i.e., SMARTs with more than two stages), and the allocation probabilities of the sth stage are defined similarly to those in Eqs. (2.7) and (2.8) for RCQ.

The proposed RCQ combines the three parts of adaptive probability in a multiplicative manner; however, other combinations can also be used. For example, one may define the allocation probability for subject i for stage s as.

$p_i^s = \varphi_1\left(p_k^{R(s)}\right)\cdot\varphi_2\left(p_{ki}^{C(s)}\right)\cdot\varphi_3\left(p_{ki}^{Q(s)}\right)$, where φ_1, φ_2 and φ_3 can be any monotonic increasing function (Yuan et al., 2011). In addition, one can also weight the three probabilities unequally in order of their importance in the treatment assignment strategy. Furthermore, the randomization probabilities can be updated as frequently as one wishes, and the group size can be adjusted to serve this purpose.

In our simulations, we updated the randomization probabilities after each group using vague priors for the model parameters in MCMC regression, which allows for independent estimation of the parameters. One can also estimate the model parameters using other priors such as distributions of historical data from similar trials or the posteriors obtained based on previous groups. However, such an approach may lead to an exclusive assignment to one of the interventions (i.e., subjects may all be assigned to one intervention) after several sequential updates, which may lead to difficulty in parameter estimation and comparison of embedded interventions in the SMART. As a remedy, a mixture of this strategy and a randomization strategy that has higher uncertainty can be applied.

In conclusion, among all the randomization methods we compared, the proposed RCQ randomization method showed the best performance in assigning fewer subjects to the inferior interventions and more subjects to the optimal Q-learning interventions than the other methods while controlling the covariate balance at an acceptable level when an appropriate group size was used.

Acknowledgements The study was funded by the National Cancer Institute (P30CA016672 to S. Shete), the Betty B. Marcus Chair in Cancer Prevention (to S. Shete), the Duncan Family Institute for Cancer Prevention and Risk Assessment (S. Shete) and the Cancer Prevention Research Institute of Texas (grant RP170259 to S. Shete).

References

Almirall, D., Lizotte, D. J., & Murphy, S. A. (2012). SMART design issues and the consideration of opposing outcomes: Discussion of "Evaluation of viable dynamic treatment regimes in a sequentially randomized trial of advanced prostate cancer" by Wang, Rotnitzky, Lin, Millikan, and Thall. *Journal of American Statistical Association, 107*(498), 509–512.

Almirall, D., Nahum-Shani, I., Sherwood, N. E., & Murphy, S. A. (2014). Introduction to SMART designs for the development of adaptive interventions: With application to weight loss research. *Translational Behavioral Medicine, 4*(3), 260–274.

Berry, D. A., & Eick, S. G. (1995). Adaptive assignment versus balanced randomization in clinical trials: A decision analysis. *Statistics in Medicine, 14*(3), 231–246.

Efron, B. (1973). International symposium on Hodgkin's disease. Session 6. Survival data and prognosis. Invited discussion: Forcing a sequential experiment to be balanced. *National Cancer Institute Monograph, 36*, 571–572.

Gail, M. H., & Green, S. B. (1976). A generalization of the one-sided two-sample Kolmogorov-Smirnov statistic for evaluating diagnostic tests. *Biometrics, 32*(3), 561–570.

Green, H., McEntegart, D. J., Byrom, B., Ghani, S., & Shepherd, S. (2001). Minimization in crossover trials with non-prognostic strata: Theory and practical application. *Journal of Clinical Pharmacy and Therapeutics, 26*(2), 121–128.

Grover, N. B. (1977). Two-sample Kolmogorov-Smirnov test for truncated data. *Computer Programs in Biomedicine, 7*(4), 247–250.

Heritier, S., Gebski, V., & Pillai, A. (2005). Dynamic balancing randomization in controlled clinical trials. *Statistics in Medicine, 24*(24), 3729–3741.

Kelleher, S. A., Dorfman, C. S., Plumb Vilardaga, J. C., Majestic, C., Winger, J., Gandhi, V., Nunez, C., Van, D. A., Shelby, R. A., Reed, S. D., Murphy, S., Davidian, M., Laber, E. B., Kimmick, G. G., Westbrook, K. W., Abernethy, A. P., & Somers, T. J. (2017). Optimizing delivery of a behavioral pain intervention in cancer patients using a sequential multiple assignment randomized trial SMART. *Contemporary Clinical Trials, 57*, 51–57.

Lei, H., Nahum-Shani, I., Lynch, K., Oslin, D., & Murphy, S. A. (2012). A "SMART" design for building individualized treatment sequences. *Annual Review of Clinical Psychology, 8*, 21–48.

Lin, Y., Zhu, M., & Su, Z. (2015). The pursuit of balance: An overview of covariate-adaptive randomization techniques in clinical trials. *Contemporary Clinical Trials, 45*(Pt A), 21–25.

Martin, A. D., Quninn, K. M., & Park, J. H. (2011). MCMCpack: Markov Chain Monte Carlo in R. *Journal of Statistical Software, 42*(9).

Moodie, E. E., Chakraborty, B., & Kramer, M. S. (2012). Q-learning for estimating optimal dynamic treatment rules from observational data. *The Canadian Journal of Statistics, 40*(4), 629–645.

Nahum-Shani, I., Qian, M., Almirall, D., Pelham, W. E., Gnagy, B., Fabiano, G. A., Waxmonsky, J. G., Yu, J., & Murphy, S. A. (2012). Q-learning: A data analysis method for constructing adaptive interventions. *Psychological Methods, 17*(4), 478–494.

Ning, J., & Huang, X. (2010). Response-adaptive randomization for clinical trials with adjustment for covariate imbalance. *Statistics in Medicine, 29*(17), 1761–1768.

Rosenberger, W. F., Stallard, N., Ivanova, A., Harper, C. N., & Ricks, M. L. (2001). Optimal adaptive designs for binary response trials. *Biometrics, 57*(3), 909–913.

Scott, N. W., McPherson, G. C., Ramsay, C. R., & Campbell, M. K. (2002). The method of minimization for allocation to clinical trials: A review. *Controlled Clinical Trials, 23*(6), 662–674.

Shao, J., & Yu, X. (2013). Validity of tests under covariate-adaptive biased coin randomization and generalized linear models. *Biometrics, 69*(4), 960–969.

Signorini, D. F., Leung, O., Simes, R. J., Beller, E., Gebski, V. J., & Callaghan, T. (1993). Dynamic balanced randomization for clinical trials. *Statistics in Medicine, 12*(24), 2343–2350.

Stuart, E. A., Lee, B. K., & Leacy, F. P. (2013). Prognostic score-based balance measures can be a useful diagnostic for propensity score methods in comparative effectiveness research. *Journal of Clinical Epidemiology, 66*(8 Suppl), S84–S90.

Sverdlov, O., Ryeznik, Y., & Wong, W. K. (2014). Efficient and ethical response-adaptive randomization designs for multi-arm clinical trials with Weibull time-to-event outcomes. *Journal of Biopharmaceutical Statistics, 24*(4), 732–754.

Thall, P. F., & Wathen, J. K. (2005). Covariate-adjusted adaptive randomization in a sarcoma trial with multi-stage treatments. *Statistics in Medicine, 24*(13), 1947–1964.

Thall, P. F., Inoue, L. Y., & Martin, T. G. (2002). Adaptive decision making in a lymphocyte infusion trial. *Biometrics, 58*(3), 560–568.

Yuan, Y., Huang, X., & Liu, S. (2011). A Bayesian response-adaptive covariate-balanced randomization design with application to a leukemia clinical trial. *Statistics in Medicine, 30*(11), 1218–1229.

Zhang, L., & Rosenberger, W. F. (2006). Response-adaptive randomization for clinical trials with continuous outcomes. *Biometrics, 62*(2), 562–569.

An Introduction to Bayesian Inference for Finite Population Characteristics

J. Sedransk

Abstract With an increased interest in using Bayesian methods for the analysis of sample survey data, it is timely to provide an introduction. We start from first principles, progressing to relatively simple parametric models. With the latter one can see how the likelihood and prior information are combined to make inferences. Assuming that the values of Y in the finite population come from a normal distribution with known variance, a sample of size n, and a conjugate prior distribution, explicit expressions for the posterior mean and variance of the *finite population* mean and variance are presented and interpreted. In a similar way, explicit expressions are given for the case where the finite population is generated from a linear regression of Y on X through the origin. This is a model typically seen in establishment surveys where Y_i and X_i represent the survey and census values for unit i. A useful extension when there is hidden cluster structure in the data is to use a Dirichlet process rather than simple parametric models such as those described in this chapter. Multiple regression with post-stratification (MRP) is based on the use of many categorical variables and specialized hierarchical priors. MRP has seen widespread application, especially when the data are from nonprobability samples or probability samples with low response rates. Finally, there is an extensive discussion of alternative (Bayesian) inferential methods when the data are categorical. This is an attractive option when one wishes to avoid postulating parametric continuous distributions.

Keywords Categorical data · Dirichlet process · Post-stratification · Survey sampling

1 Introduction

Bayesian methodology is well developed, and there are successful applications in many areas of substantive research. An important advantage of using a Bayesian approach is that it permits the use of more appropriate but complicated models.

J. Sedransk (✉)
Joint Program in Survey Methodology, University of Maryland, College Park, MD, USA
e-mail: jxs123@cwru.edu

D. D. Hanagal et al. (eds.), *Applied Statistical Methods*, Springer Proceedings in Mathematics & Statistics 380, https://doi.org/10.1007/978-981-16-7932-2_14

Typically, analyses are carried out using computational methods, notably Markov chain Monte Carlo, thus avoiding the necessity of using large-sample approximations; examples are presented in Sects. 5 and 6 of this chapter. The formal structure of a Bayesian analysis makes it easier to incorporate available prior information, common in many repeated surveys. For some, the foundational arguments for inference are compelling. In short, from the Bayesian viewpoint, inferences should be based on what we have observed, not on samples that we might have observed.

However, the use of such methods in survey sampling has been limited. The reasons for this include the presence of a well-developed methodology, i.e. design-based inference, the complexity of many survey designs, and the "observational" nature of the data. There are, though, many applications where a Bayesian approach will provide improved inferences.

In this chapter, we start from first principles, progressing to relatively simple parametric models. With the latter one can see how the likelihood and prior information are combined to make inferences. In special cases, these inferences are seen to be equivalent to those obtained by using a design-based approach.

Ericson (1969) is the first paper that presents the modern approach to Bayesian inference for finite population quantities using sample survey data. The methodology presented in this chapter assumes that the values of Y for the units in the finite population come from a model where the random variables are exchangeable. In such circumstances, one would often have selected the sample using simple random sampling or a similar design. In this chapter, it is assumed that there is neither nonresponse nor measurement error, and there are no selection effects. The approach and results given in this chapter provide the foundation for the analysis of data that are obtained using more complex survey designs or, in some cases, from observational studies.

The basic Bayesian approach to inference for finite population quantities is to make (predictive) inference for the nonsampled units, conditioning on the values for the units that have been observed. Define a finite population of N distinguishable elements labelled by the integers $1, 2, \ldots, N$. Let $\mathcal{N} = (1, 2, \ldots, N)$ define the label set and $\mathbf{Y} = (Y_1, \ldots, Y_N)$ where Y_i is the unknown value of Y for the i-th population element. The unknown Y_i can be vector valued but, for simplicity, only the scalar case is considered in this chapter. Inference concerns the N-dimensional vector $\mathbf{Y}$ and functions of $\mathbf{Y}$ such as the finite population total or median.

For a sample of size n, selected without replacement, define the statistic $(s, \mathbf{y}_s)$ to be the set of indices of distinct population elements, $s = (i_1, \ldots, i_n)$ in $\mathcal{N}$, together with the observed values y_j of Y_j, $j \in s$. Finally, define the operator $S(\mathbf{Y}) = (Y_{i_1}, \ldots, Y_{i_n})$ where, for definiteness, $i_1 < i_2 < \cdots < i_n$.

Ericson's representation of the posterior distribution of $\mathbf{Y}$ is

$$p(\mathbf{Y}|(s, \mathbf{y}_s)) \propto p(s|\mathbf{Y})p(\mathbf{Y}) \quad \text{if, and only if,} \quad S(\mathbf{Y}) = \mathbf{y}_s. \tag{1.1}$$

In Ericson's formulation, inference is for the entire finite population vector, $\mathbf{Y}$. The prior density, $p(\mathbf{Y})$, is multiplied by the selection probability, $p(\mathbf{s}|\mathbf{Y})$, and modified by requiring that $\mathbf{Y}$ be consistent with $\mathbf{y}_s$, the set of observed values in the sample.

In other words, $\mathbf{Y}$ is free to vary over $\mathcal{R}^N$, say, except that the components of $\mathbf{Y}$ in the sample must agree with $\mathbf{y}_s$, i.e. $S(\mathbf{Y}) = \mathbf{y}_s$.

If $p(s|\mathbf{Y})$ in (1.1) is independent of $\mathbf{Y}$, then

$$p(\mathbf{Y}|(s, \mathbf{y}_s)) = \frac{p(\mathbf{Y})}{\int_{S(\mathbf{Y})=\mathbf{y}_s} p(\mathbf{Y})d\mathbf{Y}} \quad \text{if, and only if,} \quad S(\mathbf{Y}) = \mathbf{y}_s. \tag{1.2}$$

For example, the expression in (1.2) holds if the selection probability depends only on a design variable Z, and the values for the entire population, $\mathbf{Z} = (Z_1, \ldots, Z_N)$, are known. If only the sample values of Z are known, it can be shown that $p(\mathbf{s}|\mathbf{Y})$ is not independent of $\mathbf{Y}$; see Sedransk (2022) for details. In this chapter, the objective is to introduce the Bayesian methodology, so the simpler formulation in (1.2) is assumed.

Note that (1.1) and (1.2) imply that one must model the entire population vector $\mathbf{Y}$. In this chapter, it is assumed that the members of $\mathbf{Y}$ are exchangeable, an assumption consistent with the notion that there are no subpopulations of $\mathcal{N}$ with important differences in distributions. Specifically, it is assumed that given a superpopulation parameter θ, the elements of $\mathbf{Y}$ are independent and identically distributed with density function $f(Y|\theta)$ and θ has a prior distribution with density $p(\theta)$.

Assuming exchangeability and $S(\mathbf{Y}) = \mathbf{y}_s$,

$$p(\mathbf{Y}|(s, \mathbf{y}_s)) \propto \int_\theta \left[\prod_{i \notin s} p(Y_i|\theta) \prod_{i \in s} p(y_i|\theta) \right] p(\theta) d\theta \tag{1.3}$$

and, alternatively,

$$p(\mathbf{Y}|(s, \mathbf{y}_s)) \propto \int_\theta \left[\prod_{i \notin s} p(Y_i|\theta) \right] p(\theta|\mathbf{y}_s) d\theta. \tag{1.4}$$

More generally, one can consider an initial sample of size m that contains repeated measurements on the same individual. However, Ericson notes that $(s, \mathbf{y}_s)$, now based on the n distinct individuals selected in the sample, is a sufficient statistic. Thus, in this case, (1.4) is the appropriate posterior distribution.

In practice, one will make inference about $\mathbf{Y}$ from (1.4) using a sampling-based method. However, it is informative to display the posterior mean and variance for several simple cases.

To introduce the methodology, a simple, straightforward model is assumed in Sect. 2.

2 Normal Distribution

Assume that Y is normally distributed with unknown mean θ and known variance v. Further, assume that the prior on θ is normal with mean m' and variance v'. This is a conjugate prior distribution, discussed in DeGroot (1970, Chap. 9).

Given a sample of n distinct units, observed values $\mathbf{y}_s = (y_1, \ldots, y_n)$, and sample mean, $\overline{y}$, the posterior distribution of θ is normal with mean

$$E(\theta|(s, \mathbf{y}_s)) \equiv m'' = \lambda\overline{y} + (1 - \lambda)m'$$

where $\lambda = v'/(v' + (v/n))$, and variance

$$Var(\theta|(s, \mathbf{y}_s)) \equiv v'' = ((n/v) + (1/v'))^{-1} = \left(\frac{v/n}{v' + (v/n)}\right) v'.$$

Thus, the posterior mean of θ is a weighted average of the sample mean, $\overline{y}$, and the prior mean, m', with weights inversely proportional to the sampling variance of $\overline{y}$ and the prior variance v'. The posterior variance of θ is the reciprocal of the sum of these weights.

From (1.4), the posterior distribution of $\mathbf{Y}$ is an $N - n$ dimensional normal distribution since the probability is concentrated on the subspace where $S(\mathbf{Y}) = \mathbf{y}_s$. Then, writing the finite population mean, $\overline{Y}$, as

$$\overline{Y} = N^{-1}(n\overline{y} + \sum_{i \notin s} Y_i),$$

the posterior distribution of $\overline{Y}$ is normal with mean

$$N^{-1}(n\overline{y} + (N - n)m'')$$

and variance

$$N^{-2}((N - n)(v + v'') + (N - n)(N - n - 1)v''). \tag{2.1}$$

These moments can also be written as

$$E(\overline{Y}|(s, \mathbf{y}_s)) = \frac{n(Nv' + v)\overline{y} + (N - n)vm'}{N(nv' + v)}$$

and

$$Var(\overline{Y}|(s, \mathbf{y}_s)) = \frac{N - n}{N} \frac{v/n}{v' + (v/n)}(v' + (v/N)). \tag{2.2}$$

The posterior mean is a weighted average of the sample mean $\overline{y}$ and prior mean m'. The weight on $\overline{y}$ is inversely proportional to the prior expectation of the conditional variance of $\overline{y}$ given $\overline{Y}$, $E_{\overline{Y}}[Var(\overline{y}|\overline{Y})]$ while the weight on m' is inversely proportional to the prior variance of $\overline{Y}$. Extensive algebra is required to obtain the posterior mean of $\overline{Y}$ and the explanatory forms of the weights. See Sect. 2.3 of Ericson (1969).

The posterior variance in (2.2) has three terms. The first is the usual finite population correction factor. From (2.1), the second term is v''/v', the ratio of the posterior variance of θ to the prior variance of θ. The third term is the prior variance of $\overline{Y}$. Extensive algebraic manipulation is required to obtain the posterior mean of $\overline{Y}$ and the explanatory forms of the weights. See Sect. 2.3 of Ericson (1969).

Note that if the prior distribution on θ is taken as locally uniform, the posterior distribution of $\overline{Y}$ is normal with mean $\overline{y}$ and variance $((N-n)/N)(v/n)$. These moments are, of course, similar to the expressions for the mean and variance from a design-based analysis assuming simple random sampling.

Inference for *any* function of **Y** can be made by noting that the posterior distribution of the *unobserved* coordinates, $\overline{S}(\mathbf{Y})$, i.e. the values of Y for the nonsampled units, is $(N-n)$ dimensional normal with common means, variances and covariances, m'', v'', $v+v''$.

The extension of these results to unknown sampling variance is straightforward, although the formulas are more cumbersome. Now, assume that Y is normally distributed with unknown mean θ and variance $1/h$. Assuming that θ and $1/h$ follow a normal-gamma distribution (DeGroot, 1970, Chap. 9), Ericson gives, in detail, the posterior distribution of $\overline{Y}$. For the special case where the prior for $(\theta, 1/h)$ is taken to be proportional to $1/h$, $\overline{Y}$ is distributed as $\overline{y} + t_{n-1}((N-n)s^2/Nn)^{1/2}$ where $\overline{y}$ and s^2 are the sample mean and variance. That is, $(\overline{Y}-\overline{y})/((N-n)s^2/Nn)^{1/2}$ has the t distribution with $(n-1)$ degrees of freedom, a familiar form.

3 Regression

A further extension assumes a simple regression model, appropriate for many establishment surveys where, for a single establishment, Y denotes the value of the variable of interest from the current survey and X denotes the value from a census. Ericson gives explicit results for the following independent sampling model for $i = 1, \ldots, N$: Given x_i,

$$Y_i \sim N(\alpha x_i, z_i/h)$$

where $z_i = g(x_i)$ for a prespecified positive function g. Three special cases are considered in some detail, i.e. $z_i = (x_i^2, x_i, 1)$. Ericson (1969) assigns a normal-gamma prior to (α, h) and gives detailed results for posterior inference about (α, h) and predictive inference for the finite population mean $\overline{Y}$. These results will be useful in situations where the prior information is in the form of data from a prior survey similar in nature to that of the main survey. Here, we summarize results obtained by using a noninformative prior distribution for (α, h), namely $p(\alpha, h) \propto h^{-1}$.

Given $(\mathbf{s}, \mathbf{y}_s)$, $\overline{Y}$ is distributed as

$$\frac{n}{N}\overline{y}_s + \overline{\alpha}^* \frac{(N-n)}{N}\overline{x_{\overline{s}}} + t_\nu^* \left[\frac{\upsilon^*(n^* z_{\overline{s}} + x_{\overline{s}})}{n^* N^2}\right]^{0.5}$$

where $\overline{y}_s$ is the sample mean of Y, $\overline{x_{\overline{s}}}$ is the mean of X for the nonsampled units, $\nu^* = n - 1$, $n^* = \sum_{i \in s} x_i^2 / z_i$,

$$\overline{\alpha}^* = \frac{\sum_{i \in s} x_i y_i / z_i}{n^*},$$

and

$$\upsilon^* \nu * = \frac{\sum_{i \in s} x_i^2/z_i \sum_{i \in s} y_i^2/z_i - (\sum_{i \in s} y_i x_i / z_i)^2}{\sum_{i \in s} x_i^2 / z_i}.$$

Thus, the posterior mean and variance of the finite population mean are

$$E(\overline{Y}|\mathbf{s}, \mathbf{y}_s) = \frac{n}{N}\overline{y}_s + \frac{N-n}{N}\overline{x_{\overline{s}}}\overline{\alpha}^*$$

and

$$V(\overline{Y}|\mathbf{s}, \mathbf{y}_s) = \frac{\upsilon^* \nu^*}{\nu^* - 2} \frac{n^* z_{\overline{s}} + x_{\overline{s}}^2}{n^* N^2}$$

where $x_{\overline{s}}$ is the sum of x for the nonsampled units.

Note that the numerator of the posterior mean of $\overline{Y}$ is the sum of the sampled y and the sum of values of Y predicted for the nonsampled units, i.e. $(N-n)\overline{\alpha}^* \overline{x_{\overline{s}}}$. For the regression model with $z_i = x_i$, i.e. the variance is proportional to x_i, the posterior mean of $\overline{Y}$ is

$$\frac{\overline{y}_s}{\overline{x}_s}\overline{y},$$

i.e. the usual ratio estimator, as one may have expected.

4 Dirichlet Process

The assumption of a normal distribution in Sect. 2 (and the extension to unknown sampling variance) may be tenuous. Nandram and Yin (2016) provide a useful alternative by assuming a (nonparametric) Dirichlet process (DP) instead. Their basic method enables predictive inference for the set of nonsampled values of Y, useful when the finite population is of moderate size. When the size of the finite population is large, they provide a good approximation to the posterior distribution of the finite population mean, $\overline{Y}$. The methodology in Nandram and Yin (2016) is an extension of work by Binder (1982) and Lo (1986). The methodology should be useful when there is hidden cluster structure in the data, and the data have ties.

Starting with the model $Y_1, \ldots, Y_N|\theta, v \overset{iid}{\sim} N(\theta, v)$, $p(\theta, v) \propto 1/v$, with $-\infty < \theta < \infty$, $v > 0$, the Bayes prediction interval for the finite population mean is $\overline{y} \pm t_{n-1,v}\sqrt{\frac{N-n}{Nn}}s$ where $t_{n-1,v}$ is the vth percentile of the t distribution with $n-1$ degrees of freedom, a result also noted in Sect. 2. While this interval is typically used in a conventional survey sampling analysis, it is not robust to non-normality, especially when the sample size is small.

Alternatively, one may use (Nandram and Yin, 2016) a DP model for the population values. The DP is

$$Y_1, \ldots, Y_N|G \overset{iid}{\sim} G \quad \text{and} \quad G|\alpha, H_\gamma(y) \sim DP(\alpha, H_\gamma(y)) \tag{4.1}$$

where the mean and variance of the DP are $E(G(y)) = H_\gamma(y)$ and $Var(G(y)) = H_\gamma(y)[1 - H_\gamma(y)]/(\alpha + 1)$.

In this context, it is natural to take $H_\gamma(y)$ to be the cdf of the normal random variable with mean θ and variance v. Note that Appendix 1 in Nandram and Yin (2016) reviews the DP.

The prior distribution that Nandram and Yin (2016) use assumes independence among θ, v and α with a Cauchy prior (two degrees of freedom in both the numerator and denominator) for α, i.e. $p(\alpha, \theta, v) \propto \frac{1}{(\alpha+1)^2}\frac{1}{v}$ with $-\infty < \theta < \infty$, $\alpha, v > 0$. With the distribution for $(Y_1, \ldots, Y_N)$ in (4.1) and this prior, the posterior distribution can be written as

$$p(\theta, v, \alpha|\mathbf{y}_s) \propto \frac{1}{v(\alpha+1)^2} g(y_1|\theta, v) \prod_{i=2}^{n} \left[\frac{1}{\alpha+i+1} \sum_{j=1}^{i-1} \delta_{y_j}(y_i) + \alpha g(y_i|\theta, v) \right] \tag{4.2}$$

where g is the pdf associated with the (DP) cdf G, and $\delta_{y_j}(y_i)$ denotes a unit point mass at $y_i = y_j$.

Inference for the nonsampled units uses the posterior distribution

$$p(\mathbf{Y}_{ns}, \alpha, \theta, v|\mathbf{y}_s) = p(\mathbf{Y}_{ns}|\alpha, \theta, v, \mathbf{y}_s) p(\alpha, \theta, v|\mathbf{y}_s). \tag{4.3}$$

First, sample from $p(\alpha, \theta, v|\mathbf{y}_s)$, and, given α, θ, v, sample $\mathbf{Y}_{ns}$ from $p(\mathbf{Y}_{ns}|\alpha, \theta, v, \mathbf{y}_s)$.

Let $\mathbf{y}_k^* = (y_1^*, \ldots, y_k^*)$ denote the k distinct sample values, $k \geq 2$. Writing $p(\mathbf{y}_s|\alpha, \theta, v) \equiv p(k, \mathbf{y}_k^*|\alpha, \theta, v)$, Antoniak (1974) has shown that

$$p(k, \mathbf{y}_k^*|\alpha, \theta, v) = p(\mathbf{y}_k^*|\theta, v) p(k|\alpha). \tag{4.4}$$

Using (4.4) and the prior, $p(\alpha, \theta, v) \propto \frac{1}{v(1+\alpha)^2}$, yields the posterior distribution, $p(\theta, v, \alpha|\mathbf{y}_s)$. Now, $y_1^*, \ldots, y_k^*|k, \theta, v \overset{iid}{\sim} N(\theta, v)$. Thus, defining $\overline{y}_*$ and s_*^2 as the sample mean and variance of the y_i^*, $\sqrt{k}(\theta - \overline{y}_*)/s_*|\mathbf{y}_k^* \sim t_{k-1}$, and $v^{-2}|s_*^2, k \sim$ Gamma $((k-1)/2, (k-2)s_*^2/2)$. Thus, it is straightforward to sample values of

θ, v. Nandram and Yin (2016) give an improved method of drawing α from $p(\alpha|k) \propto \frac{\alpha^k \Gamma(\alpha)}{\Gamma(\alpha+n)(\alpha+1)^2}$, $\alpha > 0$. Transforming α to $\rho = \frac{1}{(\alpha+1)}$, they describe a one-dimensional grid method for sampling from $p(\rho|k)$; see their Sect. 2.1.

In principle, it is easy to draw $\mathbf{Y}_{ns}$ from $p(\mathbf{Y}_{ns}|\alpha, \theta, v, \mathbf{y}_s)$ in (4.3). Writing $\mathbf{Y}_{ns} = \{Y_{n+1}, \ldots, Y_N\}$ and $\mathbf{y}_s = \{y_1, \ldots, y_n\}$,

$$y \equiv y_{n+j+1}|(\alpha, \theta, v, \mathbf{y}_s, y_1, \ldots, y_{n+j}) \sim \frac{n+j}{\alpha+n+j}\overline{F}_{n+j}(y) + \frac{\alpha}{\alpha+n+j}H_\gamma(y) \tag{4.5}$$

where $j = 1, \ldots, N-n-1$, $\overline{F}_{n+j}(y) = \sum_{i=1}^{n+j} F_{y_i}(y)/(n+j)$, $H_\gamma(y)$ is the expected value of $G(y)$ in (4.1) and $F_{y_i}(y)$ is the cdf of a point mass at y_i. For $j = 0$ in (4.5), the conditioning is limited to α, θ, v and $\mathbf{y}_s$.

It is easy to draw the nonsampled values using (4.5). One may proceed more quickly by drawing from $\overline{F}_{n+j}(y)$ using the multinomial distribution because values are repeated among those already drawn.

To obtain samples from the posterior distribution of $\mathbf{Y}_{ns}$ given $\mathbf{y}_s$ use

$$p(\mathbf{Y}_{ns}|\mathbf{y}_s) = \int p(\mathbf{Y}_{ns}|\alpha, \theta, v, \mathbf{y}_s)p(\alpha, \theta, v|\mathbf{y}_s)d\alpha d\theta dv.$$

First, sample (α, θ, v) from $p(\alpha, \theta, v|\mathbf{y}_s)$ as described below (4.4). Then for the chosen (α, θ, v), sample $\mathbf{Y}_{ns}$ from $p(\mathbf{Y}_{ns}|\alpha, \theta, v, \mathbf{y}_s)$ using (4.5). Repeating M times yields $\mathbf{Y}_{ns}^{(1)}, \ldots, \mathbf{Y}_{ns}^{(M)}$. With $\mathbf{y}_s$ known we have $\overline{Y}^{(1)}, \ldots, \overline{Y}^{(M)}$, i.e. M realizations of $\overline{Y}$ from its posterior distribution.

Order these M values and use $(\overline{Y}_{M(\alpha/2)}, \overline{Y}_{M(1-(\alpha/2))})$ as the $100(1-\alpha)\%$ prediction interval. This procedure can also be used to make inference about quantiles. For each draw of the entire finite population, compute the required quantile, Q, and then a $100(1-\alpha)\%$ credible interval is $(Q_{M\alpha/2}, Q_{M(1-(\alpha/2))})$. However, when N is much larger than n, the computation time will be longer. Typically, n is much smaller than N so the time to fit the model is negligible compared with the time to draw the $N-n$ nonsampled values from the DP model.

There is a good, approximate, Bayesian predictive interval for the finite population mean. Nandram and Yin (2016) show that, asymptotically,

$$\overline{Y}|\theta, v, \alpha, \mathbf{y}_s \sim N(E(\overline{Y}|\theta, v, \alpha, \mathbf{y}_s), Var(\overline{Y}|\theta, v, \alpha, \mathbf{y}_s)). \tag{4.6}$$

First, take M draws from the posterior density of θ, v, α, $\mathbf{y}_s$ as described above. Then for each draw use (4.6) to obtain a value of $\overline{Y}$. Order these values and use $(\overline{Y}_{M\alpha/2}, \overline{Y}_{M(1-(\alpha/2))})$ as the prediction interval.

To enhance understanding, it is useful to describe $E(\overline{Y}|\theta, v, \alpha, \mathbf{y}_s)$ and

$$Var(\overline{Y}|\theta, v, \alpha, \mathbf{y}_s),$$

given by Binder (1982). Let $\lambda = n(\alpha + N)/N(\alpha + n), \quad 0 \leq \lambda \leq 1$, a shrinkage parameter, and $\phi = 1/(\alpha + n + 1)$, the posterior correlation, Then

$E(\overline{Y}) \equiv E(\overline{Y}|\theta, v, \alpha, \mathbf{y}_s) = \lambda\overline{y} + (1-\lambda)\theta$, and

$Var(\overline{Y}) \equiv Var(\overline{Y}|\theta, v, \alpha, \mathbf{y}_s)=\lambda[(n-1)\phi(1-(n/N))s^2/n + (1-\lambda)\{\phi(\overline{y}-\theta)^2 + (1-\phi)v/n\}]$.

As $\alpha \to 0, \phi \to 1/(n+1)$ and $\lambda \to 1$. If it is assumed that the data were obtained from a simple random sample, $E(\overline{Y}) = \overline{y}$, the design-based estimator of $\overline{Y}$. Also, $Var(\overline{Y}) = [(n-1)/(n+1)](N-n)s^2/Nn$, equal to the design-based estimator of the variance of $\overline{y}$ when n is large.

As $\alpha \to \infty, \phi \to 0$ and $\lambda \to f$ where $f = n/N$. Then, $E(\overline{Y}) \to f\overline{y} + (1-f)\theta$, the (baseline) mean under normality, and $Var(\overline{Y}) \to (1-f)v/N$, the prior variance. Thus, when α is large, draws are made mostly from the normal distribution, and when α is small, draws are made mostly from the Polya posterior. See Ghosh and Meeden (1997) for a thorough description of the Polya posterior.

5 Multiple Regression with Post-stratification

Si et al. (2020) use extensive post-stratification and global local priors to make finite population inference. The covariates, X, are discretized and their cross-tabulation constructs the (post-stratification) cells, j, with population and sample sizes N_j and n_j. Si et al. (2020) take

$$Y_{ji} \overset{ind}{\sim} N(\theta_j, \sigma_y^2)$$

where $j = 1, \ldots, k$ and $i = 1, \ldots, N_j$ for the population while $i = 1, \ldots, n_j$ for the sample, the N_j assumed to be known.

Denoting $\theta = (\theta_1, \ldots, \theta_k)$ and adding covariates to (1.4)

$$p(\mathbf{Y}|(\mathbf{s}, \mathbf{y}_s)) \propto \int \prod_{j=1}^{k} \left[\prod_{i=n_j+1}^{N_j} \exp\left\{ -\frac{1}{2\sigma_y^2}(Y_{ji}-\theta_j)^2 \right\} \right] p(\theta|\mathbf{y}_s)d\theta$$

where $p(\theta|\mathbf{y}_s) \propto p(\theta) \prod_{i=1}^{n_j} \exp\{-\frac{1}{2\sigma_y^2}(y_{ji}-\theta_j)^2\}$.

An important aspect of this work is that, within cells, the selection process is assumed to be ignorable, i.e. "the post-stratification implicitly assumes that the units in each cell are included with equal probability." A key feature of Si et al. (2020) is that $p(\theta)$ is a structured prior distribution, chosen to mitigate the effects of sparse and unbalanced cell structure. Clearly, other choices of the prior density may be made.

The details of the model and prior are:

First, express the population cell mean, θ_j, as

$$\theta_j = \alpha_0 + \sum_{k \in S^{(1)}} \alpha_{jk}^{(1)} + \sum_{k \in S^{(2)}} \alpha_{jk}^{(2)} + \cdots + \sum_{k \in S^{(q)}} \alpha_{jk}^{(q)} \tag{5.1}$$

where $S^{(l)}$ is the set of all possible l-way interaction terms, and $\alpha_{jk}^{(l)}$ represents the k-th of the l-way interaction terms in the set $S^{(l)}$ for cell j. For example, the $\alpha_{jk}^{(1)}$ with $k \in S^{(1)}$ refers to the main effects, and the $\alpha_{jk}^{(2)}$ with $k \in S^{(2)}$ the two-way interaction terms for cell j. As seen in (5.1), the expression for θ_j includes all of the interactions among the q variables. The authors note that when the cell structure is sparse, variable selection is necessary.

The prior distributions, described below, are intended to handle deep interactions and account for their hierarchical structure where the high-order interaction terms will be excluded if one of the corresponding main effects is not selected. The authors note that larger main effects often lead to larger effects for the involved interaction terms. Ideally, greater shrinkage should be put on the high-order interactions than that on the main effects, and the prior setting should reflect the nested structure.

Independent prior distributions are used for the regression parameters, α, i.e.

$$\alpha_{jk}^{(l)} \sim N(0, (\lambda_k^{(l)}\sigma)^2),$$

where $\lambda_k^{(l)}$ represents the local scale and σ is the global error scale for $k \in S^{(l)}$ for $l = 1, \ldots, q$. The error scale is the same across the main effects and high-order interactions, while the local scales are different. The shrinkage effect is induced through the specification of the local scales. For the local scale of high-order interactions

$$\lambda_k^{(l)} = \delta^{(l)} \prod_{l_0 \in M^{(k)}} \lambda_{l_0}^{(1)},$$

where $\delta^{(l)}$ is the relative magnitude adjustment and $M^{(k)}$ is the collection of corresponding main effects that construct the k-th l-way interaction in the set $S^{(l)}$.

The hyperpriors on the scale parameters are taken as

$$\sigma \sim \text{Cauchy}_+(0, 1)$$

$$\lambda_k^{(1)} \sim N_+(0, 1)$$

$$\delta_l \sim N_+(0, 1)$$

where the subscript + denotes the positive part of the indicated distribution.

Finally, the authors assign a noninformative prior to α_0 and a weakly informative prior to σ_y, i.e. $\text{Cauchy}_+(0, 5)$. In Si et al. (2020), there is additional discussion of these prior assumptions, details about computation and analysis of data from the survey, "Longitudinal Study of Wellbeing."

6 Categorical Data

Assume a categorical variable Y, possibly a continuous variable that has been discretized for public release. That is, Y takes on only k different values, $y_1, \ldots, y_k$, in the finite population where $Pr(Y = y_j){=}\theta_j$ with $\sum_{j=1}^{k} \theta_j = 1$. Let $\theta = (\theta_1, \ldots, \theta_k)$ and assume that, conditional on θ, the Y_i are independent random variables. Then one may use (1.4) to make inference for $\mathbf{Y}$.

Denote by N_j and n_j the total numbers of units with $Y = y_j$ in the population and sample, and let $M_j = N_j - n_j$. Also, let $\mathbf{N} = \{N_1, \ldots, N_k\}$, $\mathbf{n} = \{n_1, \ldots, n_k\}$, and $\mathbf{M} = \{M_1, \ldots, M_k\}$. Then, from (1.3),

$$p(\mathbf{Y}|(\mathbf{s}, \mathbf{y}_s)) \propto \int_{\theta} \prod_{j=1}^{k} (\theta_j^{M_j})(\theta_j^{n_j}) p(\theta) d\theta. \tag{6.1}$$

It is clear from (6.1) that for specified $\{y_1, \ldots, y_k\}$ and observed $\mathbf{n}$, inference about $\mathbf{Y}$ only requires inference about $\mathbf{M}$. For example, the finite population total is $\sum_{j=1}^{k} y_j(M_j + n_j)$. The κ-th percentile of the finite population, ξ_κ, is such that $\xi_\kappa = y_r$ if r is the smallest integer for which $\sum_{i=1}^{r} N_i \geq N\kappa$. Since $\xi_\kappa \leq y_r$ whenever $\sum_{i=1}^{r} N_i \geq N\kappa$, the posterior distribution function of ξ_κ depends only on $\mathbf{M}$.

Rather than using (6.1), it is more convenient to represent the inference for $\mathbf{M}$ given $\mathbf{n}$ directly. Using (1.3),

$$p(\mathbf{M}|(\mathbf{s}, \mathbf{y}_s)) \propto \int_{\theta} \prod_{j=1}^{k} \left(\frac{\theta_j^{M_j}}{M_j!} \right) (\theta_j^{n_j}) p(\theta) d\theta. \tag{6.2}$$

Taking $p(\theta)$ to be the Dirichlet density, defined below in (6.3), provides a simplified, conjugate analysis. A positive feature is that a Dirichlet distribution can be thought of as representing data from a prior survey of the same type as that of the current one. Moreover, special cases of the Dirichlet distribution exhibit interesting properties. A negative aspect is that the correlation between θ_i and θ_j is negative, troublesome when there is an underlying structure in θ. Modifications, discussed below, mitigate this problem.

The Dirichlet density, $p(\theta)$, is defined on the $(k-1)$-dimensional simplex of θ such that $\theta_j > 0$ and $\sum_{j=1}^{k} \theta_j = 1$, i.e.,

$$p(\theta) = \frac{\Gamma(\sum_{j=1}^{k} \upsilon_j)}{\prod_{j=1}^{k} \Gamma(\upsilon_j)} \prod_{j=1}^{k} \theta_j^{\upsilon_j - 1} \tag{6.3}$$

with $\theta_j > 0$ and $\sum_{j=1}^{k} \theta_j = 1$.

From (6.2) and (6.3), the posterior distribution, $p(\mathbf{M}|(\mathbf{s}, \mathbf{y}_s))$, is proportional to

$$\int_{\theta} p(\mathbf{M}|\theta) p(\theta|\mathbf{n}) d\theta \propto \int_{\theta} \prod_{j=1}^{k} \frac{\theta_j^{M_j}}{M_j!} \theta_j^{n_j+\upsilon_j-1} d\theta. \quad (6.4)$$

When $\upsilon_j = 1$ for $j = 1, \ldots, k$,

$$p(\mathbf{M}|(\mathbf{s}, \mathbf{y}_s)) \propto \prod_{j=1}^{k} \binom{N_j}{n_j}. \quad (6.5)$$

Hartley and Rao (1968) use a different approach to inference. They start by assuming simple random sampling, leading to their likelihood, proportional to $\prod_{j=1}^{k} \binom{N_j}{n_j}$. Assuming a uniform prior for $\mathbf{N}$ (subject to $\sum_{j=1}^{k} N_j = N$) yields a posterior distribution for $\mathbf{M}$ that is the same as the one in (6.5). This result shows that in some situations, design-based and model-based likelihoods lead to the same posterior inference. Note that Hartley and Rao (1968) obtain a more general result by replacing the uniform prior on $\mathbf{N}$ by one attributed to work by Hoadley, i.e. the compound multinomial distribution $p(\mathbf{N}) \propto \prod_{j=1}^{k} \binom{N_j+\upsilon_j-1}{N_j}$ with $\nu_j > 0$.

Ericson (1969) gives results for the Haldane prior in (6.3), i.e. $\upsilon_j = 0$ for all j (also known as the Bayesian bootstrap, Rubin (1981)). It is notable that the posterior mean of the finite population mean, $\overline{Y}$, is $\overline{y}$, the sample mean. Moreover, the posterior variance of $\overline{Y}$ is $\frac{N-n}{N}\frac{s^2}{n}$, where s^2 is the sample variance. These results are analogous to the usual estimates of the finite population mean and sampling variance when viewed from a design-based perspective, and simple random sampling is assumed. This is not surprising because the assumptions made here are exchangeability and little prior information. Ericson (1969) also gives general expressions for moments of the posterior distribution of $\mathbf{N}$. In addition, Ericson gives explicit expressions for the posterior distribution of a percentile of the finite population.

For posterior inference, one would, in practice, sample θ from $p(\theta|\mathbf{n})$ and $\mathbf{M}$ from $p(\mathbf{M}|\theta)$ in (6.4). Then one would use the set of possible values of Y, $\{y_j : j = 1, \ldots, k\}$, the observed $\mathbf{n}$ and the sample value of $\mathbf{M}$ from (6.4) to calculate the desired $q(\mathbf{Y})$, the quantity of interest. Finally, this process is repeated B times.

Since the probabilities, $\{\theta_j : j = 1, \ldots, k\}$, are nearly independent in a Dirichlet distribution, using an alternative class of prior distributions may be beneficial. Aitchison and Shen (1980), Aitchison (1982) and Aitchison (1985) develop the logistic-normal distribution, evaluate its properties and suggest extensions. The main idea is to apply the logistic transformation to a $k-1$-dimensional normal distribution. This produces a distribution over the $k-1$-dimensional simplex $S^{k-1} = \{\theta : 0 < \theta_j, j = 1, \ldots, k-1; \sum_{j=1}^{k-1} \theta_j < 1\}$. Let $\mathbf{v}$ follow the multinormal distribution $N_{k-1}(\mu, \Sigma)$. Then the logistic transformation

$$\theta_j = \frac{e^{v_j}}{(1 + \sum_{j=1}^{k-1} e^{v_j})}$$

with $\theta_k = 1 - \sum_{j=1}^{k-1} \theta_j$ can be used to define a logistic-normal distribution over S^{k-1}.

A further alternative is to smooth the prior distribution in (6.3) by adding order restrictions, if warranted by the application. To illustrate, consider a two-phase survey (Nandram et al., 1997) of Atlantic cod whose objective is to make inference about the age distribution of the population. At the first phase, a large sample is selected and each fish is assigned to a stratum consisting of all fish having length in a specified range. At the second phase, a much smaller sample is chosen from each stratum, and the age of each of these fish is determined. Considering here only a single post-stratum, there are k age classes, i.e. $y_j = j, j = 1, \ldots, k$. With $Pr(y_j = j|\theta_j) = \theta_j$, it is well known that $\{\theta_j : j = 1, \ldots, k\}$ obey the order restriction $R_k^{(t)} = \{\theta : \theta_1 \le \cdots \le \theta_t \ge \cdots \ge \theta_k, 0 \le \theta_j \le 1, \sum_{j=1}^k \theta_j = 1\}$, where, for this illustration, the position of the maximal value is known. Then the prior distribution for $\theta \in R_k^t$ is

$$p(\theta|t) \propto \prod_{j=1}^{k} \theta_j^{v_j^{(t)}-1}, \qquad \theta \in R_k^t. \tag{6.6}$$

If the value of t is unknown, one may assign a prior distribution, $\{p(t) : t = 1, \ldots, k\}$, to the possible values of t. Then posterior inference about θ accounts for uncertainty about the value of t.

Obtaining the posterior distribution of θ from the sample likelihood

$$\frac{n!}{\prod_{j=1}^k n_j!} \prod_{j=1}^{k} \theta_j^{n_j} \tag{6.7}$$

and (6.6), Sedransk et al. (1985) describe posterior inference for θ given t, and also introduce inference for θ. Given the expression to select values of θ from its posterior distribution, it is, of course, straightforward, to select values of $\mathbf{M}$ from its posterior predictive distribution. See the left side of (6.4) where, now, $p(\theta|\mathbf{n})$ involves the order restrictions.

Sedransk et al. (1985) emphasize inference for the posterior mean of a general function, $h(\theta)$. They use importance sampling and give an efficient method to make posterior inference for this specific problem.

A general method to make inference for θ is to use the Gibbs sampler. Assuming the restricted region $R_k^{(t)}$ with t fixed, and using (6.6) and (6.7), it is straightforward to show that

$$p(\theta_j|\theta_1, \ldots, \theta_{j-1}, \theta_{j+1}, \ldots, \theta_k) \propto \kappa_j^{n_j+v_j^{(t)}} (1 - \kappa_j^{n_t+v_j^{(t)}}) \tag{6.8}$$

where $\kappa_j = p_j/(1 - (p_1 + \cdots + p_{j-1} + p_{j+1} + \cdots + p_{t-1})$ with restrictions

$\theta_{j-1} \le \theta_j \le \theta_{j+1}$ if $1 \le j \le t-1$ with $\theta_0 = 0$.

$\theta_j \ge \max(\theta_{t-1}, \theta_{t+1})$ if $j = t$.

$\theta_{j+1} \le \theta_j \le \theta_{j-1}$ if $j \ge t+1$.

That is, $p(\theta_j|\theta_1, \ldots, \theta_{j-1}, \theta_{j+1}, \ldots, \theta_k)$ is the density of a Beta random variable subject to a linear inequality. One may use the following algorithm from Devroye (1986) to select values directly from (6.8) subject to the appropriate restriction.

Let X be a random variable with distribution function F and truncated random variable Y by its distribution function G, i.e.

$$
\begin{aligned}
&= 0 \quad x < a \\
G &= \frac{F(x) - F(a)}{F(b) - F(a)} \qquad a \le x \le b \\
&= 1 \quad x > b
\end{aligned}
$$

Then Y can be generated as $F^{-1}(F(a) + U(F(b) - F(a)))$ where $U \sim U[0, 1]$.

7 Summary and Discussion

This introduction to Bayesian inference for finite population characteristics is selective, starting with a simple case, i.e. where the units in the finite population are generated from a normal distribution with known variance. This enables one to see how the sample and prior information are combined to provide inferences. The second case, regression through the origin, is a specification that is appropriate for many establishment surveys, e.g. when the outcome variable and a covariate measure the same quantity at different times. The extension to use of a Dirichlet process prior rather than a simple parametric model gives an example where only a model-based approach is feasible. The MRP, multiple regression with post-stratification, was included because it illustrates a common feature, i.e. extensive use of post-stratification, but also an innovative model and choice of prior distribution. The last section describes several inferential methods when the variables are categorical. Such methodology (including MRP) reduces the need to specify a parametric model.

As a selective introduction, many important features of survey sampling have not been described. The Bayesian literature that gives specific coverage of stratified sampling is limited and, for the most part, outdated. Rao and Ghangurde (1972) provides a good starting point. There is an extensive literature associated with multistage cluster sampling, but it is scattered. The first significant paper is Scott and Smith (1969), extended in Malec and Sedransk (1985), then further in Datta and Ghosh (1991). For categorical data, two papers that provide a starting point are Malec et al. (1997) and Ha and Sedransk (2019). For informative sampling, the review paper, Sedransk (2022), includes a discussion of probability proportional to size sampling. Inference for quantities associated with small geographical areas and populations, i.e. small area inference, is essentially a separate field. It is well established that such inferences require a model-based approach. A good place to start reading about the Bayesian approach to small area inference is Chap. 10 of Rao and Molina (2015).

With a model-based approach using complex models, analyses require using either asymptotic approximations or computational methods. Over the past twenty years, the preferred approach is computational, largely using Markov chain Monte Carlo (MCMC). For small area inference, there is an introduction in Rao and Molina (2015). However, in general, no single method is used. Even within MCMC, the approach is typically tailored to the specific application.

Acknowledgements The author is grateful to the reviewers for their comments. They have improved the focus of this chapter.

References

Aitchison, J., & Shen, S. (1980). Logistic-normal distributions: Some properties and uses. *Biometrika, 67,* 261–272.

Aitchison, J. (1982). The statistical analysis of compositional data. *JRSS-B, 44,* 139–177.

Aitchison, J. (1985). A general class of distributions on the simplex. *JRSS-B, 47,* 136–146.

Antoniak, C. (1974). Mixtures of Dirichlet processes with applications to Bayesian nonparametric problems. *Annals of Statistics, 2,* 1152–1174.

Binder, D. (1982). Nonparametric Bayesian models for samples from finite populations. *JRSS-B, 44,* 388–393.

Datta, G., & Ghosh, M. (1991). Bayesian prediction in linear models: Applications to small area estimation. *Annals of Statistics, 19*(4), 1748–1770.

DeGroot, M. (1970). *Optimal statistical decisions*. McGraw-Hill.

Devroye, L. (1986). *Non-uniform random variate generation*. Springer.

Ericson, W. (1969). Subjective Bayesian models in sampling finite populations (with discussion). *JRSS-B, 31,* 195–233.

Ghosh, M., & Meeden, G. (1997). *Bayesian methods for finite population sampling*. Cambridge University Press.

Ha, N. S., & Sedransk, J. (2019). Assessing health insurance coverage in Florida using the Behavioral Risk Factor Surveillance System. *Statistics in Medicine, 38,* 2332–2352.

Hartley, H., & Rao, J. (1968). A new estimation theory for sample surveys. *Biometrika, 55,* 547–557.

Lo, A. (1986). Bayesian statistical inference for sampling a finite population. *Annals of Statistics, 14,* 1226–1233.

Malec, D., & Sedransk, J. (1985). Bayesian methodology for predictive inference for finite population parameters in multistage cluster sampling. *Journal of the American Statistical Association, 80,* 897–902.

Malec, D., Sedransk, J., Moriarity, C., & LeClere, F. (1997). Small area inference for binary variables in the National Health Interview Survey. *Journal of the American Statistical Association, 92,* 815–826.

Nandram, B., Sedransk, J., & Smith, S. (1997). Bayesian estimation of the age composition of a fish population incorporating order restrictions. *Journal of the American Statistical Association, 92,* 33–40.

Nandram, B., & Yin, J. (2016). A nonparametric Bayesian predictive interval for a finite population mean. *Journal of Statistical Computation and Simulation, 86,* 3141–3157.

Rao, J. N. K., & Ghangurde, P. (1972). Bayesian optimization in sampling finite populations. *Journal of the American Statistical Association, 67,* 439–443.

Rao, J. N. K., & Molina, I. (2015). *Small area estimation* (2nd ed.). Wiley.

Rubin, D. (1981). The Bayesian bootstrap. *The Annals of Statistics, 9,* 130–134.

Scott, A., & Smith, T. M. F. (1969). Estimation in multi-stage surveys. *Journal of the American Statistical Association, 64,* 830–840.

Sedransk, J. (2022). Bayesian inference using survey sampling data when sampling is informative. Technical Report.

Sedransk, J., Monahan, J., & Chiu, H. Y. (1985). Bayesian estimation of finite population parameters in categorical data models incorporating order restrictions. *JRSS-B, 47,* 519–527.

Si, Y., Trangucci, R., Gabry, J., & Gelman, A. (2020). Bayesian hierarchical weighting adjustment and survey inference. *Survey Methodology, 46,* 181–214.

Reliability Measures of Repairable Systems with Arrival Time of Server

S. C. Malik

Abstract An approach to determine reliability measures of repairable systems of one or more units with arrival time of the server has been described. The use of semi-Markov process and regenerative point technique has been made to derive the expressions for mean time to system failure (MTSF) and availability of the systems in the long run. The behaviour of these measures has been examined for arbitrary values of the parameters associated with failure, repair and arrival rates. The comparison of MTSF and availability of the system models have also been made for arbitrary values of the parameters which follow negative exponential distribution. The results are numerically presented in the tables. The applications of the work can be visualized in various mechanical and electronic systems such as the system of power supply through electric transformer and generator in standby mode where components can be structured with cold standby redundancy and parallel redundancy. The system of electric transformer can be considered as an example of a single unit repairable system.

Keywords Repairable systems · Reliability measures · Exponential distribution · Semi-Markov process · Regenerative point technique

1 Introduction

Over the years, the subject reliability has gained more importance to fulfil the expectation of the users of repairable systems with minimum possible snags. In fact, the term reliability appeared in day-to-day activities after World War 2 and particularly when the failure of electronic equipment was observed at large scale in that war as a result of which several research institutions started work to know the causes of failures of the equipment. The US department of defence established a committee in 1950 known as AGREE. Robert Lusser was the first who gave the definition of reliability at a symposium in San Diego. In his words, 'Reliability is basically an ability of a system to perform the intended function'.

S. C. Malik (✉)
Department of Statistics, M.D. University, Rohtak 124001, India

D. D. Hanagal et al. (eds.), *Applied Statistical Methods*, Springer Proceedings in Mathematics & Statistics 380, https://doi.org/10.1007/978-981-16-7932-2_15

Today, the main stress of the system designers and engineers is on the development of reliable systems with least possible manufacturing and operating costs. The researchers have tried a lot to provide reliability improvement techniques for the systems, and somehow, they got success in it. There exist systems of one or more units which are being frequently used in most of the industries and management sectors. In view of these practical situations in mind, the purpose of the present chapter is to provide an approach to determine reliability measures of systems of one or more units with arrival time of the server. The reliability models for a single unit system and a system with two units in cold standby mode as well as in parallel mode are developed to see the effect of redundancy on availability of the system. The expressions for mean time to system failure (MTSF) and availability have been derived in steady state by using semi-Markov process and regenerative point technique. The behaviour of these measures has been examined for arbitrary values of various parameters. The comparison of MTSF and availability of the system models have also been made for a particular case. The results are presented numerically in Tables 1, 2, 3, 4, 5 and 6. The applications of the work can be visualized in various mechanical and electronic systems such as the system of power supply through

Table 1 MTSF versus failure rate (λ) of the unit (for a single unit system)

Failure rate (λ)	MTSF
0.001	1000.0
0.002	500.00
0.003	333.33
0.004	250.00
0.005	200.00
0.006	166.66
0.007	142.85
0.008	125.00
0.009	111.11

Table 2 Availability versus failure rate (λ) of the unit (for a single unit system)

Failure Rate (λ)	$\alpha, \beta = 2$	$\alpha = 3, \beta = 2$	$\alpha = 2, \beta = 3$
0.001	0.99900	0.99916	0.99916
0.002	0.99800	0.99833	0.99833
0.003	0.99700	0.99750	0.99750
0.004	0.99601	0.99667	0.99667
0.005	0.99502	0.99585	0.99585
0.006	0.99403	0.99502	0.99502
0.007	0.99304	0.99420	0.99420
0.008	0.99206	0.99337	0.99337
0.009	0.99108	0.99255	0.99255

Table 3 MTSF versus failure rate (λ) of the unit (for a cold standby system)

Failure Rate (λ)	α, β = 2	α = 3, β = 2	α = 2, β = 3
0.001	1,001,750	1,201,760	1,201,760
0.002	250,875	300,880	300,880
0.003	111,694	133,920	133,920
0.004	62,937	75,440	75,440
0.005	40,350	48,352	48,352
0.006	28,069	33,626	33,626
0.007	20,658	24,741	24,741
0.008	15,843	18,970	18,970
0.009	12,540	15,010	15,010

Table 4 Availability versus failure rate (λ) of the unit (for a cold standby system)

Failure Rate (λ)	α, β = 2	α = 3, β = 2	α = 2, β = 3
0.001	0.999999251	0.99999947287	0.9999994727
0.002	0.999997007	0.999997894	0.9999976714
0.003	0.999993276	0.99999526746	0.9999947661
0.004	0.999988063	0.99999159691	0.9999907048
0.005	0.999981374	0.99998688626	0.9999854911
0.006	0.999973215	0.99998113936	0.9999791286
0.007	0.999963591	0.99997436001	0.9999716208
0.008	0.999952508	0.99996655204	0.9999629713
0.009	0.999939974	0.99995771925	0.9999531836

Table 5 MTSF versus failure rate (λ) of the unit (for a parallel system)

Failure Rate (λ)	α, β = 2	α = 3, β = 2	α = 2, β = 3
0.001	501,375	601,380	601,380
0.002	125,687	150,690	150,690
0.003	56,013	67,126	67,126
0.004	31,593	37,845	37,845
0.005	20,275	24,276	24,276
0.006	14,118	16,896	16,896
0.007	10,400	12,442	12,442
0.008	7984	9547	9547
0.009	6325	7560	7560

Table 6 Availability versus failure rate (λ) of the unit (for a parallel system)

Failure Rate(λ)	α, β = 2	α = 3, β = 2	α = 2, β = 3
0.001	0.999998503	0.999998947	0.999998946
0.002	0.999994028	0.999995795	0.999995793
0.003	0.999986594	0.999990558	0.999990552
0.004	0.999976223	0.99998325	0.999983235
0.005	0.999962934	0.999973881	0.999973853
0.006	0.999946749	0.999962466	0.999962417
0.007	0.999927687	0.999949016	0.999948939
0.008	0.999905769	0.999933545	0.99993343
0.009	0.999881014	0.999916065	0.999915902

electric transformer and generator in standby mode where components can be structured with cold standby redundancy and parallel redundancy. The system of electric transformer can be considered as an example of a single unit repairable system.

2 Literature Review

The long-run availability of a parallel redundant system was obtained by Gaver (1963) by considering random failure and arbitrary distributions for repair times. Barlow and Proschan (1965) propagated the research work in reliability through the classical book 'Mathematical Theory of Reliability'. The use of semi-Markov process was considered by Branson and Shah (1971) for evaluating reliability measures of a system assuming exponential time and general repair time distribution. The regenerative point technique was adopted by Srinivasan and Gopalan (1973) to analyse a two-unit system with warm standby and single repair facility. Zuckerman (1978) has investigated semi-Markov shock models with optimal stopping. Murari and Goyal (1984) compared reliability models of a two-unit cold standby system with three types of repair facilities. Singh (1989) carried out profit of a two-unit cold standby system with a repair facility which appears and disappears randomly. Mokaddis et al. (1997) examined a two-unit warm standby system subject to degradation. Kadyan et al. (2004) analysed a system of non-identical units using the concepts of priority and different failure modes.

The idea of priority for operation and repair was introduced by Chander (2005) while studying reliability models of repairable systems. Pawar et al. (2010) discussed on operating system with repair at different level of damages subject to inspection and weather conditions. The reliability models for a computer system have been probed by Anand (2012) with the concepts of cold standby redundancy and inspection. Later, Ashish (2013) carried out cost–benefit analysis of computer systems with preventive maintenance after a maximum operation time. Gitanjali (2014) suggested reliability models for a parallel system with repair by different servers. Bhardwaj and Kaur (2014) obtained reliability and profit of a cold standby system with possible renewal of standby subject to inspection. The cost–benefit analysis of a repairable system was done by Malik and Gitanjali (2019) with alternate repair and Weibull distribution. Nandal and Malik (2019) used Lindley distribution for profit analysis of a three-unit cold standby system subject to arrival time of the server.

3 Some Fundamentals

Here, we shall describe in brief the following fundamentals.

3.1 Reliability

If 'T' is the lifetime of the system, then the system reliability is defined as follows:

$$R(t) = P_r[T > t] = \int_t^\infty f(u)du = 1 - F(t) = \overline{F}(t)$$

where $f(t)$ is a probability density function of lifetime 'T', $F(t)$ is the cumulative density function of lifetime 'T' or unreliability of the system, and $R(t)$ is the probability that the item does not fail in the time interval (0, 1] and is still functioning at time 't'.

3.2 Mean Time to System Failure (MTSF)

The expected time before the system completely fail is called mean time to system failure. Let $f(t)$ be the failure density function, then

$$MTSF = E(T) = \int_0^\infty t f(t)dt, \text{ where, T is the time to failure}$$

$$= \lim_{t\to\infty} \int_0^t R(t)dt = \lim_{s\to 0} R^*(s) \text{ where } R^*(s) \text{ is the Laplace transform of} R(t)$$

3.3 Steady-State Availability

The probability that the system is operating successfully at time 't' is called availability of the system which is given by

$$\text{Availability A(t)} = \frac{SystemUpTime}{SystemUpTime + SystemDownTime}$$

The expected fraction of time that the system operates satisfactorily in the long run is known as steady-state availability. Thus, steady-state availability is

$$A(\infty) = \lim_{t\to\infty} A(t)$$

3.4 Redundancy

Redundancy is a common approach to improve the reliability and availability of a system. The provision of parallel paths (or alternative means) in a system for performing a given task such that all means must fail before causing the system failure is called redundancy. It is mainly of two types: active redundancy and standby redundancy. The redundancy in which all spare units operate simultaneously is known as active redundancy while the standby redundancy is that in which failed unit is replaced manually or automatically by its similar spare unit, and this process will continue until all the spare units (standby) have been exhausted.

For example, the system of power supply through electric transformer and generator is a case of standby system where generator is kept as spare (called redundant) and can be switched on as and when power supply through electric transformer is interrupted.

3.5 Semi-Markov Process

The semi-Markov process is a process in which transition from one state to another is governed by the transition probabilities of a Markov process, but the time spent in each state before a transition occurs is a random variable depending upon the last transition made.

Mathematically, we assume that the process is time homogeneous, i.e.

$Pr\{X_{n+1} = \text{j}, t_{n+1} - - t_n \langle t | X_n = i\} = Q_{ij}(t), i, j \in s$ is independent of n, then there exist limiting transition probabilities.

Here, $Q_{ij}(t)$ is the CDF of passage time from regenerative state S_i to a regenerative state S_j or to a failed state S_j without visiting any other regenerative state in (0, t].

$p_{ij} = \lim Q_{ij}(t) = \Pr\{X_{n+1} = j | X_n = i\}$,

then $\{X_n,\ n = 0, 1, 2, \ldots\ldots.\}$ constitute a Markov chain with state space E and transition probability matrix (TPM)

$$P = [p_{ij}]$$

3.6 Regenerative Point Process

Regenerative stochastic process was defined by Smith (1955) and has been crucial in the analysis of complex system. In this, we take time points at which the system history prior to the time points is irrelevant to the system conditions. These points are called regenerative points. Let $X(t)$ be the state of the system of epoch. If $t_1, t_2, \ldots\ldots$ are the epochs at which the process probabilistically restarts, then these epochs are called regenerative epochs, and the process $\{X(t),\ t = t_1,\ t_2, \ldots\ldots\ldots\}$ is called **regenerative process**.

4 Common Notations

O	The system is operating
λ	The constant failure rate of the system
C_s	The unit is in cold standby
$w(t)/W(t)$	PDF/CDF of the random variable associated with arrival time of the server
$g(t)/G(t)$	PDF/CDF of the random variable associated with repair time of the unit
Fwr/FWR	The system is failed and waiting for repair/waiting for repair continuously from previous state
Fur/FUR	The system is failed and under repair/ under repair unit continuously from previous state
q_{ij}/Q_{ij}	PDF/CDF of passage time from regenerative state S_i to a regenerative state S_j or to a failed state S_j without visiting any other regenerative state in (0, t]
Ⓢ/©	Symbol for Laplace–Stieltjes convolution/Laplace convolution
*/**	Symbol for Laplace transform(LT)/Laplace–Stieltjes transform (LST)

5 Reliability Measures of Repairable Systems

The reliability measures of the following repairable systems are discussed as follows:

5.1 A single unit system model with arrival time of the server
5.2 A two-unit cold standby system model with arrival time of the server
5.3 A two-unit parallel system model with arrival time of the server

5.1 *MTSF and Availability of a Single Unit System with Arrival Time of the Server*

Let us consider a single unit system with constant failure rate (λ). The system has two modes—operative (normal mode) and completely failed. The system has complete failure from normal mode. A single server which takes some time to arrive at the system (called arrival time) is provided to carry out repair activities. The state transition diagram is shown in Fig. 1.

5.1.1 System Description

S_0	The initial state in which system is good and operating
S_1	The system is failed and waiting for repair due to non-availability of the server
S_2	The system is failed and under repair

Fig. 1 State transition diagram (for a single unit system)

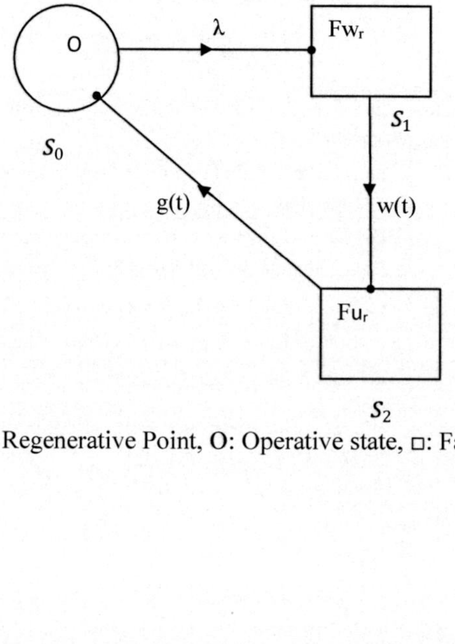

•: Regenerative Point, O: Operative state, □: Failed state

5.1.2 Transition Probabilities and Mean Sojourn Time

Simple probabilistic considerations yield the following expressions for the nonzero elements $p_{ij} = Q_{ij}(\infty) = \int_0^\infty q_{ij}(t)dt$ as

$$dQ_{01}(t) = q_{01}(t)dt = \lambda e^{-\lambda t}dt,\ dQ_{12}(t) = w(t)dt dQ_{20}(t) = g(t)dt$$

Taking Laplace–Stieltjes Transform, we get

$$Q_{01}^{**}(s) = \int_0^\infty e^{-st}[dQ_{01}(t)] = \frac{\lambda}{\lambda+s},\quad Q_{12}^{**}(s) = w^*(s),\quad Q_{20}^{**}(s) = g^*(s)$$

Taking lim $s \to 0$, we get the following transition probabilities:

$$p_{01} = 1,\ p_{12} = 1,\ p_{20} = 1$$

Mean sojourn times

$$\mu_i = \sum_j m_{ij}(i = 0, 1)$$

$$\text{But, } m_{ij} = -\frac{d}{ds}\left[Q_{ij}^{**}(s)\right]_{s=0}$$

$$\text{So, } m_{01} = -\frac{d}{ds}\left[\frac{\lambda}{\lambda+s}\right]_{s=0} = \frac{1}{\lambda},\quad m_{12} = -w^{*'}(0),\ m_{20} = -g^{*'}(0)$$

$$\text{Now, } \mu_0 = m_{01} = \frac{1}{\lambda},\quad \mu_1 = m_{12} = -w^{*'}(0), \mu_2 = m_{20} = -g^{*'}(0)$$

5.1.3 Reliability and Mean Time to System Failure (MTSF)

It is defined as the expected time for which the system is in operation before it completely fails. Let $f(t)$ be the probability density function of lifetime of the system.

We have,

$$MTSF = E(T) = \int_0^\infty tf(t)dt = \int_0^\infty R(t)dt$$

$$\text{Also, } \lim_{s\to 0} R^*(s) = \int_0^\infty R(t)dt \text{ and } MTSF = \lim_{S\to 0} R^*(s)$$

Let $\varnothing_0(t)$ be the cumulative distribution function of the first passage time from the initial state to a failed state, then

$$R^*(s) = \frac{1 - \varnothing_0^{**}(s)}{s}$$

From above equations, we have

$$\text{MTSF} = \lim_{s \to 0} R^*(s) = \lim_{S \to 0} \frac{1 - \varnothing_0^{**}(s)}{s}$$

where $R^*(s)$ and $\varnothing_0^{**}(s)$ are, respectively, the Laplace transform and Laplace–Stieltjes transform of $R(t)$ and $\varnothing_0(t)$.

Let $\varnothing_i(t)$ be the CDF of first passage time from regenerative state S_i to a failed state. Regarding the failed state as absorbing state, we have the following recursive relations for $\varnothing_i(t)$:

$$\varnothing_0(t) = Q_{01}(t)$$

Taking Laplace–Stieltjes transform, it can be written as

$$\varnothing_0^{**}(s) = Q_{01}^{**}(s)$$

Now, $$MTSF = \lim_{S \to 0} \frac{1 - \varnothing_0^{**}(s)}{s} = \lim_{S \to 0} \frac{1 - \frac{N_1}{D_1}}{D_1} = \lim_{s \to 0} \frac{[D_1 - N_1]}{sD_1} \left(\frac{0}{0}\right) \text{ Form}$$

So, by applying L' Hospital rule, we get

$$MTSF = Q_{01}^{**'}(0) = m_{01} = \mu_0 = \frac{1}{\lambda}$$

$$\text{Hence, MTSF} = \frac{1}{\lambda}$$

Reliability

We have,

$$\text{R}^*(\text{s}) = \frac{1 - \varnothing_0^{**}(s)}{s} = \frac{1 - Q_{01}^{**}(s)}{s} = \frac{1}{\lambda + S}$$

The reliability of the system model can be obtained by taking Laplace inverse of $R^*(s)$, and we get

$$\mathrm{R(t)}=L^{-1}\left(\frac{1}{\lambda+S}\right)=e^{-\lambda t},\quad \mathrm{t}>0$$

Hence,

$$R(t)=e^{-\lambda t}\quad \text{for t} > 0$$

5.1.4 Availability Analysis

Let $A_i(t)$ be the probability that the system is in upstate at instant 't' given that the system entered regenerative state $\mathrm{S_i}$ at t = 0. The recursive relations for $A_i(t)$ are given as

$$A_0(t)=\ M_0(t)+q_{01}(t)©A_1(t),\quad A_1(t)=q_{12}(t)©A_2(t),$$
$$A_2(t)=q_{20}(t)©A_0(t)$$

$$\text{Where,}\quad M_0(t)\ =\ e^{-\lambda t}$$

Taking Laplace transform of above equations, it can be written as follows:

$$A_0^*(\mathrm{s})\ =\ M_0^*(\mathrm{s})+q_{01}^*(\mathrm{s}).A_1^*(\mathrm{s}),\ A_1^*(\mathrm{s})\ =\ q_{12}^*(\mathrm{s}).A_2^*(\mathrm{s}),$$
$$A_2^*(\mathrm{s})\ =\ q_{20}^*(\mathrm{s}).A_0^*(\mathrm{s})$$

or

$$\mathrm{A_0^*(s)-q_{01}^*(s).A_1^*(s)\ =\ M_0^*(s),\ A_1^*(s)-q_{12}^*(s).A_2^*(s)\ =\ 0,}$$
$$\mathrm{A_2^*(s)-q_{20}^*(s).A_0^*(s)\ =\ 0}$$

$$\begin{bmatrix} 1 & -q_{01}^*(\mathrm{s}) & 0 \\ 0 & 1 & -q_{12}^*(\mathrm{s}) \\ -q_{20}^*(\mathrm{s}) & 0 & 1 \end{bmatrix}\begin{bmatrix} A_0^*(\mathrm{s}) \\ A_1^*(\mathrm{s}) \\ A_2^*(\mathrm{s}) \end{bmatrix} = \begin{bmatrix} M_0^*(\mathrm{s}) \\ 0 \\ 0 \end{bmatrix}$$

The above expression is of the type AX = B

$$||\ =\ \mathrm{1-q_{01}^*(s)q_{12}^*(s)q_{20}^*(s)}$$

$$|\Delta_1| = \begin{vmatrix} M_0^*(\mathrm{s}) & -q_{01}^*(\mathrm{s}) & 0 \\ 0 & 1 & -q_{12}^*(\mathrm{s}) \\ 0 & 0 & 1 \end{vmatrix},\quad ||\ =\ (\mathrm{s})$$

Applying Crammer's rule and solving for A_0^* (s), we get

$$A_0^*(s) = \frac{M_0^*(s)}{1 - q_{01}^*(s)q_{12}^*(s)q_{20}^*(s)}$$

The steady-state availability is given by

$$A(\infty) = \lim_{t\to\infty} A(t) = \lim_{s\to 0} sA_0^*(s) = \lim_{s\to 0} s[\frac{M_0^*(s)}{1 - q_{01}^*(s)q_{12}^*(s)q_{20}^*(s)} \quad \left(\frac{0}{0} form\right)$$

Using L' Hospital rule, we get

$$A(\infty) = \lim_{s\to 0} s\left[\frac{M_0^*(s)}{1 - q_{01}^*(s)q_{12}^*(s)q_{20}^*(s)}\right] = \lim_{s\to 0} \frac{1}{1 - \lambda\left[w^{*'}(0)g^*(0) + w^*(0)g^{*'}(0)\right]}$$

$$A(\infty) = \frac{1}{1 - \lambda\left[w^{*'}(0) + g^{*'}(0)\right]}$$

5.1.5 Particular Case

Suppose repair rate $g(t)$ and arrival time of the server w(t) follow negative exponential distribution, i.e. $g(t) = \alpha\, e^{-\alpha t}$ and $w(t) = \beta\, e^{-\beta t}$.

Taking Laplace transform, it can be written as follows:

$$s) = \frac{\alpha}{\alpha + s}, \quad g^{*'}(s) = -\frac{\alpha}{(\alpha + s)^2}, \quad g^{*'}(0) = -\frac{1}{\alpha}$$

$$w^*(s) = \frac{\beta}{\beta + s}, \quad w^{*'}(s) = -\frac{\beta}{(\beta + s)^2}, \quad w^{*'}(s) = -\frac{1}{\beta}$$

Hence, $A(\infty) = \frac{\alpha\beta}{\alpha\beta + \lambda[\alpha + \beta]}$.

5.1.6 Numerical and Graphical Representation of MTSF and Availability

5.2 *MTSF and Availability of a Two-Unit Cold Standby System with Arrival Time of the Server*

Let us consider a system of two identical units in which one unit is operative and the other unit is kept as spare in cold standby. There is a single server which takes some time to arrive at the system. The possible transition between states is shown in the following state transition diagram (Fig. 2):

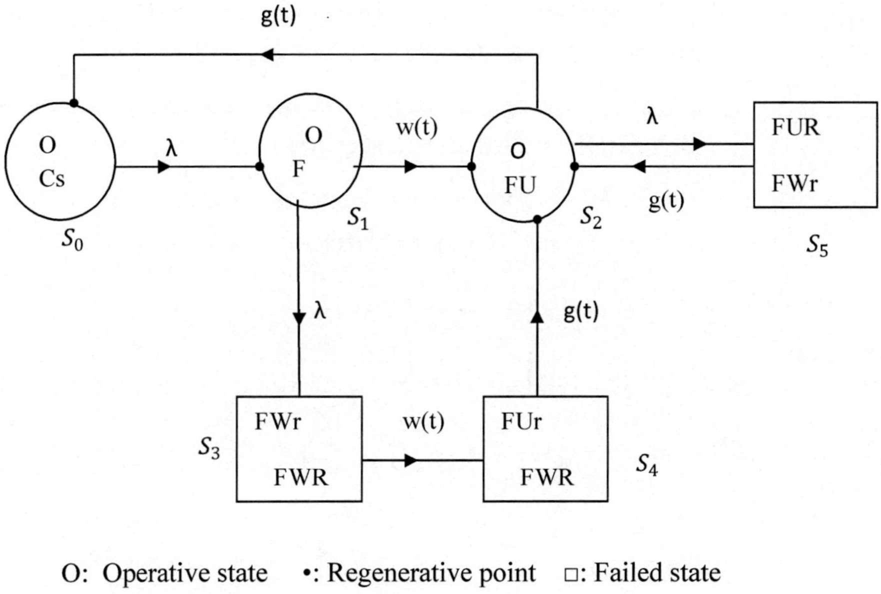

Fig. 2 State transition diagram (for a cold standby system)

5.2.1 System Description

S_0	The initial state in which one unit is operative and the other is kept as cold standby
S_1	The state in which one unit is operative and the other failed unit is waiting for repair
S_2	The state in which one unit is operative and the other unit is under repair
S_3	The state in which one failed unit is waiting for repair and the other failed unit is also waiting for repair continuously from previous state
S_4	The state in which both the units are failed, and one unit is under repair and the another unit is waiting for repair
S_5	The final state in which one unit is under repair from previous state and another unit is waiting for repair

5.2.2 Transition Probabilities and Mean Sojourn Times.

Simple probabilistic considerations yield the following expressions for the nonzero elements $p_{ij} = Q_{ij}(\infty) = \int_0^\infty q_{ij}(t)dt$ as

$$dQ_{01}(t) = q_{01}(t)dt = \lambda e^{(-\lambda t)}dt, dQ_{12}(t) = e^{-\lambda t}w(t)dt,$$
$$dQ_{13}(t) = \lambda e^{(-\lambda t)}\overline{W}(t)dt$$

$$dQ_{20}(t) = e^{-\lambda t}g(t)dt, dQ_{25}(t) = \lambda e^{-\lambda t}\overline{G}(t)dt,$$
$$dQ_{12.34}(t) = dQ_{13}(t)dQ_{34}(t)dQ_{42}(t)$$

$$dQ_{34}(t) = w(t)dt, dQ_{42}(t) = g(t)dt,$$
$$dQ_{22.5}(t) = e^{-\lambda t}\overline{G}(t)g(t)dt, dQ_{52}(t) = g(t)dt$$

Taking Laplace–Stieltjes transform, we have

$$Q_{01}^{**}(\mathrm{s}) = \int_0^\infty e^{-st}\mathrm{d}[Q_{01}(t)] = \frac{\lambda}{\lambda + S}, Q_{12}^{**}(s) = w^*(\lambda + s),$$
$$Q_{13}^{**}(s) = \frac{\lambda}{\lambda + S}[1 - w^*(\lambda + s)]$$

$$Q_{20}^{**} = g^*(\lambda + s), Q_{25}^{**}(\mathrm{s}) = \frac{\lambda}{\lambda + S}[1 - g^*(\lambda + s)]$$

$$Q_{12.34}^{**}(\mathrm{s}) = \frac{\lambda}{\lambda + S}[1 - w^*(\lambda + s)]w^*(s)g^*(s),$$
$$Q_{34}^{**}(\mathrm{s}) = w^*(s), Q_{42}^{**} = g^*(s), Q_{22.5}^{**} = \frac{\lambda}{\lambda + S}[1 - g^*(\lambda + s)]g^*(s),$$
$$Q_{52}^{**}g^*(\mathrm{s}) = g^*(s)$$

Taking limit s $\rightarrow$ 0, we get the following transition probabilities:

$$p_{01} = 1, p_{12} = w^*(\lambda), p_{13} = [1 - w^*(\lambda)], p_{20} = g^*(\lambda),$$
$$p_{25} = [1 - g^*(\lambda)]$$

$$p_{12.34} = [1 - w^*(\lambda)], p_{34} = 1, p_{42} = 1, p_{22.5} = [1 - g^*(\lambda)], p_{52} = 1$$

It can be easily verified that

$$p_{01} = p_{01} + p_{13} = p_{20} + p_{22.5} = p_{20} + p_{25} = p_{12} + p_{12.34} = p_{34} = p_{42} = p_{52} = 1$$

Mean sojourn times

$$\mu_i = \sum_j m_{ij} \ (i = 0, 1)$$

$$\text{But, } m_{ij} = -\frac{d}{ds}[Q_{ij}^{**}(\mathrm{s})]_{\mathrm{s}} = 0,$$

$$\text{So}, m_{01} = --\frac{d}{ds}\left[\frac{\lambda}{\lambda + S}\right]_{s=0} = \frac{1}{\lambda}, \ m_{12} = -w^{*'}(\lambda),$$

$$m_{13} = w^{*'}(\lambda) + \frac{1}{\lambda}\left[1 - w^{*}(\lambda)\right]$$

$$m_{20} = g^{*'}(\lambda), \ m_{25} = g^{*'}(\lambda) + \frac{1}{\lambda}\left[1 - g^{*}(\lambda)\right]$$

$$m_{12.34} = \frac{\lambda w^{*'}(\lambda) - \lambda[1 - w^{*}(\lambda)]g^{*'}(0) - \lambda[1 - w^{*}(\lambda)]w^{*'}(0) + [1 - w^{*}(\lambda)]}{\lambda}$$

$$m_{22.5} = \frac{\lambda g^{*'}(\lambda) - \lambda[1 - g^{*}(\lambda)]g^{*'}(0) + [1 - g^{*}(\lambda)]w^{*'}(0)}{\lambda}$$

$$\text{Now}, \mu_0 = m_{01} = \frac{1}{\lambda}, \ \mu_1 = m_{12} + m_{13} = \frac{1}{\lambda}\left[1 - w^{*}(\lambda)\right]$$

$$\mu_2 = m_{20} + m_{25} = \frac{1}{\lambda}\left[1 - g^{*}(\lambda)\right]$$

$$\mu_1' = m_{12} + m_{12.34} = \frac{1}{\lambda}\left\{\left[1 - w^{*}(\lambda)\right] - \lambda\left[1 - w^{*}(\lambda)\right]g^{*'}(0) - \lambda\left[1 - w^{*}(\lambda)\right]w^{*'}(0)\right\}$$

$$\mu_2' = m_{20} + m_{22.5} = \frac{1}{\lambda}\left\{\left[1 - g^{*}(\lambda)\right] - \lambda\left[1 - g^{*}(\lambda)\right]g^{*'}(0)\right\}$$

5.2.3 Reliability and Mean Time to System Failure (MTSF)

Let $\varnothing_i$(t) be the CDF of first passage time from regenerative state S_i to a failed state. Regarding the failed state as absorbing state, we have the following recursive relations for $\varnothing_i$(t):

$$\varnothing_0(t) = Q_{01}(t)\varnothing_1(t), \varnothing_1(t) = Q_{12}(t)\varnothing_2(t) + Q_{13}(t), \varnothing_2(t) = Q_{20}(t)\varnothing_0(t) + Q_{25}(t)$$

Taking Laplace–Stieltjes transform, it can be written as

$$\varnothing_0^{**}(s) = Q_{01}^{**}(s)\varnothing_1^{**}(s), \varnothing_1^{**}(s) = Q_{12}^{**}(s)\varnothing_2^{**}(s) + Q_{13}^{**}(s), \varnothing_2^{**}(s) = Q_{20}^{**}(s)\varnothing_0^{**}(s) + Q_{25}^{**}(s)$$

or

$$\begin{bmatrix} 1 & -Q_{01}^{**}(s) & 0 \\ 0 & 1 & -Q_{12}^{**}(s) \\ -Q_{20}^{**}(s) & 0 & 1 \end{bmatrix}\begin{bmatrix} \varnothing_0^{**}(s) \\ \varnothing_1^{**}(s) \\ \varnothing_2^{**}(s) \end{bmatrix} = \begin{bmatrix} 0 \\ Q_{13}^{**}(s) \\ Q_{25}^{**}(s) \end{bmatrix}$$

Or AX = B

$$D_1 = |A| = 1 - Q_{01}^{**}(s)Q_{12}^{**}(s)Q_{20}^{**}(s)$$

$$N_1 = \begin{vmatrix} 0 & -Q_{01}^{**}(s) & 0 \\ Q_{13}^{**}(s) & 1 & -Q_{12}^{**}(s) \\ Q_{25}^{**}(s) & 0 & 1 \end{vmatrix} = Q_{01}^{**}(s)Q_{13}^{**}(s) + Q_{01}^{**}(s)Q_{12}^{**}(s)Q_{25}^{**}(s)$$

On solving for $\varnothing_0^{**}(s)$ by Crammer's rule, we get

$$\text{We have } \varnothing_0^{**}(s) = \frac{Q_{01}^{**}(s)Q_{13}^{**}(s) + Q_{01}^{**}(s)Q_{12}^{**}(s)Q_{25}^{**}(s)}{1 - Q_{01}^{**}(s)Q_{12}^{**}(s)Q_{20}^{**}(s)}$$

$$\text{Also, } MTSF = \lim_{S\to 0} \frac{1 - \varnothing_0^{**}(s)}{s} = \lim_{S\to 0} \frac{1 - \frac{Q_{01}^{**}(s)Q_{13}^{**}(s) + Q_{01}^{**}(s)Q_{12}^{**}(s)Q_{25}^{**}(s)}{1 - Q_{01}^{**}(s)Q_{12}^{**}(s)Q_{20}^{**}(s)}}{s}$$

$$= \lim_{S\to 0} \frac{Q_{01}^{**}(s)Q_{13}^{**}(s) + Q_{01}^{**}(s)Q_{12}^{**}(s)Q_{25}^{**}(s)}{s\left[1 - Q_{01}^{**}(s)Q_{12}^{**}(s)Q_{20}^{**}(s)\right]} = \lim_{s\to 0} \frac{1 - \frac{N_1}{D_1}}{s} = \lim_{s\to 0} \frac{[D_1 - N_1]}{sD_1} \left(\frac{0}{0}\right) Form$$

So, by applying L' Hospital rule, we get

$$MTSF = \frac{\mu_0 + \mu_1 + \mu_2 p_{12}}{1 - p_{12}p_{20}}$$

$$\text{So, we get } R^*(s) = \frac{1 - \varnothing_0^{**}(s)}{s} = \frac{Q_{01}^{**}(s)Q_{13}^{**}(s) + Q_{01}^{**}(s)Q_{12}^{**}(s)Q_{25}^{**}(s)}{s\left[1 - Q_{01}^{**}(s)Q_{12}^{**}(s)Q_{20}^{**}(s)\right]}$$

The reliability of the system model can be obtained by taking Laplace inverse of $R^*(s)$, and we get

$$R(t) = L^{-1}\{\frac{Q_{01}^{**}(s)Q_{13}^{**}(s) + Q_{01}^{**}(s)Q_{12}^{**}(s)Q_{25}^{**}(s)}{s\left[1 - Q_{01}^{**}(s)Q_{12}^{**}(s)Q_{20}^{**}(s)\right]}$$

5.2.4 Availability Analysis

Let $A_i(t)$ be the probability that the system is in upstate at instant 't' given that the system entered regenerative state S_i at t = 0. The recursive relations for $A_i(t)$ are given as

$$\begin{aligned} A_0(t) &= M_0(t) + q_{01}(t)©A_1(t) \\ A_1(t) &= M_1(t) + q_{12}(t)©A_2(t) + q_{12.34}(t)©A_2(t) \\ A_2(t) &= M_2(t) + q_{20}(t)©tA_0(t) + q_{22.5}(t)©A_2(t) \end{aligned}$$

where $M_i(t)$ is the probability that the system is up initially in state $S_i \in E$ which is up at time t without visiting to any other regenerative state, and we have

$$M_0(t) = e^{-\lambda t}, M_1(t) = e^{-\lambda t}\overline{W}(t), M_2 = \overline{G}(t)$$

Taking Laplace transform of above equations,it can be written as

$$\begin{aligned}A_0^*(s) &= M_0^*(s) + q_{01}^*(s)A_1^*(s)\\ A_1^*(s) &= M_1^*(s) + q_{12}^*(s)A_2^*(s) + q_{12.34}^*(s)A_2^*(s)\\ A_2^*(s) &= M_2^*(s) + q_{20}^*(s)A_0^*(s) + q_{22.5}^*(s)A_2^*(s)\end{aligned}$$

Or

$$\begin{aligned}A_0^*(s) - q_{01}^*(s)A_1^*(s) &= M_0^*(s)\\ A_1^*(s) - q_{10}^*(s)A_0^*(s) - q_{11.2}^*(s)A_1^*(s) &= M_1^*(s)\\ A_2^*(s) - q_{20}^*(s)A_0^*(s) - q_{22.5}^*(s)A_2^*(s) &= M_2^*(s)\end{aligned}$$

$$\begin{bmatrix} 1 & -q_{01}^*(s) & 0\\ 0 & 1 & -\left[q_{12}^*(s) + q_{12.34}^*(s)\right]\\ -q_{20}^*(s) & 0 & 1 - q_{22.5}^*(s)\end{bmatrix}\begin{bmatrix}A_0^*(s)\\ A_1^*(s)\\ A_2^*(s)\end{bmatrix} = \begin{bmatrix}M_0^*(s)\\ M_1^*(s)\\ M_2^*(s)\end{bmatrix}$$

$$|\Delta| = \begin{vmatrix} 1 & -q_{01}^*(s) & 0\\ 0 & 1 & -\left[q_{12}^*(s) + q_{12.34}^*(s)\right]\\ -q_{20}^*(s) & 0 & 1 - q_{22.5}^*(s)\end{vmatrix}$$

$$= 1 - q_{22.5}^*(s) - q_{01}^*(s)q_{20}^*(s)q_{12}^*(s) - -q_{01}^*(s)q_{20}^*(s)q_{12.34}^*(s)$$

$$|\Delta_1| = \begin{vmatrix} M_0^*(s) & -q_{01}^*(s) & 0\\ M_1^*(s) & 1 & -\left[q_{12}^*(s) + q_{12.34}^*(s)\right]\\ M_2^*(s) & 0 & 1 - q_{22.5}^*(s)\end{vmatrix}$$

$$= M_0^*(s)[1 - q_{22.5}^*(s)] + q_{01}^*(s)M_1^*(s)[1 - q_{22.5}^*(s)] + M_2^*(s)[q_{12}^*(s) + q_{12.34}^*(s)]$$

On solving for $A_0^*(s)$ by Crammer's rule, we get
Thus,

$$A_0^*(s) = \frac{M_0^*(s)\left[1 - q_{22.5}^*(s)\right] + q_{01}^*(s)M_1^*(s)\left[1 - q_{22.5}^*(s)\right] + M_2^*(s)\left[q_{12}^*(s) + q_{12.34}^*(s)\right]}{1 - q_{22.5}^*(s) - q_{01}^*(s)q_{20}^*(s)q_{12}^*(s) - -q_{01}^*(s)q_{20}^*(s)q_{12.34}^*(s)}$$

The steady-state (long run) availability is given by

$$\begin{aligned}A(\infty) &= \lim_{t\to\infty} A(t) = \lim_{s\to 0} sA_0^*(s)\\ &= \lim_{s\to 0} s\frac{\left\{\begin{array}{l} M_0^*(s)\left[1 - q_{22.5}^*(s)\right] + q_{01}^*(s)M_1^*(s)\left[1 - q_{22.5}^*(s)\right]\\ +M_2^*(s)\left[q_{12}^*(s) + q_{12.34}^*(s)\right]\end{array}\right\}}{1 - q_{22.5}^*(s) - q_{01}^*(s)q_{20}^*(s)q_{12}^*(s) - - q_{01}^*(s)q_{20}^*(s)q_{12.34}^*(s)}\\ &= \lim_{s\to 0} s\frac{N}{D}\left(\frac{0}{0}\right)\end{aligned}$$

From,
So, by applying L' Hospital rule, we get

$$A(\infty) = \frac{\mu_0 p_{20} + \mu_1 p_{20} + \mu_2}{\mu_2' + \mu_0 p_{20} + \mu_1' p_{20}}$$

5.2.5 Particular Case

Suppose repair rate $g(t)$ and arrival time of the server $w(t)$ follow negative exponential distribution, i.e.

$$g(t) = \alpha e^{-\alpha t} and w(t) = \beta e^{-\beta t}$$

Taking Laplace transform, it can be written as

$$g^*(s) = \frac{\alpha}{\alpha + s}, g^{*'}(s) = -\frac{\alpha}{(\alpha + s)^2}, g^{*'}(0) = -\frac{1}{\alpha}$$

$$w^*(s) = \frac{\beta}{\beta + s}, w^{*'}(s) = -\frac{\beta}{(\beta + s)^2}, w^{*'}(0) = -\frac{1}{\beta}$$

$$MTSF = \frac{\mu_0 + \mu_1 + \mu_2 p_{12}}{1 - p_{12} p_{20}} = \frac{(\alpha + \lambda)(\alpha + \lambda) + (\alpha + \lambda)\lambda + \beta\lambda}{\lambda[(\beta + \lambda)(\alpha + \lambda) - \alpha\beta]}$$

$$A(\infty) = \frac{\mu_0 p_{20} + \mu_1 p_{20} + \mu_2}{\mu_2' + \mu_0 p_{20} + \mu_1' p_{20}} = \frac{\alpha\beta[\alpha(\beta + \lambda) + \alpha\lambda + \lambda(\beta + \lambda)]}{\beta\lambda(\beta + \lambda)(\alpha + \lambda) + \alpha^2\beta(\beta + \lambda) + \lambda^2\alpha\beta + \lambda^2\alpha^2 + \lambda\alpha^2\beta}$$

5.2.6 Numerical and Graphical Representation of MTSF and Availability

5.3 MTSF and Availability of a Two-Unit Parallel System with Arrival Time of the Server

Let us consider a system of two identical units in which both units are operative in parallel mode. There is a single server which is allowed to take some time to arrive at the system (called arrival time of the server). The possible transition between states is shown in the following state transition diagram (Fig. 3):

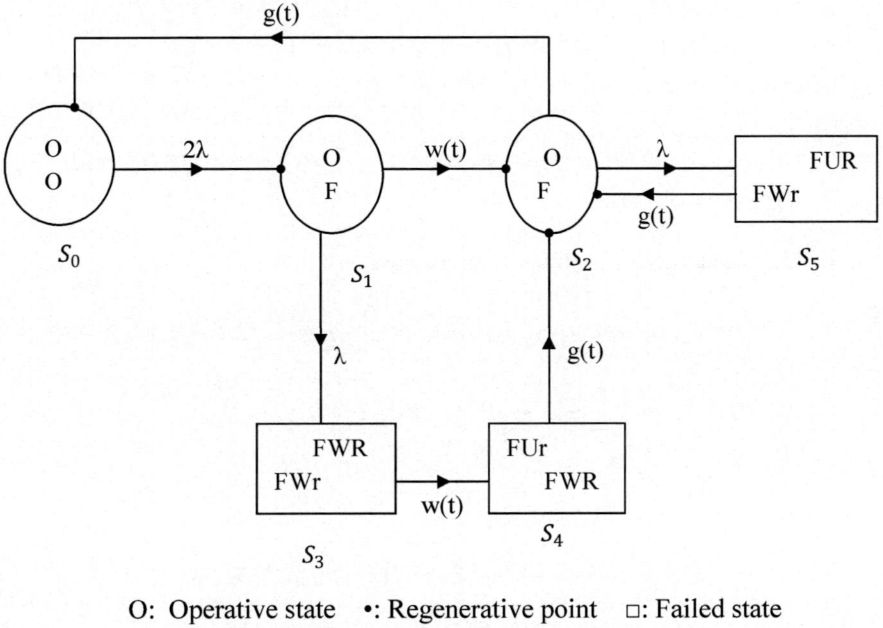

Fig. 3 State transition diagram (for a parallel system)

5.3.1 System Description

S_0	The initial state in which one unit is operative and the other one unit in parallel
S_1	The state in which one unit is operative and the other failed unit is waiting for repair
S_2	The state in which one unit is operative and the other unit is under repair
S_3	The state in which one failed unit is waiting for repair and the other failed unit is continuously waiting for repair from previous state
S_4	The state in which one failed unit is under repair and the other failed unit is continuously waiting for repair from previous state
S_5	The final state in which one unit is under repair from previous state and other unit is waiting for repair

5.3.2 Transition Probabilities and Mean Sojourn Times

Simple probabilistic considerations yield the following expressions for the nonzero elements $p_{ij} = Q_{ij}(\infty) = \int_0^\infty q_{ij}(t)dt$ as

$$dQ_{01}(t) = q_{01}(t)dt = 2\lambda e^{-2\lambda t}dt, dQ_{12}(t) = e^{-\lambda t}w(t)dt,$$
$$dQ_{13}(t) = \lambda e^{-\lambda t}\overline{W}(t)dt$$
$$dQ_{20}(t) = e^{-\lambda t}g(t)dt, dQ_{25}(t) = \lambda e^{-\lambda t}\overline{G}(t)dt, dQ_{12.34}(t) = dQ_{13}(t)©dQ_{34}(t)©dQ_{42}(t)$$

$$dQ_{34}(t) = w(t)dt, dQ_{42}(t) = g(t)dt, dQ_{22.5}(t) = \lambda e^{-\lambda t}\overline{G}(t)g(t)dt,$$
$$dQ_{52}(t) = g(t)dt$$

Taking Laplace–Stieltjes transform, we have

$$Q_{01}^{**}(s) = \int_0^\infty e^{-st}[Q_{01}(t)] = \frac{2\lambda}{2\lambda + S}, Q_{12}^{**}(s) = w^*(\lambda + s),$$
$$Q_{13}^{**}(s) = \frac{\lambda}{\lambda + S}\left[1 - w^*(\lambda + s)\right]$$

$$Q_{20}^{**}(s) = g^*(\lambda + s), Q_{25}^{**}(s) = \frac{\lambda}{\lambda + S}\left[1 - g^*(\lambda + s)\right],$$

$$Q_{12.34}^{**}(s) = \frac{\lambda}{\lambda + S}\left[1 - w^*(\lambda + s)\right]w^*(s)g^*(s),$$
$$Q_{34}^{**}(s) = w^*(s), Q_{42}^{**}(s) = g^*(s),$$

$$Q_{22.5}^{**}(s) = \frac{\lambda}{\lambda + S}\left[1 - g^*(\lambda + s)\right]g^*(s), Q_{52}^{**}(s) = g^*(s)$$

Taking $lims \to 0$, we get the following transition probabilities:

$$p_{01} = 1, p_{12} = w^*(\lambda), p_{13} = \left[1 - w^*(\lambda)\right], p_{20} = g^*(\lambda)$$

$$p_{25} = \left[1 - g^*(\lambda)\right], p_{12.34} = \left[1 - w^*(\lambda)\right], p_{34} = 1, p_{42} = 1,$$

$$p_{22.5} = [1 - g^*(\lambda)], p_{52} = 1$$

It can be easily verified that

$$p_{01} = p_{01} + p_{13} = p_{20} + p_{22.5} = p_{20} + p_{25} = p_{12} + p_{12.34} = p_{34} = p_{42} = p_{52} = 1$$

Mean sojourn times

$$\mu_i = \sum_j m_{ij}(i = 0, 1)$$

$$But, m_{ij} = -\frac{d}{ds}[Q_{ij}^{**}(s)]_{s=0}, m_{01} = -\frac{d}{ds}[]_{s=0} = \frac{1}{2\lambda}, m_{12} = -w^{*'}(\lambda)$$

$$m_{13} = w^{*'}(\lambda) + \frac{1}{\lambda}\left[1 - w^*(\lambda)\right], m_{20} = g^{*'}(\lambda),$$

$$m_{25} = g^{*'}(\lambda) + \frac{1}{\lambda}\left[1 - g^*(\lambda)\right]$$

$$m_{12.34} = \frac{\lambda w^{*'}(\lambda) - \lambda[1 - w^*(\lambda)]g^{*'}(0) - \lambda[1 - w^*(\lambda)]w^{*'}(0) + [1 - w^*(\lambda)]}{\lambda}$$

$$m_{22.5} = \frac{\lambda g^{*'}(\lambda) - \lambda[1 - g^*(\lambda)]g^{*'}(0) + [1 - g^*(\lambda)]w^{*'}(0)}{\lambda}$$

$$Now, \mu_0 = m_{01} = \frac{1}{2\lambda}, \mu_1 = m_{12} + m_{13} = \frac{1}{\lambda}\left[1 - w^*(\lambda)\right]$$

$$\mu_2 = m_{20} + m_{25} = \frac{1}{\lambda}\left[1 - g^*(\lambda)\right]$$

$$\mu_1' = m_{12} + m_{12.34} = \frac{1}{\lambda}\left\{\left[1 - w^*(\lambda)\right] - \lambda\left[1 - w^*(\lambda)\right]g^{*'}(0) - \lambda\left[1 - w^*(\lambda)\right]w^{*'}(0)\right\}$$

$$\mu_2' = m_{20} + m_{22.5} = \frac{1}{\lambda}\left\{\left[1 - g^*(\lambda)\right] - \lambda\left[1 - g^*(\lambda)\right]g^{*'}(0)\right\}$$

5.3.3 Reliability and Mean Time to System Failure (MTSF)

Let $\emptyset_i(t)$ be the CDF of first passage time from regenerative state S_i to a failed state. Regarding the failed state as absorbing state, we have the following recursive relations for $\emptyset_i(t)$:

$$\begin{aligned}
\emptyset_0(t) &= Q_{01}(t)\emptyset_1(t) \\
\emptyset_1(t) &= Q_{12}(t)\emptyset_2(t) + Q_{13}(t) \\
\emptyset_2(t) &= Q_{20}(t)\emptyset_0(t) + Q_{25}(t)
\end{aligned}$$

Taking Laplace–Stieltjes transform, it can be written as

$$\begin{aligned}
\emptyset_0^{**}(s) &= Q_{01}^{**}(s)\emptyset_1^{**}(s) \\
\emptyset_1^{**}(s) &= Q_{12}^{**}(s)\emptyset_2^{**}(s) + Q_{13}^{**}(s) \\
\emptyset_2^{**}(s) &= Q_{20}^{**}(s)\emptyset_0^{**}(s) + Q_{25}^{**}(s)
\end{aligned}$$

Or

$$\begin{bmatrix} 1 & -Q_{01}^{**}(s) & 0 \\ 0 & 1 & -Q_{12}^{**}(s) \\ -Q_{20}^{**}(s) & 0 & 1 \end{bmatrix} \begin{bmatrix} \emptyset_0^{**}(s) \\ \emptyset_1^{**}(s) \\ \emptyset_2^{**}(s) \end{bmatrix} = \begin{bmatrix} 0 \\ Q_{13}^{**}(s) \\ Q_{25}^{**}(s) \end{bmatrix}$$

Or AX = B

$$|\Delta| = 1 - Q_{01}^{**}(s)Q_{12}^{**}(s)Q_{20}^{**}(s)$$

$$|\Delta_1| = \begin{vmatrix} 0 & -Q_{01}^{**}(s) & 0 \\ Q_{13}^{**}(s) & 1 & -Q_{12}^{**}(s) \\ Q_{25}^{**}(s) & 0 & 1 \end{vmatrix} = Q_{01}^{**}(s)Q_{13}^{**}(s) + Q_{01}^{**}(s)Q_{12}^{**}(s)Q_{25}^{**}(s)$$

On solving for $\emptyset_0^{**}(s)$ by Crammer's rule, we get

$$\emptyset_0^{**}(s) = \frac{Q_{01}^{**}(s)Q_{13}^{**}(s) + Q_{01}^{**}(s)Q_{12}^{**}(s)Q_{25}^{**}(s)}{1 - Q_{01}^{**}(s)Q_{12}^{**}(s)Q_{20}^{**}(s)}$$

Now,

$$\text{MTSF} = \lim_{S\to 0} \frac{1 - \emptyset_0^{**}(s)}{s} = \lim_{S\to 0} \frac{1 - \frac{Q_{01}^{**}(s)Q_{13}^{**}(s) + Q_{01}^{**}(s)Q_{12}^{**}(s)Q_{25}^{**}(s)}{1 - Q_{01}^{**}(s)Q_{12}^{**}(s)Q_{20}^{**}(s)}}{s}$$

$$= \lim_{S\to 0} \frac{Q_{01}^{**}(s)Q_{13}^{**}(s) + Q_{01}^{**}(s)Q_{12}^{**}(s)Q_{25}^{**}(s)}{s\left[1 - Q_{01}^{**}(s)Q_{12}^{**}(s)Q_{20}^{**}(s)\right]} = \lim_{S\to 0} \frac{1 - \frac{N_1}{D_1}}{S} = \lim_{S\to 0} \frac{[D_1 - N_1]}{SD_1}\left(\frac{0}{0}\right)$$

Indeterminant Form.
So, by applying L' Hospital rule, we get

$$\text{MTSF} = \frac{\mu_0 + \mu_1 + \mu_2 p_{12}}{1 - p_{12}p_{20}}$$

So, we have

$$R^*(s) = \frac{1 - \emptyset_0^{**}(s)}{s} = \frac{Q_{01}^{**}(s)Q_{13}^{**}(s) + Q_{01}^{**}(s)Q_{12}^{**}(s)Q_{25}^{**}(s)}{s\left[1 - Q_{01}^{**}(s)Q_{12}^{**}(s)Q_{20}^{**}(s)\right]}$$

The reliability of the system model can be obtained by taking Laplace inverse of R*(s), and we get

$$R(t) = L^{-1}\left\{\frac{Q_{01}^{**}(s)Q_{13}^{**}(s) + Q_{01}^{**}(s)Q_{12}^{**}(s)Q_{25}^{**}(s)}{s\left[1 - Q_{01}^{**}(s)Q_{12}^{**}(s)Q_{20}^{**}(s)\right]}\right\}$$

5.3.4 Availability Analysis

Let $A_i(t)$ be the probability that the system is in upstate at instant 't' given that the system entered regenerative state S_i at t = 0. The recursive relations for $A_i(t)$ are given as

$$\begin{aligned} A_0(t) &= M_0(t) + q_{01}(t)©A_1(t) \\ A_1(t) &= M_1(t) + q_{12}(t)©A_2(t) + q_{12.34}(t)©A_2(t) \\ A_2(t) &= M_2(t) + q_{20}(t)©A_0(t) + q_{22.5}(t)©A_2(t) \end{aligned}$$

where $M_i(t)$ is the probability that the system is up initially in state $S_i \epsilon E$ which is up at time t without visiting to any other regenerative state, and we have

$$M_0(t) = e^{-2\lambda t}, \; M_1(t) = e^{-\lambda t}\overline{W}(t), \; M_2(t) = \overline{G}(t)$$

Taking Laplace transform of above equations, it can be written as
or

$$\begin{aligned} A_0^*(s) &= M_0^*(s) + q_{01}^*(s) + A_1^*(s) = (s) + (s) \\ A_1^*(s) &= M_1^*(s) + q_{12}^*(s) + A_2^*(s) + q_{12.34}^*(s) + A_2^*(s) \\ A_2^*(s) &= M_2^*(s) + q_{20}^*(s) + A_0^*(s) + q_{22.5}^*(s) + A_2^*(s) \\ &\mathit{or} \\ A_0^*(s) - q_{01}^*(s)A_1^*(s) &= M_0^*(s) \\ A_1^*(s) - q_{10}^*(s)A_0^*(s) - q_{11.2}^*(s)A_1^*(s) &= M_1^*(s) \\ A_2^*(s) - q_{20}^*(s)A_0^*(s) - q_{22.5}^*(s)A_2^*(s) &= M_2^*(s) \end{aligned}$$

$$\begin{bmatrix} 1 & -q_{01}^*(\mathrm{s}) & 0 \\ 0 & 1 & -[q_{12}^*(\mathrm{s}) + q_{12.34}^*(\mathrm{s})] \\ -q_{20}^*(\mathrm{s}) & 0 & 1 - q_{22.5}^*(\mathrm{s}) \end{bmatrix} \begin{bmatrix} A_0^*(s) \\ A_1^*(s) \\ A_2^*(s) \end{bmatrix} = \begin{bmatrix} M_0^*(\mathrm{s}) \\ M_1^*(\mathrm{s}) \\ M_2^*(\mathrm{s}) \end{bmatrix}$$

On solving for $A_0^*(s)$ by Crammer's rule, we get

$$|\Delta| = \begin{vmatrix} 1 & -q_{01}^*(s) & 0 \\ 0 & 1 & -[q_{12}^*(s) + q_{12.34}^*(s)] \\ -q_{20}^*(s) & 0 & 1 - q_{22.5}^*(s) \end{vmatrix}$$
$$= 1 - q_{22.5}^*(s) - q_{01}^*(s)q_{20}^*(s)q_{12}^*(s) - q_{01}^*(s)q_{20}^*(s)q_{12.34}^*(s)$$

$$|\Delta_1| = \begin{vmatrix} M_0^*(s) & -q_{01}^*(s) & 0 \\ M_1^*(s) & 1 & -[q_{12}^*(s) + q_{12.34}^*(s)] \\ M_2^*(s) & 0 & 1 - q_{22.5}^*(s) \end{vmatrix}$$
$$= M_0^*(s)[1 - q_{22.5}^*(s)] + q_{01}^*(s)M_1^*(s)[1 - q_{22.5}^*(s)] + M_2^*(s)[q_{12}^*(s) + q_{12.34}^*(s)]$$

On solving for $A_0^*(s)$ by Crammer's rule, we get
Thus,

$$A_0^*(s) = \frac{M_0^*(\mathrm{s})[1 - q_{22.5}^*(\mathrm{s})] + q_{01}^*(\mathrm{s})M_1^*(\mathrm{s})[1 - q_{22.5}^*(\mathrm{s})] + M_2^*(\mathrm{s})[q_{12}^*(\mathrm{s}) + q_{12.34}^*(\mathrm{s})]}{1 - q_{22.5}^*(\mathrm{s}) - q_{01}^*(\mathrm{s})q_{20}^*(\mathrm{s})q_{12}^*(\mathrm{s}) - -q_{01}^*(\mathrm{s})q_{20}^*(\mathrm{s})q_{12.34}^*(\mathrm{s})}$$

The steady-state (long run) availability is given by

$$A(\infty) = \lim_{t\to\infty} A(t) = \lim_{s\to 0} s\, A_0^*(s)$$

$$= \lim_{s\to 0} s \frac{\{M_0^*(s)[1 - q_{22.5}^*(s)] + q_{01}^*(s)M_1^*(s)[1 - q_{22.5}^*(s)] + M_2^*(s)\,[q_{12}^*(s) + q_{12.34}^*(s)]\}}{1 - q_{22.5}^*(s) - q_{01}^*(s)q_{20}^*(s)q_{12}^*(s) -- q_{01}^*(s)q_{20}^*(s)q_{12.34}^*(s)} \left(\frac{0}{0}\right) \text{From}$$

So, by applying L' Hospital rule, we get

$$A(\infty) = \frac{\mu_0 p_{20} + \mu_1 p_{20} + \mu_2}{\mu_2' + \mu_0 p_{20} + \mu_1' p_{20}}$$

5.3.5 Particular Case

Suppose repair rate $g(t)$ and arrival time of the server $w(t)$ follow negative exponential distribution, i.e.

$$g(t) = \alpha e^{-\alpha t} \text{ and } w(t) = \beta e^{-\beta t}$$

Taking Laplace transform, it can be written as

$$g^*(s) = \frac{\alpha}{\alpha + s},\ g^{*'}(s) = -\frac{\alpha}{(\alpha + s)^2},\ g^{*'}(0) = -\frac{1}{\alpha}$$

$$w^*(s) = \frac{\beta}{\beta + s},\ w^{*'}(s) = -\frac{\beta}{(\beta + s)^2},\ w^{*'}(0) = -\frac{1}{\beta}$$

$$MTSF = \frac{\mu_0 + \mu_1 + \mu_2 p_{12}}{1 - p_{12}p_{20}} = \frac{(\alpha + \lambda)(\beta + \lambda) + (\alpha + \lambda)2\lambda + 2\beta\lambda}{2\lambda[(\beta + \lambda)(\alpha + \lambda) - \alpha\beta]}$$

$$A(\infty) = \frac{\mu_0 p_{20} + \mu_1 p_{20} + \mu_2}{\mu_2' + \mu_0 p_{20} + \mu_1' p_{20}} = \frac{\alpha\beta[\alpha(\beta + 3\lambda) + 2\lambda(\beta + \lambda)]}{\alpha[\alpha\beta(\beta + \lambda) + 2\lambda(\alpha\lambda + \beta\lambda + \alpha\beta)] + 2\beta\lambda(\alpha + \lambda)(\beta + \lambda)}$$

5.3.6 Numerical and Graphical Representation of MTSF and Availability

See Tables 5 and 6.

6 Discussion and Conclusion

An approach to determine mean time to system failure (MTSF) and availability of repairable systems has been given in this chapter. The systems with single unit and two units are considered to know the effect of redundancy (cold standby and parallel standby) on reliability measures. The reliability measures are obtained in steady state using semi-Markov process and regenerative point technique. The random variables associated with failure time of the unit, repair time of the unit and arrival time of the server follow negative exponential distribution. A single repair facility has been taken up for all types of repair activities, and it takes some time to attend the system (called arrival time of the server). For arbitrary values of the parameters, it is observed that the mean time to system failure (MTSF) and availability decrease with the increase of failure rate of the unit while the values of these measures keep on increasing with the increase of repair rate of the unit and arrival rate of the server. The mean time to system failure and availability of a single unit system are less than that of cold standby system followed by the parallel system. Hence, the provision of cold standby redundancy is more helpful in improving the system performance and thus reliability.

References

Anand, J. (2012). *Reliability modelling and economic analysis of computer systems with H/W and S/W failures*. Ph.D. Thesis, M.D. University, Rohtak.

Ashish. (2013). *Reliability and cost–benefit analysis of computer systems subject to maximum operation and repair time*. Ph.D Thesis, M.D.University, Rohtak.

Balagurusamy, E. (1984). *Reliability engineering*. Tata McGraw Hill Publishing Co., Ltd.

Barlow, R. E., & Prochan, P. (1965). *Mathematical theory of reliability*. John Wiley & Sons.

Bhardwaj, R. K., & Kaur, K. (2014). Reliability and profit analysis of a redundant system with possible renewal of standby subject to inspection. *International Journal of Statistics and Reliability Engineering, 1*(1), 36–46.

Branson, M. H., & Shah, B. (1971). Reliability analysis of system comprises of units with arbitrary repair time distributions. *IEEE Transactions on Reliability, 20*, 217–223.

Chander, S. (2005). Reliability models with priority for operation and repair with arrival time of server. *Pure and Applied Mathematika Sciences, 61*(1–2), 9–22.

Gaver, D. P. (1963). Time to failure and availability of paralleled systems with repair. *IEEE Transactions on Reliability, 12*(2), 30–38.

Gitanjali. (2014). Cost-Benefit analysis of parallel unit system reliability models with maximum repair times. Ph.D. Thesis, M.D. University, Rohtak.

Kadyan, M. S., Chander, S., & Grewal, A. S. (2004). Stochastic analysis of non-identical units reliability models with priority and different modes of failure. *Decision and Mathematical Sciences, 9*(1–3), 59–82.

Malik, S. C., & Gitanjali, G. (2019). Cost-Benefit evaluation of a repairable system with alternate repair and Weibull distribution. *International Journal of Agricultural and Statistical Sciences, 15*(2):741–749.

Mokaddis, G. S., Labib, S. W., & Ahmed, A. M. (1997). Analysis of a two-unit warm standby system subject to degradation. *Microelectronics Reliability, 37*, 641–648.

Murari, K., & Goyal, V. (1984). Comparison of two-unit cold standby reliability models with three types of repair facilities. *Journal of Microelectronics Reliability, 24*(1), 35–49.
Nandal, N., & Malik, S. C. (2019). On use of Lindley distribution for profit analysis of a three unit cold standby system subject to arrival time of the server. *International Journal of Statistics and Reliability Engineering, 6*(2), 145–151.
Pawar, D., Malik, S. C., & Bahl, S. (2010). Steady state analysis of an operating system with repair at different levels of damages subject to inspection and weather conditions. *International Journal of Agriculture and Statistical Sciences, 6*(1), 225–234.
Singh, S. K. (1989). Profit evaluation of a two-unit cold standby system with random appearance and disappearance time of the service facility. *Microelectronics Reliability, 29*, 705–709.
Smith, W. L. (1955). Regenerative stochastic processes. *Proceedings of the Royal Society of London, Series A: Mathematical and Physical Sciences, 232*(1188), 6–31.
Srinath, L. S. (1985). *Concept in reliability engineering*. Affiliated East-West Press (P) Ltd.
Srinivasan, S. K., & Gopalan, M. N. (1973). Probabilistic analysis of two-unit system with warm standby and single repair facility. *Operations Research, 21*(3), 748–754.
Zuckerman, D. (1978). Optimal stopping in a semi-markov shock model. *Journal of Applied Probability, 15*(3), 629–634.

Stress-strength Reliability Estimation for Multi-component System Based on Upper Record Values Under New Weibull-Pareto Distribution

Parameshwar V. Pandit and Shubhashree Joshi

Abstract In this chapter, estimation of multi-component system reliability, particularly, s-out-of-k system is considered, where the system has k independent and similar components under stress-strength setup. The system reliability is evaluated assuming new Weibull-Pareto distribution for strength and stress variables. Likelihood ratio test is also constructed to test the equality of scale parameters on which the reliability of the system depends. Maximum likelihood estimator of the system reliability is obtained based on upper records. Approximate Bayes estimators are also obtained using Lindley's approximation technique. The estimators are compared based on mean squares error criteria using simulation. The estimation procedures are illustrated using real data.

Keywords Bayes estimator · Maximum likelihood estimator · New Weibull-Pareto distribution · Stress-strength reliability · Upper record values

1 Introduction

The study of reliability is related to know the ability of a component, equipment, subsystem or a system to perform a required function, under given environmental and operational conditions for a specified period. Hence, reliability can be defined as probability that a system will satisfactorily perform its intended function under given circumstances. Stress-strength model is one of the important models studied in the literature. The term stress-strength can be described as the assessment of reliability of a component in terms of random variables representing 'stress' of the component and that representing 'strength' of the component available to overcome the possible stress. If the stress exceeds the strength, then the system fails. The stress is a function of the environment in which the component is located and can be estimated from the available technological knowledge about the relevant conditions of the system and the manner in which they interact. Stress can be treated as a random variable based on a priori considerations and can only be estimated by means of statistical

P. V. Pandit (✉) · S. Joshi
Department of Statistics, Bangalore University, Bengaluru 560056, India

D. D. Hanagal et al. (eds.), *Applied Statistical Methods*, Springer Proceedings in Mathematics & Statistics 380, https://doi.org/10.1007/978-981-16-7932-2_16

methods from the results of the tests specifically geared for this purpose. However, considerations regarding the nature of stress and strength do not require that they are related in any way. Hence, many authors studied inference about $P(X > Y)$ under the assumption that strength X and stress Y are independent variables.

The stress-strength model was originated in a non-parametric setup which can be found in the works of Birnbaum (1956), Birnbaum and McCarty (1958), Govindrajulu (1967), among others. By the end of seventies, the estimation of $P(X < Y)$ was carried out for the major distributions such as exponential (Kelley et al., 1976, Tong, 1974), normal (Church & Harris, 1970) and exponential families (Tong, 1977). Further, after the introduction of variety of bivariate exponential distribution by Freund (1961), Marshall and Olkin (1967), Block and Basu (1974) and others, it became possible to study dependent exponential random variables with various types of dependence. The estimation of reliability for bivariate exponential distribution was considered by Hanagal (1995). Hanagal (1997a) addressed a similar problem for bivariate Pareto and exponential distributions. The estimation of reliability for a two-parameter inverse Chen distribution was considered by Pandit and Joshi (2019a). Pandit and Joshi (2019b) and Joshi and Pandit 2018) studied the estimation of multi-component system reliability for a bivariate generalized Rayleigh distribution and inverse Chen distribution, respectively. Hanagal (1997b) considered the estimation of stress-strength reliability by censoring stress at strength where stress and strength follow bivariate exponential (BVE) model of Marshall and Olkin (1967). Huizhen and Krishnamoorthy (2004) considered the problem of testing and interval estimation of the reliability parameter $P(Y_1 > Y_2)$, where Y_1 and Y_2 are independent normal random variables with unknown means and variances.

Recently, many authors are interested in estimating multi-component stress-strength reliability for various distributions like log-logistic, generalized exponential, generalized inverted exponential, Rayleigh, Burr Type XII and generalized Rayleigh distributions, respectively, by Rao and Kantam (2010), Rao (2012a, b, 2014) and Rao et al., (2013, 2015). Pandit and Kantu (2013) and Pandit and Joshi (2018) developed procedures for estimating multi-component stress-strength reliability when strength and stress variables follow exponential distribution and generalized Pareto distribution, respectively. Hassan et al. (2015) studied the estimation of R when X and Y are independently distributed Burr XII random variables based on different ranked set sampling schemes. Nadar and Kizilaslan (2015, 2016) considered estimation of multi-component stress-strength reliability using both classical and Bayesian approaches when underlying distribution is Weibull and bivariate Kumarswamy distributions. Joshi and Pandit (2018) studied the estimation of system reliability in s-out-of-k system under stress-strength setup for inverse Chen distribution.

The modelling of systems in a stress-strength setup has attracted many researchers in the field of statistics, particularly, who are interested in application of statistics in engineering and technology. The problem of estimation of the stress-strength reliability of s-out-of-k system has received a considerable attention of many researchers. The researchers like Rao and Kantam (2010), Rao (2014), Nadar and Kizilaslan (2015, 2016) studied the problem of estimating stress-strength reliability of an s-out-of-k system when the underlying distributions, respectively, are log-logistic, Rayleigh distribution, Weibull, bivariate Kumarswamy, etc. Estimation of stress-strength reliability for parallel and series systems was considered by the application of s-out-of-k system which can be seen in many real-life situations, particularly in industry and military (refer Kuo & Zuo, 2003).

In this chapter, estimation of system reliability is considered using upper record values. The main idea of record values was introduced by Chandler (1952). This record values have become important field of research in recent years due to its vast applications in the sports, hydrology and life tests. The advantage of using record values is the reduction in number of measurements to be made as compared to that in complete sample, and the measurement saving can be done when the measurements of the experiments are expensive in case of destructive experiments. We give below the definition of upper record values as in the literature.

Definition 1.1: for a sequence of independent and identically distributed random variables $\{X_1, X_2, ...\}$, a random variable X_j is said to be an upper record if $X_j > X_i$ for every $i < j$. An analogous definition can be given for lower records.

Use of record values in estimating system reliability in a stress-strength setup has been studied in the literature by many authors for various distributions to mention a few Wang and Shi (2013) and Tarvirdizade and Ahmadpour (2016) when underlying distributions are exponential and two-parameter bathtub-shaped.

This chapter discusses stress-strength reliability of s-out-of-k (s $\leq$ k) system when the underlying distribution is new Weibull-Pareto (NWP) due to Tahir et al. (2016), and it is extensively discussed by Nasiru and Luguterah (2015). Cumulative distribution function and probability density function of NWP distribution are given below

$$F(x) = 1 - e^{-\delta\left(\frac{x}{\theta}\right)^{\beta}},\ x > 0,\ \delta, \theta,\ \beta > 0$$

and

$$f(x) = \frac{\beta\delta}{\theta}\left(\frac{x}{\theta}\right)^{\beta-1} e^{-\delta\left(\frac{x}{\theta}\right)^{\beta}},\ x > 0,\ \delta, \theta,\ \beta > 0$$

This distribution has constant failure rate when $\beta = 1$ and increasing (decreasing) failure rates when $\beta > 1$ ($\beta < 1$).

In Sect. 2, the reliability of s-out-of-k system ($R_{s,k}$) under stress-strength setup is derived. Section 3 consists of maximum likelihood estimation of the parameters of NWP distribution. The likelihood ratio test is developed for testing equality of scale parameters in Sect. 4. The maximum likelihood estimator and Bayes estimators of Rs,k are given in Sect. 5. In Sect. 6, the mean square errors (MSEs) are obtained using a simulation study, and real-life example is given in Sect. 7. Section 8 gives summary and conclusions.

2 System Reliability

Bhattacharyya and Johnson (1974) derived the general expression for the reliability of s-out-of-k system in a stress-strength setup which is given by

$$R_{s,k} = P(\text{atleast s of the } (X_1, X_2, \ldots, X_k) \text{ exceed } Y)$$

$$= \sum_{i=s}^{k} \binom{k}{i} \int_0^\infty [1 - F(y)]^i \, [F(y)]^{k-i} \, dG(y) \tag{2.1}$$

where the strengths $X_1, X_2, \ldots, X_k$ are independent random variables with common distribution function F(x) and stress Y is a random variable with distribution function G(y).

Here, our focus is on studying s-out-of-k ($s \leq k$) system reliability when component strengths $X_1, X_2, \ldots, X_k$ follow NWP distribution with parameters $(\delta_1, \theta, \beta)$ and stress Y follow NWP distribution with parameters $(\delta_2, \theta, \beta)$. For this setup, the system reliability, $R_{s,k}$ is derived using (2.1) and is given by

$$R_{s,k} = \sum_{i=s}^{k} \binom{k}{i} \int_0^\infty \frac{\delta_2 \beta}{\theta} \left(1 - \left(1 - e^{-\delta_1 (\frac{y}{\theta})^\beta}\right)\right)^i \left(1 - e^{-\delta_1 (\frac{y}{\theta})^\beta}\right)^{k-i} \left(\frac{y}{\theta}\right)^{\beta - 1} e^{-\delta_2 (\frac{y}{\theta})^\beta} dy$$

$$= \begin{cases} \sum_{i=s}^{k} \sum_{j=0}^{k-i} \binom{k}{i} \binom{k-i}{j} (-1)^j \frac{\delta_2}{[\delta_1 (i+j) + \delta_2]} & if \quad \delta_1 \neq \delta_2 \\ \sum_{i=s}^{k} \sum_{j=0}^{k-i} \binom{k}{i} \binom{k-i}{j} (-1)^j \frac{1}{(i+j+1)} & if \quad \delta_1 = \delta_2 = \delta \end{cases}$$

(Using binomial expansion)

It is to be noted that the case when stress and strengths are identically distributed according to NWP distribution with parameters (δ, θ, β), the reliability of s-out-of-k (s≤k) system does not depend on the parameters (δ, θ, β). Hence, estimation of reliability of the system suffices to estimating Rs,k only when stress and strength variables are not identically distributed (i.e. δ1 ≠ δ2). However, to check whether δ1 = δ2 or not, we conduct likelihood ratio (LR) test. For that, first, the maximum likelihood estimators of the parameters when δ1 = δ2 and δ1 ≠ δ2 are derived.

3 Maximum Likelihood Estimators (MLE) of Parameters

Let $\underset{\sim}{u} = (u_1, \ldots, u_n)$ and $\underset{\sim}{v} = (v_1, \ldots, v_m)$ be the sets of upper records from distributions with pdfs of f and g, respectively. Let F and G be the cdfs of f and g, respectively. Then, the likelihood given records are obtained as product of L_1 and L_2 ($L = L_1 L_2$), where L_1 and L_2 are given by

$$L_1(\delta_1,\ \theta,\ \beta) = f(u_n)\prod_{i=1}^{n-1}\left(\frac{f(u_i)}{1 - F(u_i)}\right),\ 0 < u_1 < \ldots < u_n < \infty$$

$$L_2(\delta_2,\ \theta,\ \beta) = g(v_m)\prod_{j=1}^{m-1}\left(\frac{g(v_j)}{1 - G(v_j)}\right),\ 0 < v_1 < \ldots < v_m < \infty$$

(see Chandler (1952) for details).

Now, the parameters are estimated based on records when.

(i) stress and strength variables are identically distributed with NWP (i.e. $\delta_1 = \delta_2$).

(ii) stress and strength variables are not identically distributed with NWP which differ only with respect to scale parameter δ (i.e. δ1≠δ2).

Case 1: Let $u_1, \ldots, u_n$ and $v_1, \ldots, v_m$be two sets of records from NWP distribution with parameters $(\delta,\ \theta,\ \beta)$ (i.e. for the case $\delta_1 = \delta_2$). Then, likelihood function is given by.

$$L(\delta,\ \theta,\ \beta) = \left(\frac{\delta\beta}{\theta}\right)^n e^{-\delta\left(\frac{u_n}{\theta}\right)^\beta}\prod_{i=1}^{n}\left(\frac{u_i}{\theta}\right)^{\beta-1}\left(\frac{\delta\beta}{\theta}\right)^m e^{-\delta\left(\frac{v_m}{\theta}\right)^\beta}\prod_{j=1}^{m}\left(\frac{v_j}{\theta}\right)^{\beta-1}$$

Thus, the log-likelihood function of δ,θ and β is

$$\log L(\delta,\ \theta,\ \beta\ ) = (n+m)\log\delta + (n+m)\log\beta - (n+m)\log\theta + (\beta-1)\sum_{i=1}^{n}\log\left(\frac{u_i}{\theta}\right)$$
$$+ (\beta-1)\sum_{j=1}^{m}\log\left(\frac{v_j}{\theta}\right) - \delta\left[\left(\frac{u_n}{\theta}\right)^\beta + \left(\frac{v_m}{\theta}\right)^\beta\right]$$

The likelihood equations are

$$\frac{\partial\log L}{\partial\delta} = \frac{n+m}{\delta} - \left[\left(\frac{u_n}{\theta}\right)^\beta + \left(\frac{v_m}{\theta}\right)^\beta\right] = 0;$$

$$\frac{\partial\log L}{\partial\theta} = -\left(\frac{n+m}{\theta}\right) + \frac{\delta\beta(u_n)^\beta}{\theta^{\beta+1}} + \frac{\delta\beta(v_m)^\beta}{\theta^{\beta+1}} - \frac{n}{\theta}(\beta-1) - \frac{m}{\theta}(\beta-1) = 0 \text{ and}$$

$$\frac{\partial\log L}{\partial\beta} = \frac{n+m}{\beta} - \delta\left(\frac{u_n}{\theta}\right)^\beta\log\left(\frac{u_n}{\theta}\right) - \delta\left(\frac{v_m}{\theta}\right)^\beta\log\left(\frac{v_m}{\theta}\right)$$
$$+ \sum_{i=1}^{n}\log\left(\frac{u_i}{\theta}\right) + \sum_{j=1}^{m}\log\left(\frac{v_j}{\theta}\right) = 0$$

We obtain the MLE of δ as a function of β and θ.

The maximum likelihood estimators are

$$\hat{\delta} = \frac{n+m}{\left[\left(\frac{u_n}{\theta}\right)^{\beta} + \left(\frac{v_m}{\theta}\right)^{\beta}\right]}$$

where $\hat{\beta}$ and $\hat{\theta}$ are the solution of the nonlinear equation of the form,

$$\left(\frac{n+m}{\theta}\right) + \frac{\delta\beta(u_n)^{\beta}}{\theta^{\beta+1}} + \frac{\delta\beta(v_m)^{\beta}}{\theta^{\beta+1}} - \frac{n}{\theta}(\beta-1) - \frac{m}{\theta}(\beta-1) = 0 \text{ and}$$

$$\frac{n+m}{\beta} - \delta\left(\frac{u_n}{\theta}\right)^{\beta}\log\left(\frac{u_n}{\theta}\right) - \delta\left(\frac{v_m}{\theta}\right)^{\beta}\log\left(\frac{v_m}{\theta}\right) + \sum_{i=1}^{n}\log\left(\frac{u_i}{\theta}\right) + \sum_{j=1}^{m}\log\left(\frac{v_j}{\theta}\right) = 0.$$

Here, $\hat{\beta}$ and $\hat{\theta}$ can be obtained by using any iterative scheme, in particular, using Newton–Raphson method.

Case 2: Let $u_1, \ldots, u_n$ be the records from NWP distribution with parameters $(\delta_1, \theta, \beta)$ and $v_1, \ldots, v_m$ be the records from NWP distribution with parameters $(\delta_2, \theta, \beta)$ (i.e. $\delta_1 \neq \delta_2$).

Now, the likelihood function is given by

$$L_1(\delta_1, \delta_2, \theta, \beta) = \left\{\left(\frac{\delta_1\beta}{\theta}\right)^{n} e^{-\delta_1\left(\frac{u_n}{\theta}\right)^{\beta}} \prod_{i=1}^{n}\left(\frac{u_i}{\theta}\right)^{\beta-1}\right\}\left\{\left(\frac{\delta_2\beta}{\theta}\right)^{m} e^{-\delta_2\left(\frac{v_m}{\theta}\right)^{\beta}} \prod_{j=1}^{m}\left(\frac{v_j}{\theta}\right)^{\beta-1}\right\}$$

Thus, the log-likelihood function of δ_1,δ_2,θ and β is

$$\log L(\delta_1, \delta_2, \theta, \beta) = n\log\delta_1 + m\log\delta_2 + (n+m)\log\beta - (n+m)\beta\log\theta + (\beta-1)\sum_{i=1}^{n}\log\left(\frac{u_i}{\theta}\right)$$
$$+ (\beta-1)\sum_{j=1}^{m}\log\left(\frac{v_j}{\theta}\right) - \delta_1\left(\frac{u_n}{\theta}\right)^{\beta} - \delta_2\left(\frac{v_m}{\theta}\right)^{\beta}.$$

The likelihood equations are

$$\frac{\partial\log L}{\partial\delta_1} = \frac{n}{\delta_1} - \left(\frac{u_n}{\theta}\right)^{\beta} = 0; \quad \frac{\partial\log L}{\partial\delta_2} = \frac{m}{\delta_2} - \left(\frac{v_m}{\theta}\right)^{\beta} = 0$$

$$\frac{\partial\log L}{\partial\theta} = -\left(\frac{n+m}{\theta}\right) + \frac{\delta_1\beta(u_n)^{\beta}}{\theta^{\beta+1}} + \frac{\delta_2\beta(v_m)^{\beta}}{\theta^{\beta+1}} - \frac{n}{\theta}(\beta-1) - \frac{m}{\theta}(\beta-1) = 0 \text{ and}$$

$$\frac{\partial\log L}{\partial\beta} = \frac{n+m}{\beta} - \delta_1\left(\frac{u_n}{\theta}\right)^{\beta}\log\left(\frac{u_n}{\theta}\right) - \delta_2\left(\frac{v_m}{\theta}\right)^{\beta}\log\left(\frac{v_m}{\theta}\right)$$
$$+ \sum_{i=1}^{n}\log\left(\frac{u_i}{\theta}\right) + \sum_{j=1}^{m}\log\left(\frac{v_j}{\theta}\right) = 0$$

The MLEs of δ_1,δ_2 are obtained as functions of β and θ.

The maximum likelihood estimators are.

$$\hat{\delta}_1 = \frac{n}{\left(\frac{u_n}{\theta}\right)^{\beta}} \; and \; \hat{\delta}_2 = \frac{m}{\left(\frac{v_m}{\theta}\right)^{\beta}}$$

where $\hat{\beta}$ and $\hat{\theta}$ are the solution of the nonlinear equation of the form,

$$\left(\frac{n+m}{\theta}\right) + \frac{\delta_1 \beta (u_n)^{\beta}}{\theta^{\beta+1}} + \frac{\delta_2 \beta (v_m)^{\beta}}{\theta^{\beta+1}} - \frac{n}{\theta}(\beta - 1) - \frac{m}{\theta}(\beta - 1) = 0 \, and$$

$$\frac{n+m}{\beta} - \delta_1 \left(\frac{u_n}{\theta}\right)^{\beta} \log\left(\frac{u_n}{\theta}\right) - \delta_2 \left(\frac{v_m}{\theta}\right)^{\beta} \log\left(\frac{v_m}{\theta}\right) + \sum_{i=1}^{n} \log\left(\frac{u_i}{\theta}\right) + \sum_{j=1}^{m} \log\left(\frac{v_j}{\theta}\right) = 0.$$

Here, $\hat{\beta}$ and $\hat{\theta}$ can be obtained by using any iterative scheme, namely Newton–Raphson method.

The estimators $\hat{\beta}$ and $\hat{\theta}$ of β and θ are used to obtain MLEs of δ_1 and δ_2.

4 Likelihood Ratio (LR) Test for Equality of Scale Parameters

Suppose, $\underset{\sim}{u} = (u_1, \ldots, u_n)$ and $\underset{\sim}{v} = (v_1, \ldots, v_m)$ are the sets of upper records from distributions with pdfs f and g, respectively.

The likelihood function of $\underline{\mu} = (\delta_1, \delta_2, \theta, \beta)$ given $\underset{\sim}{u}$,$\underset{\sim}{v}$ is

$$L(\underline{\mu}) = L_1(\delta_1, \theta, \beta) \, L_2(\delta_2, \theta, \beta)$$

The LR test is constructed for the problem of testing of testing $H_0 : \delta_1 = \delta_2 = \delta$ against $H_1 : \delta_1 \neq \delta_2$.

For that, the whole parameter space is $\Theta = \{\underline{\mu} = (\delta_1, \delta_2, \theta, \beta) : \delta_1, \delta_2, \theta, \beta > 0\}$, and the parameter space under H_0 is $\Theta_0 = \{\underline{\mu} = (\delta_1, \delta_2, \theta, \beta) : \delta_1 = \delta_2 = \delta, \theta, \beta > 0\}$.

Then, the likelihood ratio statistic is given by

$$\lambda_{n,m} = \frac{\sup\limits_{\underline{\mu} \in \Theta_0} L(\underline{\mu}, \underline{x}, \underline{y})}{\sup\limits_{\underline{\mu} \in \Theta} L(\underline{\mu}, \underline{x}, \underline{y})}$$

If $\underline{\hat{\mu}_0} = (\hat{\delta}, \hat{\theta}, \hat{\beta})$ is MLE of $\underline{\theta}$ under Θ_0 and $\underline{\hat{\mu}} = (\hat{\delta}_1, \hat{\delta}_2, \hat{\theta}, \hat{\beta})$ is MLE of $\underline{\theta}$ under Θ, then

$$\lambda_{n,m} = \frac{L(\underline{\hat{\mu}_0}, \underline{x}, \underline{y})}{L(\underline{\hat{\mu}}, \underline{x}, \underline{y})}.$$

The test criterion is to reject H_0 if $\lambda_{n,m} > c$. The null distribution of $\lambda_{n,m}$ is tedious. However, the asymptotic distribution of $-2\log\lambda_{n,m}$ is chi-square distribution with one degree of freedom. Hence, the test criterion for large n, m is to reject H_0 if $-2\log\lambda_{n,m} > \chi_1^2(1-\alpha)$ where $\chi_1^2(1-\alpha)$ is determined such that $P_{H_0}(-2\log\lambda_{n,m} > \chi_1^2(1-\alpha)) = \alpha$.

The behaviour of LR test in terms empirical power is evaluated for $n, m =$ *25,30,40,50.* For different combinations of (δ_1, δ_2) with (θ, β) as (1,1), (0.4,0.7), (0.5,0.7) and (1,0.6) are considered for evaluating empirical powers.

Tables for empirical powers are not reported to limit the number of pages. However, the observations on the power of the test are given below:

1. The power of LRTs increases as the difference in the parameter values increases.
2. The power of the tests increases as the sample size increases.
3. Both the above observations reveal that the LR test constructed here possesses the desirable property of any reasonable test such as
   (a) The chance of rejection is more when the value of the parameter moves away from the parameters under null hypothesis.
   (b) The LR test is consistent.

5 Estimation of $R_{s,k}$ Using Maximum Likelihood and Bayesian Methods

Based on the decision of the LR test, the estimators of $R_{s,k}$ are obtained. If the LR test rejects H_0 $(\delta_1 \neq \delta_2)$, then using invariance property, the MLE of $R_{s,k}$ is given by,

$$\hat{R}_{s,k} = \sum_{i=s}^{k}\sum_{j=0}^{k-i}\binom{k}{i}\binom{k-i}{j}(-1)^j \frac{\hat{\delta}_2}{\left[\hat{\delta}_1(i+j)+\hat{\delta}_2\right]}$$

(Using the results of Sect. 3).

Next, the Bayes estimator of $R_{s,k}$ is obtained for s-out-of-k ($s \leq k$) system under stress-strength reliability assuming prior distribution for δ_1,δ_2,θ and β as below. Here, the

prior distribution for δ_1,δ_2,θ and β is assumed to be gamma with hyperparameter $((c_1, d_1), (c_2, d_2), (c_3, d_3)$ and $(c_4, d_4))$, respectively.

The joint pdf of $(\delta_1, \delta_2, \theta, \beta)$ is

$$g(\delta_1, \delta_2, \theta, \beta) = g(\delta_1)\, g(\delta_2)\, g(\theta) g(\beta),$$

where

$$g(\delta_1) = \frac{d_1^{c_1}}{\Gamma c_1}\delta_1^{c_1-1}e^{-\delta_1 d_1};\ \delta_1 > 0,\ c_1,\ d_1 > 0$$

$$g(\delta_2) = \frac{d_2^{c_2}}{\Gamma c_{2\,\delta}} \delta_2^{c_2-1} e^{-\delta_2 d_2};\ \delta_2 > 0,\ c_2,\ d_2 > 0$$
$$g(\theta) = \frac{d_3^{c_3}}{\Gamma c_3} \theta^{c_1-1} e^{-\theta d_1};\ \theta > 0,\ c_3,\ d_3 > 0$$
$$g(\beta) = \frac{d_4^{c_4}}{\Gamma c_4} \beta^{c_4-1} e^{-\beta d_4};\ \beta > 0,\ c_4,\ d_4 > 0$$

The corresponding joint posterior density function of δ_1,δ_2,θ and β is

$$\pi(\delta_1, \delta_2, \theta, \beta) = \frac{g(\delta_1, \delta_2, \theta, \beta) L_s(\delta_1, \delta_2, \theta, \beta)}{\int_0^\infty \int_0^\infty \int_0^\infty \int_0^\infty g(\delta_1, \delta_2, \theta, \beta) L_s(\delta_1, \delta_2, \theta, \beta)\, d\delta_1\, d\delta_2 d\theta\, d\beta}$$
$$= \frac{A}{\int_0^\infty \int_0^\infty \int_0^\infty \int_0^\infty A\ d\delta_1\, d\delta_2 d\theta\, d\beta}$$

where

$$A = \delta_1^{n+c_1-1} \delta_2^{m+c_2-1} \theta^{-(n+m)+c_3-1} \beta^{n+m+c_4-1} \prod_{i=1}^{n} \frac{u_i}{\theta}^{\beta-1} \prod_{j=1}^{m} \frac{v_j}{\theta}^{\beta-1} e^{-\delta_1\left(\left(\frac{u_n}{\theta}\right)^\beta + d_1\right)}$$
$$e^{-\delta_2\left(\left(\frac{v_m}{\theta}\right)^\beta + d_2\right)} e^{-d_3\theta} e^{-d_4\beta}$$

Here, it is to be noted that under squared error (SE) loss function, Bayes estimator of $R_{s,k}$ is given by

$$\hat{R}^B_{s,k} = \int_0^\infty \int_0^\infty \int_0^\infty \int_0^\infty R_{s,k}\, \pi(\delta_1, \delta_2, \theta, \beta)\, d\delta_1\, d\delta_2 d\theta\, d\beta$$

The evaluation of posterior mean is not tractable. However, approximate posterior mean is obtained using Lindley's approximation method Lindley (1980).

The Bayes estimator under squared error loss function is

$$\hat{R}_B = u + (u_1 a_1 + u_2 a_2 + a_4) + \frac{1}{2}\left[\begin{array}{l} A(u_1\sigma_{11}) + B(u_2\sigma_{22}) \\ +C(u_1\sigma_{31} + u_2\sigma_{32}) + D(u_1\sigma_{41} + u_2\sigma_{42}) \end{array}\right]$$

where,

$$a_1 = \rho_1\sigma_{11} + \rho_3\sigma_{13} + \rho_4\sigma_{14},\ a_2 = \rho_2\sigma_{22} + \rho_3\sigma_{23} + \rho_4\sigma_{24},\ a_4 = \frac{1}{2}(u_{11}\sigma_{11} + u_{22}\sigma_{22}).$$
$$A = \sigma_{11}L_{111} + \sigma_{33}L_{331} + 2\sigma_{34}L_{341} + \sigma_{44}L_{441},$$
$$B = 2\sigma_{23}L_{232} + \sigma_{22}L_{222} + \sigma_{33}L_{332} + 2\sigma_{34}L_{342} + 2\sigma_{24}L_{242} + \sigma_{44}L_{442} + 2\sigma_{14}L_{142},$$
$$C = 2\sigma_{13}L_{133} + 2\sigma_{23}L_{233} + \sigma_{33}L_{333} + 2\sigma_{34}L_{343} + 2\sigma_{24}L_{243} + \sigma_{44}L_{443} + 2\sigma_{14}L_{143},$$
$$D = 2\sigma_{13}L_{134} + 2\sigma_{23}L_{234} + \sigma_{33}L_{334} + 2\sigma_{34}L_{344} + 2\sigma_{24}L_{244} + \sigma_{44}L_{444} + 2\sigma_{14}L_{144},$$

Here,

$$\rho_1 = \frac{(c_1-1)}{\delta_1} - d_1,\ \rho_2 = \frac{(c_2-1)}{\delta_2} - d_2,\ \rho_3 = \frac{(c_3-1)}{\theta} - d_3 \text{ and } \rho_4 = \frac{(c_4-1)}{\beta} - d_4.$$

The values of L_{ij} can be obtained as follows for i, j = 1,2,3,4.

$$L_{11} = -\frac{n}{\delta_1^2},\ L_{13} = L_{31} = \beta\frac{u_n^\beta}{\theta^{\beta+1}},\ L_{14} = L_{41} = -\left(\frac{u_n}{\theta}\right)^\beta \ln\left(\frac{u_n}{\theta}\right),\ L_{22} = -\frac{m}{\delta_2^2}.$$

$$L_{23} = L_{32} = \beta\frac{v_m^\beta}{\theta^{\beta+1}},\ L_{24} = L_{42} = -\left(\frac{v_m}{\theta}\right)^\beta \ln\left(\frac{v_m}{\theta}\right),$$

$$L_{33} = -\frac{n+m}{\theta^2} - \delta_1\frac{\beta\, u_n^\beta(\beta+1)}{\theta^{\beta+2}} - \delta_2\frac{\beta\, v_m^\beta(\beta+1)}{\theta^{\beta+2}} - \frac{n(\beta-1)}{\theta^2} - \frac{m(\beta-1)}{\theta^2}$$

$$L_{34} = n\theta + m\theta + \delta_1\frac{\beta\, u_n^\beta(\beta[\ln u_n - \ln\theta]+1)}{\theta^{\beta+1}} + \delta_2\frac{\beta\, v_m^\beta(\beta[\ln v_m - \ln\theta]+1)}{\theta^{\beta+1}}$$

$$L_{44} = -\frac{(n+m)}{\beta^2} - \delta_1\left(\frac{u_n}{\theta}\right)^\beta\left[\ln\left(\frac{u_n}{\theta}\right)\right]^2 - \delta_2\left(\frac{v_m}{\theta}\right)^\beta\left[\ln\left(\frac{v_m}{\theta}\right)\right]^2,$$

and the values of L_{ijk} for i, j, k = 1, 2, 3,4 are

$$L_{111} = \frac{2n}{\delta_1^3},\ L_{222} = \frac{2m}{\delta_2^3},\ L_{331} = L_{133} = L_{313} = -\frac{\beta u_n^\beta(\beta+1)}{\theta^{\beta+2}},$$

$$L_{441} = -\left(\frac{u_n}{\theta}\right)^\beta\left[\log\left(\frac{u_n}{\theta}\right)\right]^2,\ L_{332} = L_{233} = L_{323} = -\frac{\beta v_m^\beta(\beta+1)}{\theta^{\beta+2}},$$

$$L_{342} = \frac{v_m^\beta(\beta(\ln v_m - \ln\theta)+1)}{\theta^{\beta+1}},\ L_{442} = -\left(\frac{v_m}{\theta}\right)^\beta\left[\log\left(\frac{v_m}{\theta}\right)\right]^2,$$

$$L_{333} = \frac{2(n+m)}{\theta^3} + \frac{\delta_1\beta(\beta+1)(\beta+2)}{\theta^{\beta+3}} + \frac{\delta_2\beta(\beta+1)(\beta+2)}{\theta^{\beta+3}}$$

$$L_{444} = \frac{2(n+m)}{\beta^3} - \delta_1\left(\frac{u_n}{\theta}\right)^\beta\left[\log\left(\frac{u_n}{\theta}\right)\right]^3 - \delta_2\left(\frac{v_m}{\theta}\right)^\beta\left[\log\left(\frac{v_m}{\theta}\right)\right]^3$$

$$L_{343} = n + m - \frac{\left[\beta\theta^\beta + (\beta+1)\theta^\beta(\beta\ln u_n - \beta\ln\theta + 1)\right]}{\theta^{2(\beta+1)}} - \frac{\left[\beta\theta^\beta + (\beta+1)\theta^\beta(\beta\ln v_m - \beta\ln\theta + 1)\right]}{\theta^{2(\beta+1)}}$$

$$L_{443} = -\delta_1\left[u_n^\beta\ln\left(\frac{u_n}{\theta}\right)\left(2\theta^{-\beta+1} - \ln\left(\frac{u_n}{\theta}\right)\beta\theta^{-(\beta+1)}\right)\right] - \delta_2\left[v_m^\beta\ln\left(\frac{v_m}{\theta}\right)\left(2\theta^{-\beta+1} - \ln\left(\frac{v_m}{\theta}\right)\beta\theta^{-(\beta+1)}\right)\right]$$

since $u = u(\delta_1, \delta_2, \theta, \beta) = R_{s,k}$,

$$u_1 = \sum_{i=s}^{k}\sum_{j=o}^{k-i}\binom{k}{i}\binom{k-i}{j}(-1)^j \frac{-\delta_2(i+j)}{(\delta_1(i+j)+\delta_2)^2},$$

$$u_2 = \sum_{i=s}^{k}\sum_{j=o}^{k-i}\binom{k}{i}\binom{k-i}{j}(-1)^j \frac{\delta_1(i+j)}{(\delta_1(i+j)+\delta_2)^2}.$$

$$u_{11} = \sum_{i=s}^{k}\sum_{j=o}^{k-i}\binom{k}{i}\binom{k-i}{j}(-1)^j \frac{-2\delta_2(i+j)}{(\delta_1(i+j)+\delta_2)^3},$$

$$u_{22} = \sum_{i=s}^{k}\sum_{j=o}^{k-i}\binom{k}{i}\binom{k-i}{j}(-1)^j \frac{-2\delta_1(i+j)}{(\delta_1(i+j)+\delta_2)^3}.$$

$$u_{12} = \sum_{i=s}^{k}\sum_{j=o}^{k-i}\binom{k}{i}\binom{k-i}{j}(-1)^j \frac{(i+j)(\delta_1(i+j)-\delta_2)}{(\delta_1(i+j)+\delta_2)^3}$$

6 Simulation Study

A simulation study is conducted to compute mean square error of the estimators by generating samples from new Weibull-Pareto distribution for comparison of estimates based on 100,000 random samples of size n and m. We have evaluated empirical mean square errors for different sets of values for $(\delta_1, \delta_2, \theta, \beta)$ for an s-out-of-k system. The results are shown for particular sets of values of $\delta_1, \delta_2, \theta, \beta$ in Tables 1, 2, 3 and 4.

For the present study, the values $(\delta_1, \delta_2, \theta, \beta)$ are (0.2, 0.1, 1, 1), (1.4, 1.2, 0.4, 0.7), (0.5, 0.6, 0.5, 0.7) and (0.2, 0.7, 1, 0.6). The corresponding true values of stress-strength reliability for s-out-of-k system with (s, k) = (1, 3) are 0.5428, 0.7068, 0.7970 and 0.9627 that for (s, k) = (2, 4) are 0.3904, 0.5516, 0.6565 and 0.9104.

The Bayesian estimators under squared error loss function using gamma prior are $c_1 = 2$, $c_2 = 7$, $c_3 = 4$, $d_1 = 3$, $d_2 = 5$, $d_3 = 2$ (Prior 1) and $c_1 = 1$, $c_2 = 1$, $c_3 = 1$, $d_1 = 1$, $d_2 = 1$, $d_3 = 1$ (Prior 2).

7 Real Data Analysis

In this section, we present a real data sets which was reported by (Saracoglu et al., 2012). For the data sets, to check the goodness of fit of new Weibull-Pareto model, the Kolmogorov–Smirnov (K-S) test is used. The estimated parameters with corresponding standard errors, K-S distances and the corresponding p-values are presented in table given below.

Table 1 MLE and Bayes estimates of reliability and corresponding mean square errors

$\delta_1 = 0.5,\ \delta_2 = 0.1,\ \theta = 1,\ \beta = 1.2$, Prior 1

(s, k)	$R_{s,k}$	n = m	$\hat{R}^M_{s,k}$	$\hat{R}^B_{s,k}$	MSE($\hat{R}^M_{s,k}$)	MSE($\hat{R}^B_{s,k}$)
(1, 3)	0.5428	10	0.5500	0.5497	0.0081	0.0073
		15	0.5490	0.5521	0.0077	0.0071
		20	0.5514	0.5471	0.0072	0.0059
		30	0.5489	0.5473	0.0065	0.0034
		35	0.5446	0.5456	0.0051	0.0025
		40	0.5423	0.5420	0.0022	0.0012
		50	0.5416	0.5412	0.0013	0.0008
(2, 4)	0.3904	10	0.3987	0.3976	0.0086	0.0075
		15	0.3990	0.3962	0.0072	0.0081
		20	0.3978	0.3951	0.0054	0.0043
		30	0.3964	0.3934	0.0037	0.0026
		35	0.3945	0.3929	0.0028	0.0018
		40	0.3921	0.3913	0.0016	0.0010
		50	0.3911	0.3900	0.0009	0.0005

Table 2 MLE and Bayes estimates of reliability and corresponding mean square errors

$\delta_1 = 1.4,\ \delta_2 = 1.2,\ \theta = 0.7,\ \beta = 0.4$, prior2

(s, k)	$R_{s,k}$	n = m	$\hat{R}^M_{s,k}$	$\hat{R}^B_{s,k}$	MSE($\hat{R}^M_{s,k}$)	MSE($\hat{R}^B_{s,k}$)
(1, 3)	0.7068	10	0.7122	0.7081	0.0088	0.0071
		15	0.7097	0.7088	0.0076	0.0057
		20	0.7084	0.7077	0.0064	0.0042
		30	0.7076	0.7065	0.0051	0.0025
		35	0.7071	0.7054	0.0032	0.0016
		40	0.7067	0.7061	0.0012	0.0010
		50	0.7059	0.7055	0.0007	0.0005
(2, 4)	0.5516	10	0.5612	0.5608	0.0077	0.0063
		15	0.5609	0.5597	0.0053	0.0042
		20	0.5591	0.5581	0.0042	0.0037
		30	0.5574	0.5557	0.0039	0.0024
		35	0.5537	0.5543	0.0025	0.0014
		40	0.5524	0.5512	0.0014	0.0011
		50	0.5510	0.5504	0.0009	0.0004

Table 3 MLE and Bayes estimates of reliability and corresponding mean square errors

$\delta_1 = 0.5,\ \delta_2 = 0.6,\ \theta = 0.7,\ \beta = 0.5$, Prior 1

(s, k)	$R_{s,k}$	n = m	$\hat{R}^M_{s,k}$	$\hat{R}^B_{s,k}$	MSE($\hat{R}^M_{s,k}$)	MSE($\hat{R}^B_{s,k}$)
(1, 3)	0.7970	10	0.7984	0.7997	0.0087	0.0081
		15	0.7981	0.7972	0.0074	0.0062
		20	0.7976	0.7971	0.0065	0.0058
		30	0.7971	0.7964	0.0047	0.0035
		35	0.7969	0.7961	0.0032	0.0024
		40	0.7965	0.7954	0.0017	0.0011
		50	0.7963	0.7957	0.0005	0.0003
(2, 4)	0.6565	10	0.6587	0.6577	0.0079	0.0064
		15	0.6584	0.6572	0.0067	0.0058
		20	0.6571	0.6584	0.0066	0.0047
		30	0.6569	0.6564	0.0057	0.0039
		35	0.6564	0.6559	0.0041	0.0024
		40	0.6557	0.6543	0.0038	0.0012
		50	0.6551	0.6544	0.0015	0.0009

Table 4 MLE and Bayes estimates of reliability and corresponding mean square errors

$\delta_1 = 0.2,\ \delta_2 = 0.7,\ \theta = 0.6,\ \beta = 1$, Prior 2

(s, k)	$R_{s,k}$	n = m	$\hat{R}^M_{s,k}$	$\hat{R}^B_{s,k}$	MSE($\hat{R}^M_{s,k}$)	MSE($\hat{R}^B_{s,k}$)
(1, 3)	0.9627	10	0.9681	0.9696	0.0086	0.0089
		15	0.9675	0.9671	0.0078	0.0071
		20	0.9668	0.9664	0.0065	0.0047
		30	0.9657	0.9645	0.0057	0.0034
		35	0.9644	0.9637	0.0051	0.0027
		40	0.9634	0.9624	0.0034	0.0016
		50	0.9631	0.9614	0.0019	0.0004
(2, 4)	0.9104	10	0.9194	0.9188	0.0085	0.0077
		15	0.9182	0.9176	0.0071	0.0066
		20	0.9174	0.9163	0.0062	0.0045
		30	0.9166	0.9154	0.0058	0.0037
		35	0.9148	0.9129	0.0043	0.0025
		40	0.9127	0.9117	0.0026	0.0014
		50	0.9111	0.9109	0.0016	0.0007

Data set	$\hat{\delta}$	$\hat{\beta}$	$\hat{\theta}$	K-S distances	P-value
X	0.06808 (0.1613)	1.56534 (0.1038)	2.75988 (4.1515)	$1.8041e^{-16}$	0.9599
Y	14.9301 (1.0026)	1.36069 (0.0461)	22.794 (1.5281)	$6.9389e^{-18}$	0.9634

For the data considered here, upper record values are observed as follows.
u: 21.8, 70.7, 138.6, 151.9.
v: 71.46, 419.02, 585.57, 688.16, 756.70, 765.14.
From the result, it can be seen that new Weibull-Pareto distribution fits better for the data sets. When the estimated parameter values are $\delta_1 = 1.5654$, $\delta_2 = 1.3610$, $\theta = 3.1769$, $\beta = 0.1526$, the maximum likelihood and Bayes estimate are obtained as $\hat{R}^M_{(1,3)} = 0.7656$, $\hat{R}^B = 0.7741$ under Prior 1 and $\hat{R}^M_{(2,4)} = 0.6115$, $\hat{R}^B = 0.6224$ under Prior 2.

8 Summary and Conclusions

The estimation of s-out-of-k ($s \leq k$) system reliability under stress-strength setup is considered in this chapter when the underlying distributions for stress and strength variables are new Weibull-Pareto. The maximum likelihood estimators of the parameters of NWP distribution based on record values are derived. The likelihood ratio test is constructed to check whether the strength and stress variables are identically distributed or not with reference to scale parameters. The estimation of system reliability depends on the decision taken using LR test. The system reliability using maximum likelihood and Bayes methods is obtained based on the decision taken using LR test. Bayes estimator is obtained using Lindley's approximation under squared error loss function with conjugate priors. Simulation study results indicate that MSEs of both MLE and Bayes estimators of system reliability are inversely proportional to the sample sizes. Bayes estimator with the gamma priors performs better than maximum likelihood estimator in terms of MSE. A real data analysis is conducted to illustrate the estimation procedures studied in this chapter.

Acknowledgements Authors gratefully thank the editor and reviewers for their comments which lead to the improvement of the contents of the chapter.

References

Bhattacharyya, G. K., & Johnson, R. A. (1974). Estimation of reliability in multicomponent stress-strength model. *Journal of American Statistical Association, 69*, 966–970.

Birnbaum, Z. W. (1956). On a use of Mann-Whitney statistics. *Proceeding THird Berkley Symposium on Mathematical Statistics and Probability, 1*, 13–17.

Birnbaum, Z. W., & McCarty, B. C. (1958). A distribution-free upper confidence bounds for Pr(Y < X) based on independent samples of X and Y. *The Annals of Mathematical Statistics, 29*(2), 558–562.

Block, H. W., & Basu, A. P. (1974). A continuous bivariate exponential extension. *Journal of American Statistical Association, 69*, 1031–1037.

Chandler, N. K. (1952). The distribution and frequency of record values. *Journal of Royal Statistical Society, 14*, 220–228.

Chruch, J. D., & Harris, B. (1970). The estimation of Reliability from Stress-Strength relationships. *Technometrics,* 1249–54.

Freund, J. E. (1961). A bivariate extension of the exponential distribution. *Journal of the American Statistical Association, 56*, 971–977.

Govindarajulu, Z. (1967). Two sided confidence interval for P(Y < X) based on normal samples of X and Y. *Sankhya, B, 29*, 35–40.

Hanagal, D. D. (1995). Testing reliability in a bivariate exponential stress-strength model. *Journal of the Indian Statistical Association, 33*, 41–45.

Hanagal, D. D. (1997a). Estimation of Reliability when stress is censored at strength. *Communications in Statistics-Theory and Methods, 26*(4), 911–919.

Hanagal, D. D. (1997b). Note on estimation of reliability under bivariate pareto stress- strength model. *Statistical Papers, 38*, 453–459.

Huizhen, G., & Krishnamoorthy, K. (2004). New approximate inferential methods for the reliability parameter in a stress-strength model: The normal case. *Communication Statistics Theory and Methods, 33*, 1715–1731.

Joshi, S., & Pandit, P. V. (2018). Estimation of stress-strength reliability in s-out-of-k system for a two-parameter inverse Chen distribution. *Journal of Computer and Mathematical Sciences, 9*(12A), 1898–1906.

Kelley, G. D., Kelley, J. A., & Schucany, W. R. (1976). Efficient estimation of P[Y < X] in the exponential case. *Technometrics, 18*, 359–360.

Kuo, W., & Zuo, M. J. (2003). *Optimal reliability modelling, principles and applications.* Wiley.

Lindley, D. V. (1980). Approximate Bayes method. *Trabajos De Estadistica, 3*, 281–288.

Marshall, A. W., & Olkin, I. (1967). A multivariate exponential distribution. *Journal of the American Statistical Association, 62*, 30–44.

Nadar, M., & Kizilaslan, F. (2016). Estimation of reliability in multicomponent stress-strength model based on bivariate Kumarswamy distribution. *Statistical Papers,* 1–34.

Nadar, M., & Kizilaslan, F. (2015). Classical and Bayesian estimation of Reliabilityin Multicomponent Stress-Strength Model based on Weibull distribution. *Revista Clombiana De Estadistica, 38*(2), 467–484.

Nasiru, S., & Luguterah, A. (2015). The new Weibull Pareto distribution. *Pakistan Journal of Statistics and Operational Research, 11*(1), 103–114.

Pandit, P. V., & Joshi, S. (2018). Reliability estimation in multicomponent stress-strength model based on generalized Pareto distribution. *American Journal of Applied Mathematics and Statistics, 6*(5), 210–217.

Pandit, P. V., & Joshi, S. (2019a). Estimation of reliability for a two-parameter inverse Chen distribution. *Research and Reviews: Journal of Statistics, 8*(1), 17–25.

Pandit, P. V., & Joshi, S. (2019b). Estimation of multicomponent system reliability for a bivariate generalized Rayleigh distribution. *International Journal of Computational and Theoretical Statistics, 6*(1), 51–63.

Pandit, P. V., & Kantu, K. J. (2013). System reliability estimation in multicomponent exponential stress-strength models. *International Journal of Reliability and Applications, 14*(2), 97–105.

Rao, G. S. (2012a). Estimation of reliability in multicomponent stress-strength model based on Rayleigh distribution. *Probability Statistics Forum, 5*, 150–161.

Rao, G. S. (2012b). Estimation of reliability in multicomponent stress-strength model based on generalized exponential distribution. *Colombian Journal of Statistics, 35*(1), 67–76.

Rao, G. S. (2014). Estimation of reliability in multicomponent stress-strength model based on generalized Rayleigh distribution. *Journal of Modern Applied Statistical Methods, 13*(1), 367–379.

Rao, G. S., & Kantam, R. R. L. (2010). Estimation of reliability in multicomponent stress-strength model: Log-logistic distribution. *Electronic Journal of Applied Statistical Analysis, 3*(2), 75–84.

Rao, G. S., Kantam, R. R. L., Rosaiah, K., & Reddy, J. P. (2013). Estimation of reliability in multicomponent stress-strength model based on inverse Rayleigh distribution. *Journal of Statistics and Applied Probability, 3*, 261–267.

Rao, G. S., Aslam, M., & Kundu, D. (2015). Burr Type XII distribution parametric estimation and estimation of reliability in multicomponent stress-strength. *Communications in Statistics-Theory and Methods, 44*(23), 4953–4961.

Saracoglu, B., Kinaci, I., & Kundu, D. (2012). On estimation of R = P(Y < X) for exponential distribution under progressive type-II censoring. *Journal of Statistical Computation and Simulation, 82*(5), 729–744.

Tahir, M. H., Gauss, M., Ayman, A., Mansoor, M., & Zabair, M. (2016). A new Weibull Pareto distribution: Properties and Applications. *Communications in Statistics- Simulation and Computation, 45*(10), 3548–3567.

Tarvirdizade, B., & Ahmadpour, M. (2016). Estimation of the stress- strength reliability for the two-parameter bathtub-shaped lifetime distribution based on upper record values. *Statistical Methodology, 31*, 58–72.

Tong, H. (1974). A note on the estimation of P[Y < X] in exponential case. *Technometrics, 16*, 625.

Tong, H. (1977). On the estimation of P[Y < X]of exponential families. *IEEE Transactions on Reliability, 26*, 54–56.

Wang, L., & Shi, Y. (2013). Reliability analysis of a class of exponential distribution under record values. *Journal of Computational and Applied Mathematics, 239*, 367–379.

Record Values and Associated Inference on Muth Distribution

V. S. Vaidyanathan and Hassan Bakouch

Abstract In this chapter, specific distributional results associated with upper record values from Muth distribution, namely survival function, joint and conditional densities, and expression for the kth moment, are derived. Parameter estimation based on upper record values is carried out using the method of moments, maximum likelihood, and Bayesian approach. Further, prediction of future upper record values is made using both frequentist and Bayesian methods. Numerical illustration is provided through simulated and real-life data sets.

Keywords Likelihood function · Metropolis-Hastings algorithm · Monte Carlo integration · Moments · Muth distribution · Record values

1 Introduction

Record values have been used in the literature to characterize various distributions and to make inference on the parameters involved. The primary advantage of using record values is that modelling and inference can be done using only a handful of 'record' observations rather than the entire sample observations. This may save time and cost associated with measuring and recording all the observations. Further, record values and associated statistics are of great interest in many real-life situations involving data relating to sports, economics, weather, and life testing. Let $X_1, X_2, ...$ be a sequence of independent and identically distributed (iid) random variables with cumulative distribution function (cdf) $F(x; \theta)$ and probability density function (pdf) $f(x; \theta)$. The jth upper (lower) record value denoted by R_j of this sequence is defined as

$$R_j = Y_m \quad if \quad Y_m > (<) R_{j-1}, j = 2, 3, ... \tag{1.1}$$

V. S. Vaidyanathan (✉)
Department of Statistics, Pondicherry University, Puducherry, India

H. Bakouch
Department of Mathematics, Faculty of Science, Tanta University, Tanta, Egypt

D. D. Hanagal et al. (eds.), *Applied Statistical Methods*, Springer Proceedings in Mathematics & Statistics 380, https://doi.org/10.1007/978-981-16-7932-2_17

where Y_m = max (min)$\{X_1, X_2, ..., X_m\}$, $m > 1$. R_1, the first record value, is usually taken as X_1. Thus, $R_1, R_2, ...$ denote the sequence of record values based on $X_1, X_2, ...$ A comprehensive study on record values together with various results and applications can be found in Arnold et al. (1998). Statistical inference on life-time distributions using record values has been attempted by various researchers in the last decade. Some references include Asgharzadeh et al. (2016), Kumar (2015), MirMostaface et al. (2016), Teimouri and Gupta (2012). Among the many life-time distributions available in literature, Muth distribution introduced by Muth (1977) has strict positive memory, and its mean residual life function corresponds to that of exponential distribution. It can be used to model positively skewed data. The cdf and pdf of a random variable X having Muth distribution are given by

$$F(x;\alpha) = 1 - exp(\alpha x - \frac{1}{\alpha}(e^{\alpha x} - 1)), x > 0, \tag{1.2}$$

$$f(x;\alpha) = (e^{\alpha x} - \alpha) exp(\alpha x - \frac{1}{\alpha}(e^{\alpha x} - 1)), x > 0, \tag{1.3}$$

where $\alpha \in (0, 1]$ denotes the shape parameter. This distribution has been overlooked in the literature until recently when Pedro Jodrá et al. (2015) studied its various mathematical and statistical properties. It is a unimodal distribution with limiting distribution as standard exponential when the parameter $\alpha \downarrow 0$. One interesting property of this distribution is that it has less tail probability when compared with common unimodal distributions like gamma, Weibull. Moreover, we provide the next new remarks on this distribution.

Remark 1 Let X have the cdf defined by (1.2) and consider the transformation $Y = e^{\alpha X} - 1$. Then, the distribution of Y has the cdf

$$F(y;\alpha) = 1 - (1 + y)\, e^{-\frac{1}{\alpha}y}, \;\; y > 0, \alpha \in (0, 1].$$

This form of cdf is not available in the literature. However, Dara (2012) has proposed a similar cdf of the form

$$F(y;\lambda) = 1 - (1 + \lambda y)\, e^{-\lambda y}, \;\; y > 0, \lambda > 0.$$

Remark 2 The pdf of the Muth distribution given in (1.3) can be represented as a weighted version of the Gompertz distribution as follows. The classical Gompertz distribution has the pdf

$$g(x;\lambda,\sigma) = \lambda\sigma e^{\lambda x - \sigma(e^{\lambda x} - 1)}, \;\; x > 0, \lambda, \sigma > 0.$$

Reparametrizing λ and σ to α and $\frac{1}{\alpha}$, respectively, we get

$$g(x;\alpha) = e^{\alpha x - \frac{1}{\alpha}(e^{\alpha x} - 1)}.$$

Applying the weight function $w(x) = e^{\alpha x} - \alpha, x > 0, \alpha \in (0, 1]$, to $g(x; \alpha)$, we get the pdf of the Muth distribution given by (1.3), noting that $E[w(x)] = 1$.

Considering the various properties of Muth distribution mentioned above and motivated by the fact that not much work has been done on this distribution, in the present article, certain distributional results based on record values are derived. In addition, estimation of parameter using record values and prediction of future records are addressed. Apart from the introduction section, the chapter contains six sections organized as follows. In Sect. 2, expression for survival function, joint and conditional densities, and kth moment based on upper record values from Muth distribution is derived. Parameter estimation based on upper record values using moments, likelihood and Bayesian approach is discussed in Sect. 3. Section 4 contains numerical illustration through simulated data sets and Sect. 5 presents results from real-life application. Section 6 details prediction of future upper records using frequentist and Bayesian approaches. Concluding remarks are given in Sect. 7.

2 Survival Function, Joint and Conditional Densities, and Moments of Upper Records from Muth Distribution

Let $R_1, R_2, ...$ denote the sequence of upper record values from a continuous distribution with pdf $f(x; \theta)$ and cdf $F(x; \theta)$. Then the following results are established (see Arnold et al. (1998)).

1. The survival function of nth record is given by

$$S(r_n) = P(R_n > r_n) = [1 - F(r_n)] \sum_{k=0}^{n} \frac{[-log(1 - F(r_n))]^k}{k!}. \tag{2.1}$$

The corresponding cdf is

$$P(R_n \leq r_n) = \int_0^{[-log(1-F(r_n))]} \frac{w^n e^{-w}}{n!} dw. \tag{2.2}$$

If F is absolutely continuous, then the pdf of R_n is given by

$$f_{R_n}(r_n) = \frac{f(r_n)[-log(1 - F(r_n))]^n}{n!}. \tag{2.3}$$

2. The joint pdf of the set of records $R_1, R_2, ..., R_n$ is given by

$$f_{R_1,R_2,...,R_n}(r_1, r_2, ..., r_n) = f(r_n) \prod_{i=1}^{n-1} h(r_i), \tag{2.4}$$

where $h(r) = \frac{f(r)}{1-F(r)}$ is the failure rate function.
3. The joint pdf of the records R_m, R_n is given by

$$f_{R_m,R_n}(r_m, r_n) = \frac{[-log(1-F(r_m))]^m}{m!} \frac{[-log(\frac{1-F(r_n)}{1-F(r_m)})]^{n-m-1}}{(n-m-1)!} \frac{f(r_m)f(r_n)}{[1-F(r_m)]}, r_m < r_n. \quad (2.5)$$

4. The conditional density of (n-1) records given n^{th} record is given by

$$f(r_1, r_2, ..., r_{n-1}|r_n) = n! \frac{\prod_{i=1}^{n-1} h(r_i)}{[-log(1-F(r_n))]^n}. \quad (2.6)$$

5. The kth moment of R_n is given by

$$E(R_n^k) = \int_0^{\infty} r_n^k f_{R_n}(r_n) dr_n. \quad (2.7)$$

Using Equations (2.1) to (2.7) and the cdf and pdf given in (1.2) and (1.3), the corresponding expressions for Muth distribution based on upper record values are obtained as below.
1. The pdf of R_n is

$$f_{R_n}(r_n) = (e^{\alpha r_n} - \alpha) e^{(\alpha r_n - \frac{(e^{\alpha r_n}-1)}{\alpha})} \frac{[\frac{(e^{\alpha r_n}-1)}{\alpha} - \alpha r_n]^n}{n!}, 0 \leq r_n < \infty,$$

and the survival function is

$$S(r_n) = e^{(\alpha r_n - [\frac{e^{\alpha r_n}-1}{\alpha}])} \sum_{k=0}^{n} \frac{[(\frac{e^{\alpha r_n}-1}{\alpha}) - \alpha r_n]^k}{k!}, 0 \leq r_n < \infty.$$

2. The joint pdf of $R_1, R_2, ..., R_n$ is

$$f_{R_1,R_2,...,R_n}(r_1, r_2, ..., r_n) = (e^{\alpha r_n} - \alpha) e^{(\alpha r_n - \frac{(e^{\alpha r_n}-1)}{\alpha})} \prod_{i=1}^{n-1} (e^{\alpha r_i} - \alpha).$$

3. The joint pdf of the records R_m, R_n is given by

$$f_{R_m,R_n}(r_m, r_n) = T_1 T_2 T_3,$$

where

$$T_1 = [\frac{(e^{\alpha r_m} - 1)}{\alpha} - \alpha r_m]^m [\frac{(e^{\alpha r_n} - e^{\alpha r_m})}{\alpha} - \alpha(r_n - r_m)]^{n-m-1},$$
$$T_2 = e^{(\alpha r_n - \alpha)} e^{(\alpha r_n - \frac{e^{\alpha r_n}-1}{\alpha})} [\frac{(e^{\alpha r_n} - 1)}{\alpha} - \alpha]^n,$$

$$T_3 = e^{(\alpha r_m - \alpha)}[\frac{(e^{\alpha r_m} - 1)}{\alpha} - \alpha]^m \frac{1}{n!(m!)^2(n-m-1)!}$$

4. The conditional density of (n-1) records given n^{th} record is

$$f(r_1, r_2, ..., r_{n-1}|r_n) = \frac{n!}{[\frac{(e^{\alpha r_n}-1)}{\alpha} - \alpha r_n]^n} \prod_{i=1}^{n-1}(e^{\alpha r_i} - \alpha).$$

5. To obtain the expression for the kth moment based of R_n, we proceed as follows. Consider

$$E(R_n^k) = \int_0^{\infty} r_n^k (e^{\alpha r_n} - \alpha) exp(\alpha r_n - [\frac{e^{\alpha r_n} - 1}{\alpha}]) \frac{[(\frac{e^{\alpha r_n}-1}{\alpha}) - \alpha r_n]^n}{n!} dr_n \quad (2.8)$$

Let

$$I = \int_0^{\infty} r_n^k (e^{\alpha r_n} - \alpha) exp(\alpha r_n - [\frac{e^{\alpha r_n} - 1}{\alpha}]) \frac{[(\frac{e^{\alpha r_n}-1}{\alpha}) - \alpha r_n]^n}{n!} dr_n.$$

Put $z = \frac{e^{\alpha r_n}-1}{\alpha} \Rightarrow r_n = \frac{1}{\alpha} ln(1+\alpha z) \Rightarrow dr_n = \frac{1}{1+\alpha z} dz$ so

$$I = \alpha^{-k} \int_0^{\infty} (ln(1+\alpha z))^k (1 + \alpha z - \alpha) e^{ln(1+\alpha z) - z} \frac{[(z - ln(1+\alpha z)]^n}{n!} \frac{1}{1+\alpha z} dz$$

$$= \frac{\alpha^{-k}}{n!} \int_0^{\infty} (ln(1+\alpha z))^k (1 - \alpha + \alpha z) e^{-z} z^n [(1 - \frac{ln(1+\alpha z)}{z}]^n dz$$

$$= \frac{\alpha^{-k}}{n!} \sum_{m=0}^{n} \binom{n}{m} (-1)^m \int_0^{\infty} (ln(1+\alpha z))^{k+m} z^{n-m} (1 - \alpha + \alpha z) e^{-z} dz.$$

Put $1 + \alpha z = g \Rightarrow z = \frac{g-1}{\alpha} \Rightarrow dz = \frac{dg}{\alpha}$ so

$$I = \frac{\alpha^{-k-1}}{n!} \sum_{m=0}^{n} \binom{n}{m} (-1)^m \int_1^{\infty} ln(g)^{k+m} (\frac{g-1}{\alpha})^{n-m} (g - \alpha) e^{-\frac{g}{\alpha} + \frac{1}{\alpha}} dg$$

$$= \frac{e^{\frac{1}{\alpha}}}{n!} \sum_{m=0}^{n} \binom{n}{m} \frac{(-1)^m}{\alpha^{n+k+1-m}} \int_1^{\infty} ln(g)^{k+m} (g-1)^{n-m} (g - \alpha) e^{-\frac{g}{\alpha}} dg$$

$$= \frac{e^{\frac{1}{\alpha}}}{n!} \sum_{m=0}^{n} \sum_{h=0}^{n-m} \binom{n}{m} \binom{n-m}{h} \frac{(-1)^{m+n-m-h}}{\alpha^{n+k+1-m}} \int_1^{\infty} ln(g)^{k+m} g^h (g - \alpha) e^{-\frac{g}{\alpha}} dg.$$

By using the integral as given in Page number 576 (4.358) of Gradshteyn and Ryzhik (2007), we have

$$\int_1^{\infty} x^{v-1}e^{-ux}ln(x)^n dx = \frac{\partial^n u^{-v}\Gamma(v,u)}{\partial v^n}, n = 0, 1, 2, 3, \ldots; v > 0, u > 0.$$

Now, replacing n by $k+m$, u by $\frac{1}{\alpha}$ and v by $h+2, h+1$, respectively, we get,

$$I = \frac{e^{\frac{1}{\alpha}}}{n!}\sum_{m=0}^{n}\sum_{h=0}^{n-m}\binom{n}{m}\binom{n-m}{h}\frac{(-1)^{m+n-m-h}}{\alpha^{n+k+1-m}}\{\frac{\partial^{k+m}\alpha^{h+2}\Gamma(h+2,\frac{1}{\alpha})}{\partial(h+2)^{k+m}} - \alpha(\frac{\partial^{k+m}\alpha^{h+1}\Gamma(h+1,\frac{1}{\alpha})}{\partial(h+1)^{k+m}})\}$$

$$= \frac{e^{\frac{1}{\alpha}}}{n!\alpha^{n+k}}\sum_{m=0}^{n}\sum_{h=0}^{n-m}\binom{n}{m}\binom{n-m}{h}(-1)^{n-h}\{\frac{\partial^{k+m}\alpha^{h+m+1}\Gamma(h+2,\frac{1}{\alpha})}{\partial(h+2)^{k+m}} - (\frac{\partial^{k+m}\alpha^{h+m+1}\Gamma(h+1,\frac{1}{\alpha})}{\partial(h+1)^{k+m}})\} \quad (2.9)$$

Using (2.9), moments of R_n can be determined. However, obtaining numerical values for moments using (2.9) is difficult due to the presence of partial derivatives. Towards this, we use Monte Carlo integration. Note that, (2.8) can be expressed as

$$E(R_n^k) = \int_0^{\infty} g(r_n)f(r_n;\alpha)dr_n,$$

where $g(r_n) = [r_n{}^k]\frac{[(\frac{e^{\alpha r_n}-1}{\alpha})-\alpha r_n]^n}{n!}$ and $f(r_n;\alpha)$ is the density function of Muth distribution evaluated at $x = r_n$. Using Monte Carlo integration, the above integral can be approximated by $\bar{g_n} = \frac{\sum_{j=1}^{m} g(r_{nj})}{m}$, where r_{nj} is the simulated record values from $f(r_n;\alpha)$ and m denotes the number of samples generated. Method of simulating observations from Muth distribution is explained in Sect. 4. The variance of $\bar{g_n}$ can be estimated from the sample values using $V(\bar{g_n}) = \frac{\sum_{j=1}^{m}[g(r_{nj})-\bar{g_n}]^2}{m^2}$ (see Robert and Casella (2004)). Thus, for large m, $\frac{\bar{g_n}-E(R_n^k)}{\sqrt{(V(\bar{g_n}))}}$ is approximately distributed as $N(0,1)$ variable, and hence, confidence intervals (CI) for moments of R_n can be constructed. Numerical values of first order moments, their estimated variances, and width of 95% CI's are displayed in Table 1 for first five records. Values inside parenthesis denote estimated variance and those inside square bracket denote width of CI. Scientific notation of decimal number has been used wherever necessary. For example 1.33E-03 denotes $1.33 * 10^{-3}$ and 2.37E+02 denotes $2.37 * 10^2$.

Table 1 Values of $E(R_n)$, estimated variance, and width of 95% CI for different records and α

Record	α					
	0.15	0.30	0.45	0.60	0.75	0.90
R_1	1.88 (1.33E-03) [0.143]	1.7 (9.38E-04) [0.120]	1.66 (7.20E-04) [0.106]	1.56 (5.91E-04) [0.095]	1.5 (4.90E-04) [0.087]	1.44 (4.30E-04) [0.082]
R_2	9.83 (0.045) [0.836]	8.53 (0.038) [0.769]	7.65 (0.025) [0.628]	6.77 (0.017) [0.521]	6.34 (0.015) [0.482]	5.63 (0.011) [0.420]
R_3	44.14 (1.880) [5.379]	36 (0.875) [3.668]	31.51 (0.643) [3.144]	26.36 (0.389) [2.445]	24.03 (0.313) [2.194]	21.39 (0.265) [2.021]
R_4	158.79 (23.632) [19.057]	126.37 (14.323) [14.836]	105.71 (9.375) [12.003]	94.33 (6.700) [10.148]	85.23 (5.344) [9.062]	75.14 (3.981) [7.821]
R_5	514.76 (2.37E+02) [60.374]	387.38 (1.47E+02) [47.506]	323.41 (9.30E+01) [37.797]	291.75 (7.61E+01) [34.206]	267.91 (7.12E+01) [33.196]	221.7 (4.46E+01) [26.181]

3 Parameter Estimation Based on Upper Records Using Moment, Likelihood, and Bayesian Approaches

3.1 Moment Estimation of α

In this section, we find the moment estimate of α based on upper records through inverse transformation (Asgharzadeh et al. (2016)). Defining $V_i = -log[1 - F(x_i)]$, we have $V_i^* = -log[1 - F(r_i)]$, where V_i^* is the ith upper record arising from a sequence of iid standard exponential random variables V_i. Set $Y_1 = V_1^*$, $Y_i = V_i^* - V_{i-1}^*$, $i = 2, 3, ..., n$. From Arnold et al. (1998), we have Y_i as iid random variables from a standard exponential distribution. Note that, $S = \sum_{i=1}^n V_i^*$ can be considered as a weighted sum of Y_i as $S = \sum_{i=1}^n (n - i + 1)Y_i$. It is easy to verify that $E(S) = \frac{n(n+1)}{2}$ and $Var(S) = \frac{n(n+1)(2n+1)}{6}$. Using Markov inequality, we get $\frac{S}{\frac{n(n+1)}{2}} \to 1$ in probability as $n \to \infty$, i.e. $S \to \frac{n(n+1)}{2}$ in probability as $n \to \infty$, where $S = \sum_{i=1}^n V_i^* = \sum_{i=1}^n [\frac{e^{\alpha R_i} - 1}{\alpha} - \alpha R_i]$. Thus, a moment estimate of α can be obtained by solving the nonlinear equation

$$\sum_{i=1}^{n} [\frac{e^{\alpha R_i} - 1}{\alpha} - \alpha R_i] = \frac{n(n+1)}{2}.$$

A $100(1-p)\%$ confidence interval for α can be constructed using

$$P[s_{(\frac{p}{2})} < S < s_{(1-\frac{p}{2})}] = 1 - p, p \in (0, 1),$$

where $s_{(\frac{p}{2})}$ and $s_{(1-\frac{p}{2})}$ denote the lower and upper $(\frac{p}{2})^{th}$ percentage points of S. These points can be computed using the pdf of S given by

$$f_S(t) = \sum_{i=1}^{n} C_{i,n}\lambda_i e^{\lambda_i t}, t > 0,$$

where $C_{i,n} = \prod_{j=1, j\neq i}^{n} \frac{\lambda_j}{\lambda_j - \lambda_i}$. (See Asgharzadeh et al. (2016)).
Thus, a $100(1-p)\%$ confidence interval for α is given by (L,U) where L and U are the solutions of the equations

$$\sum_{i=1}^{n}[\frac{e^{\alpha R_i}-1}{\alpha} - \alpha R_i] = s_{(\frac{p}{2})}$$

and

$$\sum_{i=1}^{n}[\frac{e^{\alpha R_i}-1}{\alpha} - \alpha R_i] = s_{(1-\frac{p}{2})},$$

respectively.

3.2 *Maximum Likelihood Estimation*

From (1.3), the likelihood function (L) based on the upper records is

$$L(\alpha|r_1, r_2, ..., r_n) = (e^{\alpha r_n} - \alpha)e^{(\alpha r_n - \frac{(e^{\alpha r_n}-1)}{\alpha})} \prod_{i=1}^{n-1}(e^{\alpha r_i} - \alpha). \tag{3.1}$$

The corresponding log likelihood function is

$$lnL = \sum_{i=1}^{n} ln(e^{\alpha r_i} - \alpha) + \alpha r_n - \frac{(e^{\alpha r_n}-1)}{\alpha}.$$

The ML estimator $\hat{\alpha}$ of α is the solution of the log likelihood equation $\frac{dlnL}{d\alpha} = 0$,where

$$\frac{dlnL}{d\alpha} = \sum_{i=1}^{n} \frac{(r_i e^{\alpha r_i} - 1)}{e^{\alpha r_i} - \alpha} + r_n - \frac{r_n e^{\alpha r_n}}{\alpha} + \frac{(e^{\alpha r_n} - 1)}{\alpha^2}.$$

Since the log likelihood equation is nonlinear in α, one has to use numerical methods or gradient-based search algorithms to obtain the estimate $\hat{\alpha}$ of α. An estimate of the variance of $\hat{\alpha}$ can then be obtained from the sample Fisher information measure using

$$\frac{d^2 lnL}{d\alpha^2} = \sum_{i=1}^{n} [\frac{(e^{\alpha r_i} - \alpha)(r_i^2 e^{\alpha r_i}) - (r_i e^{\alpha r_i} - 1)^2}{(e^{\alpha r_i} - \alpha)^2}] - \frac{r_n^2 e^{\alpha r_n}}{\alpha} + \frac{r_n e^{\alpha r_n}}{\alpha^2} - \frac{(\alpha^2 r_n e^{\alpha r_n} - (e^{\alpha r_n} - 1)2\alpha)}{\alpha^4}$$

evaluated at $\hat{\alpha}$.

3.3 Bayesian Estimation

In this section, we discuss the method of estimating α using Bayesian approach. Since α takes values between 0 and 1, a natural choice for the prior distribution of α is beta distribution of first kind with density function of the form

$$\pi(\alpha) = \frac{\alpha^{a-1}(1-\alpha)^{b-1}}{B(a,b)}, 0 < \alpha \leq 1; a, b > 0, \tag{3.2}$$

where a and b are the hyperparameters and $B(a,b) = \frac{\Gamma(a)\Gamma(b)}{\Gamma(a+b)}$.

Using (3.1) and (3.2), the posterior density of α is obtained as

$$\pi(\alpha|r_1, r_2, ..., r_n) \propto L(\alpha|r_1, r_2, ..., r_n)\pi(\alpha) = (e^{\alpha r_n} - \alpha)e^{(\alpha r_n - \frac{(e^{\alpha r_n}-1)}{\alpha})} \prod_{i=1}^{n-1} (e^{\alpha r_i} - \alpha) \frac{\alpha^{a-1}(1-\alpha)^{b-1}}{B(a,b)}. \tag{3.3}$$

The proportionality constant is defined as $m = \int_0^1 L(\alpha|r_1, r_2, ..., r_n)\pi(\alpha)d\alpha$. Bayes estimators in general are obtained under squared error loss (SEL) function or linear-exponential (LINEX) loss function. Let δ be a decision rule for estimating the parameter α. Then, SEL function associated with δ is defined as $L_{SEL}(\delta, \alpha) = (\delta - \alpha)^2$. The corresponding LINEX loss function is given by $L_{LI}(\delta, \alpha) \propto e^{c(\delta-\alpha)} - c(\delta - \alpha) - 1, c \neq 0$. Here, c denotes the shape parameter associated with the loss function. It should be noted that LINEX loss function is an asymmetric loss function, and the shape parameter controls the degree and direction of symmetry. For more details on these loss functions, one may refer to Casella and Berger (2002). The Bayes estimator of α under SEL, say $\hat{\alpha}_{SEL}$, is defined as the posterior mean of α, i.e.

$$
\begin{aligned}
\hat{\alpha}_{SEL} &= E(\alpha|r_1, r_2, ..., r_n) \\
&= \int_0^1 \alpha\pi(\alpha|r_1, r_2, ..., r_n)d\alpha \\
&= \frac{1}{m}\int_0^1 (e^{\alpha r_n} - \alpha)e^{(\alpha r_n - \frac{(e^{\alpha r_n}-1)}{\alpha})}\prod_{i=1}^{n-1}(e^{\alpha r_i} - \alpha)\frac{\alpha^a(1-\alpha)^{b-1}}{B(a,b)}d\alpha. \qquad (3.4)
\end{aligned}
$$

Under LINEX loss function, Bayes estimator of α, say $\hat{\alpha}_{LI}$, is given by

$$
\begin{aligned}
\hat{\alpha}_{LI} &= \frac{-1}{c}ln[E(e^{-c\alpha}|r_1, r_2, ..., r_n)], c \neq o \\
&= \frac{-1}{c}ln[\frac{1}{m}\int_0^1 e^{-c\alpha}(e^{\alpha r_n} - \alpha)e^{(\alpha r_n - \frac{(e^{\alpha r_n}-1)}{\alpha})}\prod_{i=1}^{n-1}(e^{\alpha r_i} - \alpha)\frac{\alpha^{a-1}(1-\alpha)^{b-1}}{B(a,b)}d\alpha].
\end{aligned}
\qquad (3.5)
$$

The above integrals are complex to evaluate, and hence, to estimate α, we make use of Monte Carlo Markov Chain (MCMC) technique. Since the posterior density of α given in (3.3) is not in closed form, Metropolis-Hastings (MH) algorithm is used to simulate samples. MH algorithm is an iterative procedure to generate samples from the posterior density using the proposal (candidate) density function. At each step, the algorithm generates sample based on the proposal distribution and makes use of acceptance probability to either accept or reject the generated sample value. More details on the algorithm and its working principle can be found in Robert and Casella (2004). The proposal distribution can be either symmetric or asymmetric. In the former case, the acceptance probability is independent of the candidate distribution, and this makes computation easy. In the present work, normal distribution is taken as the candidate distribution. The parameters of the candidate distribution are estimated using MLE and the inverse of sample Fisher information value. Based on the samples generated from posterior distribution, Bayes estimate of α under SEL and LINEX loss function are obtained as

$$\hat{\alpha}_{SEL} = \frac{1}{N}\sum_{i=1}^{N}\alpha_i. \qquad (3.6)$$

$$\hat{\alpha}_{LI} = \frac{-1}{c}ln(\frac{1}{N}\sum_{i=1}^{N}e^{(-c\alpha_i)}), c \neq 0. \qquad (3.7)$$

Here, α_i, $i = 1, 2, ..., N$ denote the posterior sample observations, and N denotes the sample size.

4 Numerical Illustration

In this section, comparison of the estimates of α obtained from record values of Muth distribution based on ML and Bayesian approach is presented using simulated data sets. To simulate samples from Muth distribution, the inverse transformation approach is used. Let X have Muth distribution with cdf as given in (1.2). Then, equating $F(x; \alpha) = u$ for $u \in (0, 1)$ and solving for x, we get,

$$x = \frac{1}{\alpha} log(1-u) - \frac{1}{\alpha} W_{-1}(\frac{u-1}{\alpha e^{\frac{1}{\alpha}}}) - \frac{1}{\alpha^2}. \tag{4.1}$$

Here, W_{-1} denotes the negative branch of Lambert function.
In the simulation study, random samples from Muth distribution are generated using (4.1) for three choices of parameter values, namely $\alpha = 0.4, 0.6$ and 0.8. The number of record values (n) considered under each choice is 5 and 8 respectively. ML estimates are obtained by implementing the maxLik function available in R using the likelihood function given in (3.1). Bayes estimate of α is obtained using the SEL and LINEX loss functions (with c = 1 and -2) under two choices of the hyperparameters of the prior distribution, namely (1) a=1,b=2 and (2) a=2,b=4. These choices of the hyperparameters are taken so that the prior distribution has equal means but unequal variances. Samples of size 50000 is generated from the posterior distribution using MH algorithm with a burn-in period of 10000 samples that are discarded. Estimates of the parameter based on the remaining 40000 samples are computed using (3.6) and (3.7). The total number of simulation runs performed is 100. Table 2 presents the mean value of the estimates (estimate), average bias (bias), and mean square error (MSE) under ML and Bayesian approach.

From the simulation results presented in Table 2, the following can be established.

1. The average bias associated with estimates obtained by ML is large when compared with Bayesian approach.
2. As the number of record observation increases, the estimates become closer to the true parameter values. This is reflected by the diminishing MSE values associated with the estimates.
3. For large α, the estimates obtained are 'good' in terms of small bias and MSE.
4. The MSE values of the estimates obtained under Bayesian approach are smaller when compared with ML. The choice of taking normal distribution as proposal distribution in MH algorithm for generating samples and thereby determining the estimate is justified by the high acceptance rate of the algorithm as given in Table 3.

Table 2 Estimate, Bias, and MSE based on simulated data

α	n	Measure	ML	Prior 1			Prior 2		
				SEL	c=1	c=-2	SEL	c=1	c=-2
0.4	5	Estimate	0.6938	0.5682	0.5538	0.5921	0.5414	0.5329	0.5580
		Bias	0.2938	0.1682	0.1538	0.1921	0.1414	0.1329	0.1580
		MSE	0.1567	0.0678	0.0754	0.0665	0.0465	0.0425	0.0553
	8	Estimate	0.5207	0.4810	0.4671	0.4962	0.4788	0.4751	0.4863
		Bias	0.1207	0.0810	0.0671	0.0962	0.0788	0.0751	0.0863
		MSE	0.0323	0.0302	0.0422	0.0268	0.0192	0.0188	0.0202
0.6	5	Estimate	0.7666	0.6394	0.6266	0.6629	0.6092	0.6022	0.6234
		Bias	0.1666	0.0394	0.0266	0.0629	0.0922	0.0022	0.0234
		MSE	0.0795	0.0502	0.0639	0.0371	0.0288	0.0283	0.0293
	8	Estimate	0.6871	0.6412	0.6369	0.6494	0.6099	0.6064	0.6169
		Bias	0.0871	0.0412	0.0369	0.0494	0.0099	0.0064	0.0169
		MSE	0.0407	0.0242	0.0241	0.0246	0.0197	0.0198	0.0196
0.8	5	Estimate	0.8996	0.7138	0.7058	0.7292	0.6626	0.6567	0.6751
		Bias	0.0996	-0.086	-0.094	-0.707	-0.137	-0.143	-0.124
		MSE	0.0825	0.0261	0.0293	0.0209	0.0341	0.0363	0.0299
	8	Estimate	0.8814	0.7881	0.7844	0.7954	0.7385	0.7349	0.7455
		Bias	0.0814	-0.011	-0.015	-0.004	-0.061	-0.065	-0.054
		MSE	0.0373	0.0103	0.0102	0.0106	0.0020	0.0020	0.0219

Table 3 Mean acceptance rate

α	n	Prior 1	Prior 2
0.4	5	0.6151	0.4795
	8	0.7217	0.6512
0.6	5	0.5874	0.4379
	8	0.7145	0.5945
0.8	5	0.5167	0.3464
	8	0.5770	0.3962

5 Real-life Application

In this section, a real-life application of the estimating procedure is illustrated through Carrol data set. The data set contains 83 observations (with n=5 upper record values) on the monthly total rainfall (in mm) during the period from January 2000 to February 2007 in the rain gauge station of Carrol, Australia. The suitability of Muth distribution to model the data is verified by Pedro Jodrá et al. (2015) through several goodness of fit tests. The ML estimate for the parameter α as reported by them is $\hat{\alpha} = 0.4608$. Here, we attempt to estimate α through record observations using ML and Bayesian

Table 4 Estimates for real-life data

ML	Prior 1			Prior 2		
	SEL	c=1	c=-2	SEL	c=1	c=-2
0.5281	0.4676	0.4571	0.4879	0.4468	0.4387	0.4628

approach. The choice of prior and hyperparameters is same as indicated in Sect. 4. Also, normal distribution is used as proposal distribution in MH algorithm. The resulting estimates for α using likelihood and Bayesian approach are presented in Table 4.

From the results, it is observed that the estimates obtained using Bayesian approach are in close agreement with the estimate provided by Jodrá et al. In particular, estimate obtained using Prior 2 under LINEX loss function (with c=-2) turns out be fairly close when compared with the other estimates. Also, the acceptance rate of the MH algorithm under the two priors is found to be 0.7442 and 0.6491, respectively, which is more than the rule of thumb cutoff 0.4.

6 Prediction of Future Records

The prediction of future record value based on current records is dealt in this section using (i) frequentist approach and (ii) Bayesian approach.

6.1 Frequentist Approach

Let $R_1, R_2, ..., R_m$ denote 'm' record values observed from a population with density function $f(x;\alpha)$, and let $Z = R_n$ denote the future record. Since record values satisfy Markovian property, the conditional density of R_n given $R_1, R_2, ..., R_m$ depends only on R_m. The conditional density of Z given R_m called the predictive probability density is given by

$$f_{Z|R_m=r_m}(z) = \frac{[H(z) - H(r_m)]^{n-m-1}}{\Gamma(n-m)} \frac{f(z;\alpha)}{1 - F(r_m;\alpha)}, \; z > r_m, \tag{6.1}$$

where H(.) is the cumulative hazard function defined as $H(.) = -ln[1 - F(.)]$. For Muth distribution, we have $H(x) = \frac{e^{\alpha x}-1}{\alpha} - \alpha x$, and therefore,

$$H(z) - H(r_m) = \frac{e^{\alpha z} - e^{\alpha r_m}}{\alpha} + \alpha(r_m - z). \tag{6.2}$$

Also,

$$\frac{f(z;\alpha)}{1-F(r_m;\alpha)}=\frac{(e^{\alpha z}-\alpha)e^{[\alpha z-\frac{e^{\alpha z}-1}{\alpha}]}}{e^{[\alpha r_m-\frac{e^{\alpha r_m}-1}{\alpha}]}}. \tag{6.3}$$

Substituting (6.2),(6.3) in (6.1), we get the predictive probability density of Z given R_m as

$$f_{Z|R_m=r_m}(z)=\frac{[[\frac{(e^{\alpha z}-e^{\alpha r_m})}{\alpha}]+\alpha(r_m-z)]^{n-m-1}}{(n-m-1)!}(e^{\alpha z-\alpha})e^{\alpha(z-r_m)-\frac{(e^{\alpha z}-e^{\alpha r_m})}{\alpha}}, z>r_m. \tag{6.4}$$

Using (6.4), a $100(1-p)\%$ prediction interval for z, namely (z_L, z_U), can be obtained by solving

$$\int_{r_m}^{z_L} f_{Z|R_m=r_m}(z)dz=\frac{p}{2} \tag{6.5}$$

and

$$\int_{z_U}^{\infty} f_{Z|R_m=r_m}(z)dz=\frac{p}{2}, \tag{6.6}$$

where $p\in(0,1)$.

6.2 Bayesian Approach

The posterior predictive density of the future record $Z=R_n$ given the records $\mathbf{R}=(R_1, R_2, ..., R_m)$ is given by

$$p(z|\mathbf{r})=\int_0^1 f_{Z|R_m=r_m}(z)\pi(\alpha|\mathbf{r})d\alpha. \tag{6.7}$$

Using (3.3) and (6.4) in (6.7), we get

$$\begin{aligned}p(z|\mathbf{r})=\int_0^1\{&\frac{[\frac{(e^{\alpha z}-e^{\alpha r_m})}{\alpha}+\alpha(r_m-z)]^{n-m-1}}{(n-m-1)!}\\&(e^{\alpha z}-\alpha)e^{\alpha(z-r_m)-\frac{(e^{\alpha z}-e^{\alpha r_m})}{\alpha}}\\&m(\mathbf{r})\frac{\alpha^{a-1}(1-\alpha)^{b-1}}{B(a,b)}e^{[\alpha r_m-\frac{(e^{\alpha r_m})-1}{\alpha}]}\\&\prod_{i=1}^{m}(e^{\alpha r_i}-\alpha)\}d\alpha,\end{aligned} \tag{6.8}$$

where $m(\mathbf{r})$ is the proportionality constant so that $\int_0^1 \pi(\alpha|\mathbf{r})d\alpha = 1$. Upon simplifying (6.8), we get

$$p(z|\mathbf{r}) = \frac{m(\mathbf{r})}{B(a,b)} \int_0^1 (e^{\alpha z} - \alpha)^2 \prod_{i=1}^{m} (e^{\alpha r_i} - \alpha) \frac{[\frac{(e^{\alpha z} - e^{\alpha r_m})}{\alpha} + \alpha(r_m - z)]^{n-m-1}}{(n-m-1)!} e^{\alpha z - \frac{(e^{\alpha z}-1)}{\alpha}} \alpha^{a-1}(1-\alpha)^{b-1} d\alpha. \tag{6.9}$$

It can be seen that the above integral cannot be solved analytically to get a nice form of the posterior predictive density. When n=m+1, (6.9) reduces to

$$p(z|\mathbf{r}) = \frac{m(\mathbf{r})}{B(a,b)} \int_0^1 (e^{\alpha z} - \alpha)^2 \prod_{i=1}^{m} (e^{\alpha r_i} - \alpha) e^{\alpha z - \frac{(e^{\alpha z}-1)}{\alpha}} \alpha^{a-1}(1-\alpha)^{b-1} d\alpha. \tag{6.10}$$

To compute $E(Z|\mathbf{r})$, one can use MH algorithm to generate samples from the posterior predictive density. Suppose $\alpha_1, \alpha_2, ..., \alpha_N$ denote the generated samples, then a consistent estimator of $p(z|\mathbf{r})$ is given by

$$\hat{p}(z|\mathbf{r}) = \frac{1}{N} \sum_{j=1}^{N} f_{Z|R_m=r_m}(z; \alpha_j), \tag{6.11}$$

where $f_{Z|R_m=r_m}(z; \alpha_j)$ is obtained from (6.4) at $\alpha = \alpha_j$. Under squared error and LINEX loss functions, the Bayes predictor of $Z = R_m$ is given by

$$\hat{Z}_{SEL} = E(Z|\mathbf{r}) = \int_{r_m}^{\infty} z p(z|\mathbf{r}) dz$$

and

$$\hat{Z}_{LI} = \frac{-1}{c} ln[E(e^{-cZ}|\mathbf{r})] = \frac{-1}{c} ln[\int_{r_m}^{\infty} e^{-cZ} p(z|\mathbf{r})] dz,$$

respectively. Thus, the sample based Bayes predictor of Z will be

$$\hat{Z}_{SEL} = \frac{1}{N} \sum_{j=1}^{N} \int_{r_m}^{\infty} z f_{Z|R_m=r_m}(z; \alpha_j) dz$$

and

$$\hat{Z}_{LI} = \frac{1}{N} \sum_{j=1}^{N} \int_{r_m}^{\infty} e^{-cz} f_{Z|R_m=r_m}(z; \alpha_j) dz,$$

respectively. A $100(1-p)\%$ prediction interval for the future upper record $Z = R_n$ is obtained by solving

$$\int_{r_m}^{Z_L} p(z|\mathbf{r}) dz = \frac{p}{2}$$

and

$$\int_{Z_U}^{\infty} p(z|\mathbf{r}) dz = \frac{p}{2},$$

respectively, with respect to the lower and upper limits Z_L and Z_U.

7 Concluding Remarks

Studies on Muth distribution are scarce in the literature, and the present work has focused on deriving expression for survival function, joint cum conditional densities, moments, and estimating the parameter using record values. Parameter estimation is done using moment, likelihood, and Bayesian approach with beta distribution as prior. Due to the complex nature of integrals involved in evaluating the posterior distribution, the Metropolis-Hastings algorithm is used with normal distribution as the proposal distribution to generate random samples. The simulation study and real-life application suggest that the Bayesian approach yields better estimates when compared with the likelihood approach.

References

Arnold, Barry C., Balakrishnan, N., & Nagaraja, H. N. (1998). *Records*. New York: John Wiley and Sons Inc.

Asgharzadeh, A., Fallah, A., Raqab, M. Z., & Valiollahi, R. (2016). Statistical Inference based on Lindley record data. *Statistical Papers, 59*(2), 759–779.

Casella, G., & Berger, R. L. (2002). *Statistical Inference* (2nd ed.). United States: Duxbury.

Dara, S.T. (2012). *Recent Advances in Moment Distributions and their Hazard Rate*. Ph.D. Thesis. National College of Business Administration and Economics, Lahore, Pakistan.

Gradshteyn, I. S., & Ryzhik, I. M. (2007). *Table of Integrals, series and products* (7th ed.). Amsterdam: Academic Press.

Kumar, D. (2015). Explicit expressions and statistical inference of Generalized Rayleigh distribution based on lower record values. *Mathematical Methods of Statistics, 24*(3), 225–241.

MirMostaface, S. M. T. K., Asgharzadeh, A., & Fallah, A. (2016). Record values from NH distribution and associated inference. *Metron, 74*, 37–59.
Muth, J. E. (1977). Reliability models with positive memory derived from the mean residual function. In: C. P. Tsokos & L. Shimi (Eds.), *The Theory and Applications of Reliability, 2* (pp. 401–435). New York: Academic Press Inc.
Jodrá, Pedro, Jiménez-Gamero, María Dolores., & Alba-Fernández, María Virtudes. (2015). On the Muth distribution. *Mathematical Modelling and Analysis, 20*(3), 291–310.
Robert, C. P., & Casella, G. (2004). *Monte Carlo Statistical Methods* (2nd ed.). New York: Springer.
Teimouri, M., & Gupta, A. K. (2012). On the Weibull record statistics and associated inferences. *STATISTICA, anno LXXII, 2*, 145–162.

Statistical Linear Calibration in Data with Measurement Errors

Shalabh

Abstract The direct and inverse regression-based estimators are used for linear statistical calibration. The direct regression finds the relationship between the dependent and independent variables, and inverse regression uses it for calibration. When both the dependent and independent variables in linear calibration are subjected to measurement errors, the resultant estimators become biased and inconsistent. Assuming the availability of replicated observations on the independent variable, a calibration estimator is presented which has zero large sample asymptotic bias. The large sample efficiency properties of the calibration estimators are derived and analyzed assuming the random errors are not necessarily normally distributed.

Keywords Measurement errors · Calibration · Direct regression · Inverse regression · Non-normal errors

1 Introduction

Statistical calibration is an important aspect in many real applications, see, e.g., Sun et al. (2014) Salter et al. (2016), Joseph et al. (2015), Sansó et al. (2009), Gregory et al. (1993), etc. for a variety of application of statistical calibration. It is used in the creation of a scale on any measuring device which improves the precision of the instruments. The process of calibration involves two steps. In the first step, the readings from two different instruments or measuring devices are obtained. Then in the second step, a statistical relationship between the two sets of observations is established. In the next step, the thus obtained relationship is used for making the prediction of measurements of one device on the basis of readings obtained from the other device. Generally, it is expensive, tedious, complex and may be destructive to get accurate observations from one of the instrument. This instrument is considered as a standard instrument, and the obtained observations from it are termed as true

Shalabh (✉)
Department of Mathematics and Statistics, Indian Institute of Technology Kanpur, Kanpur 208016, India
e-mail: shalab@iitk.ac.in

D. D. Hanagal et al. (eds.), *Applied Statistical Methods*, Springer Proceedings in Mathematics & Statistics 380, https://doi.org/10.1007/978-981-16-7932-2_18

values. On the other hand, another device yields observations which are relatively much cheaper, easy to get, simple and nondestructive. Such obtained observations are termed as readings. This type of setup of calibration is used in many areas such as medical science, engineering science, physical science, etc. In spite of being so useful, such an approach poses several challenges in real data analysis in drawing statistical inference; see, e.g., Osborne (1991) for an interesting review of statistical developments on calibration.

The direct or classical regression and inverse regression estimation methods under linear regression model are utilized in controlled calibration experiments assuming that the observations are accurately recorded without any error; see Scheffé (1973) for a detailed discussion on statistical calibration. Such an assumption may not always hold in practice, and instead, the observations are contaminated by measurement errors. The difference between the observed and true values of the variable is termed as measurement error. The measurement errors enter into the data due to many reasons. For example, suppose someone wants to measure the content of nitrogen in the soil after sprinkling a known quantity of fertilizer over a given field. When the nitrogen content is to be measured, a sample of soil is taken, and the chemical analysis determines the concentration of true level of nitrogen. Since the content of nitrogen will be varying in soil from one point to another in the entire agricultural field/forestry land and a single value is reported based on chemical analysis, so the recorded values will have measurement errors. Thus if we wish to analyze the relationship between the soil nitrogen and the yield of a certain crop, it may have issues arising due to measurement errors in the observations. So assuming the measurement error-free observations may not be tenable in many practical situations, and the values of working standards may be subjected to measurement errors as pointed out by Lwin and Spiegelman (1986). In such circumstances, it is imperative to employ the framework of measurement error models; see, e.g., Cheng and Van Ness (1999), Fuller (1987) and Rao (2008) for an interesting exposition. The presence of measurement errors in the data disturbs the optimal properties of direct and inverse regression estimators, which become biased as well as inconsistent estimators of the regression coefficient.

The classical or direct regression technique in a linear model is employed to obtain the true relationship between the dependent and independent variables. Then the inverse regression is used based on the determined relationship to know the value of the variable. The performance properties of the classical and inverse calibration estimators along with others have been analyzed by Yum and Lee (1991) under the framework of the functional variant of the measurement error model; see also Lee and Yum (1989), Srivastava and Shalabh (1997). An interesting observation emerging from such investigations is that both the calibration estimators generally possess nonzero asymptotic bias which may often be substantial even when there are replicated observations on the working standards.

Another common assumption in the statistical analysis of calibration estimators is that the random errors follow a normal distribution. How the properties of the calibration estimators deviate when there is a departure from the normal distribution of measurement errors is a question which has been treated neglectfully in the literature. Most of the work in calibration estimators are available under the normally distributed random errors and in a no measurement error situation. The performance properties of the classical and inverse calibration estimators along with their modified and extended forms under the normality of errors have been studied by, e.g., Osborne (1991). Later, Brown (1993, Chap. 2) discusses the aspect of controlled calibration with classical and inverse calibration estimators but under the assumption of the normal distribution of errors.

We have attempted and present a modest effort in finding a calibration estimator in this chapter which is asymptotically unbiased in the sense of having zero large sample asymptotic bias and discuss the effect of non-normally distributed measurement errors.

The organization of this chapter is as follows. In Sect. 2, we describe the model and present the calibration estimators. Their approximate asymptotic bias and variance are derived using the large sample asymptotic approximation theory and are discussed in Sect. 3. An illustrative example is presented in Sect. 4, and some conclusions are placed in Sect. 5. Lastly, the derivations of the main results are provided in the Appendix.

2 Development of Calibration Estimators

Consider a calibration experiment in which $Y_1, Y_2, \ldots, Y_n$ denote the true values on the dependent or study variables corresponding to the true standard values $X_1, X_2, \ldots, X_n$, respectively. Thus, we postulate the following linear relationship between the true values Y_i and X_i

$$Y_i = \alpha + \beta X_i \quad , i = 1, 2, \ldots, n \tag{2.1}$$

where α is the intercept term and β is the slope parameter. Due to the presence of measurement errors in the data, the true values on dependent variables are obtained as $y_1, y_2, \ldots, y_n$ and we can postulate the relationship with additive measurement errors as

$$y_i = Y_i + u_i \quad , i = 1, 2, \ldots, n$$

where u_i is the measurement error associated with the ith observed value of the dependent variable y_i. Without loss of generality, we assume that the usual random error in the model (2.1) is subsumed in u_i. Further, the standard values on independent variables X_i are also subjected to the measurement errors, and we assume that p replicated observations are available corresponding to each standard value which are observed with measurement errors. Accordingly, let $x_{i1}, x_{i2}, \ldots, x_{ip}$ be the p replicated observed values of X_i. The jth observation on the ith value of the independent variable X_i is obtained as x_{ij} which is affected by the measurement errors v_{ij}. Assuming the measurement errors to be additive, we can express

$$x_{ij} = X_i + v_{ij} \quad j = 1, 2, \ldots, p$$

where v_{ij} is the measurement error associated with the jth observation on the ith value of the independent variable. Further, u_i and v_{ij}'s ($i = 1, 2, \ldots, n$; $j = 1, 2, \ldots, p$) are assumed to be independent of each other.

Now a value Y of the dependent variable is recorded corresponding to an unknown standard value, X, and the problem is to predict or estimate X.

We can thus express

$$Y = \alpha + \beta X + U \tag{2.2}$$

where U denotes the random error.

If we write

$$\bar{x}_i = \frac{1}{p}\sum_j x_{ij}, \quad \bar{\bar{x}} = \frac{1}{np}\sum_i\sum_j x_{ij}, \quad \bar{y} = \frac{1}{n}\sum_i y_i,$$

the classical and inverse regression estimators of β, denoted as b_c and b_I respectively, are given by

$$b_c = \frac{p\sum(\bar{x}_i - \bar{\bar{x}})(y_i - \bar{y})}{\sum\sum(x_{ij} - \bar{\bar{x}})^2} \tag{2.3}$$

$$b_I = \frac{\sum(y_i - \bar{y})^2}{\sum(\bar{x}_i - \bar{\bar{x}})(y_i - \bar{y})} \tag{2.4}$$

which are inconsistent for β in the presence of measurement errors in the data; see, e.g., Cheng and Van Ness (1999) and Fuller (1987). The inconsistency of these estimators can be corrected either by using some information from outside the sample or using the replicated observations, provided they are available.

So we assume the availability of replicated observations on independent variables which are also measurement error ridden. Then it is possible to construct a consistent estimator of β as follows:

$$b_c^* = \frac{p\sum(\bar{x}_i - \bar{\bar{x}})(y_i - \bar{y})}{\sum\sum(x_{ij} - \bar{\bar{x}})^2 - (\frac{p}{p-1})\sum\sum(x_{ij} - \bar{x}_i)^2}. \tag{2.5}$$

Utilizing these estimators (2.3)–(2.5), we get the following calibration estimator of X:

$$\hat{X}_c = \bar{\bar{x}} + \frac{\sum\sum(x_{ij} - \bar{\bar{x}})^2}{p\sum\bar{x}_i - \bar{\bar{x}})(y_i - \bar{y})}(Y - \bar{y}) \tag{2.6}$$

$$\hat{X}_c^* = \bar{\bar{x}} + \frac{\sum\sum(x_{ij} - \bar{\bar{x}})^2 - (\frac{p}{p-1})\sum\sum(x_{ij} - \bar{x}_i)^2}{p\sum(\bar{x}_i - \bar{\bar{x}})(y_i - \bar{y})}(Y - \bar{y}) \tag{2.7}$$

$$\hat{X}_I = \bar{\bar{x}} + \frac{\sum(\bar{x}_i - \bar{\bar{x}})(y_i - \bar{y})}{\sum(y_i - \bar{y})^2}(Y - \bar{y}). \tag{2.8}$$

Here, $\hat{X}_c$ and $\hat{X}_I$ are the conventional classical and inverse calibration estimators respectively, while the calibration estimator $\hat{X}_c^*$ is a modified version of $\hat{X}_c$.

The area of calibration estimators has remained a popular choice of research among researchers due to its enormous utility in various applications. Krutchkoff (1967, 1969) considered the calibration estimators based on direct and inverse regression estimators. Using the findings from Monte Carlo simulation experiment, it was concluded that the estimators based on inverse calibration approach yield better calibration estimators than direct regression approach under the criterion of mean squared error in the range of calibration but have a larger mean squared error in extrapolation. Then Halperin (1970) and Williams (1969) investigated and concluded that the classical calibration estimator is superior even within the range of calibration. Shalabh and Toutenburg (2006) derived the efficiency properties of the direct and inverse calibration estimators using small error asymptotic theory. Krutchkoff (1971) numerically investigated the issue of dominance of direct and inverse calibration estimators under the criterion of Pitman closeness and concluded that the inverse calibration estimator is superior or equivalent to classical calibration estimator and the ranges where classical calibration estimator is superior, it is only mildly superior. Scheffé (1973) presented a detailed discussion of the calibration problem. Tallis (1969) discussed the theory of identifiability of mixtures of distribution in calibration. Pepper (1973) discussed the calibration issue from the random walk perspective. Dunsmore (1968), Williford et al. (1979) and Hunter and Lamboy (1981) considered the Bayesian framework and discussed the calibration problem; see also Gray et al. (2019). Friedland (1977) and Brown (1982) discussed various issues concerned with multivariate calibration; see also Brown and Sundberg (1987, 1989). Gregory et al. (1993) studied the calibration in macroeconomics. Misquitta and Ruymgaart (2005) considered the statistical calibration in a nonparametric setup. Sansó et al. (2009), Sansó and Forest (2009) considered statistical calibration of the climate system. Yu et al. (2010) considered the regression calibration in semiparametric accelerated failure time models. Spiegelman et al. (2011) considered the regression calibration with heteroscedastic error variance. Spiegelman (2013) and Strand et al. (2015) used regression calibration with instrumental variables for longitudinal models and used it to study air pollution. Blas et al. (2013) generalized the controlled calibration model by assuming replication on both variables and used the likelihood-based methodology for the point and interval estimation of the parameters. Han et al. (2016) studied the calibration and simultaneous tolerance intervals in polynomial regression. Bartlett et al. (2018) examined the Bayesian correction for covariate measurement error in regression calibration. Some other aspects of calibration issues are considered in Aitchison (1977), Berkson (1969), Minder and Whitney (1975), Lwin and Maritz (1980), Wang et al. (1997), Huang (2005) and Skrondal et al. (2012). All such works were considered in a no measurement error situation except Gray et al. (2019).

3 Performance Properties

In order to analyze the performance properties of calibrations estimators, we assume that $u_1, u_2, \ldots, u_n, U$ are independently and identically distributed random variables following a distribution having mean 0, variance σ_u^2, third moment $\sigma_u^3\gamma_{1u}$ and fourth moment $\sigma_u^4(\gamma_{2u} + 3)$. Similarly, $v_{11}, v_{12}, \ldots, v_{np}$ are independently and identically distributed random variables following a distribution having mean 0, variance σ_v^2, third moment $\sigma_v^3\gamma_{1v}$ and fourth moment $\sigma_v^4(\gamma_{2v} + 3)$. Here $\gamma_{1\cdot}$ and $\gamma_{2\cdot}$ are the coefficients of skewness and kurtosis respectively associated with the respective distribution. Further, both the sets of random variables on u's and v's are stochastically independent. Thus we do not assume any specific functional form like normality for the error distributions. Only the existence of the first four finite moments of the corresponding random variables is assumed.

Let us now introduce the following notations:

$$
\begin{aligned}
d &= \left(\frac{\bar{X} - X}{\sigma_v}\right), \\
s^2 &= \frac{1}{n}\sum(X_i - \bar{X})^2, \\
\lambda_x &= \frac{s^2}{s^2 + \sigma_v^2}, \quad 0 \le \lambda_x \le 1, \\
\lambda_y &= \frac{\beta^2 s^2}{\beta^2 s^2 + \sigma_u^2}, \quad 0 \le \lambda_x \le 1, \\
\theta &= \left(\frac{1 - \lambda_y}{\lambda_y}\right) + \left(\frac{1 - \lambda_x}{\lambda_x}\right) d^2
\end{aligned}
$$

where $\bar{X}$ is the mean of $X_1, X_2, \ldots, X_n$.

The quantities λ_y and λ_x are the reliability ratios associated with the dependent and independent variables, respectively. The reliability ratio is defined as the ratio of variances of true and observed values of the variables. The concept of reliability ratios is popular in psychometrics, and their values can be found using various psychological tools. The information on reliability ratio also helps in obtaining the consistent estimators of β in measurement error models when the replicated observations are not available; see Gleser (1992, 1993). Deriving the exact sample properties of these calibration estimators is complicated and cumbersome. Even if obtained, the expressions may be so complicated that they may not provide any clear inference. So we have employed the large sample asymptotic approximation theory to derive the properties of these calibration estimators.

3.1 Large Sample Asymptotic Bias (LSAB)

Theorem 1 *The large sample asymptotic approximation of the bias (LSAB) of the calibration estimator* $\hat{X}_c$, $\hat{X}_I$ *and* $\hat{X}_c^*$ *to order* $O(n^{-1/2})$ *is given by*

$$LSAB(\hat{X}_c) = E(\hat{X}_c - X) = -\left(\frac{1-\lambda_x}{\lambda_x}\right)\sigma_v d \tag{3.1}$$

$$LSAB(\hat{X}_I) = E(\hat{X}_I - X) = (1-\lambda_y)\sigma_v d \tag{3.2}$$

$$LSAB(\hat{X}_c^*) = E(\hat{X}_c^* - X) = 0. \tag{3.3}$$

Proof See Appendix.

It is interesting to observe from the above results that the calibration estimator $\hat{X}_c^*$ is asymptotically unbiased in the sense of having zero LSAB. On the other hand, $\hat{X}_c$ and $\hat{X}_I$ are asymptotically biased in the sense of having nonzero LSAB, and their biases are in opposite directions. Further, the calibration estimator $\hat{X}_c$ has smaller magnitude of LSAB in comparison to $\hat{X}_I$ when

$$\left(\lambda_y + \frac{1}{\lambda_x}\right) < 2 \tag{3.4}$$

where it may be noticed that λ_x and λ_y lie between 0 and 1.

The opposite is true, i.e., the calibration estimator $\hat{X}_I$ has a smaller magnitude of LSAB than $\hat{X}_c$ when the condition (3.4) holds with a reversed inequality sign.

3.2 Large Sample Asymptotic Variance (LSAV)

As the calibration estimators $\hat{X}_c$ and $\hat{X}_I$ have nonzero LSAB while $\hat{X}_c^*$ has zero LSAB, it is more appropriate to compare them with respect to the criterion of large sample asymptotic approximation of variance (LSAV) rather than the corresponding mean squared error.

Theorem 2 *The large sample asymptotic approximation of variance (LSAV) of the calibration estimators* $\hat{X}_c$, $\hat{X}_I$ *and* $\hat{X}_c^*$ *to order* $O(n^{-1})$ *is given by*

$$\begin{aligned} LSAV(\hat{X}_c) &= E[\hat{X}_c - E(\hat{X}_c)]^2 \\ &= \frac{\sigma_v^2}{np}\left[1 + \frac{p(1-\lambda_y)}{\lambda_y\lambda_x(1-\lambda_x)}\right. \\ &\quad \left. + \frac{\theta}{\lambda_x^2}\left\{\frac{1}{\lambda_y} - 2\lambda_x(1-\lambda_x) + \frac{p\lambda_x(1-\lambda_y)}{\lambda_y(1-\lambda_x)}\right\} - N_c\right] \end{aligned} \tag{3.5}$$

$$LSAV(\hat{X}_I) = E[\hat{X}_I - E(\hat{X}_I)]^2$$

$$= \frac{\sigma_v^2}{np}\left[1 + \frac{p\lambda_x\lambda_y(1-\lambda_y)}{(1-\lambda_x)} + \theta\lambda_y\left\{1 + p\lambda_x\left(\frac{1-\lambda_y}{1-\lambda_x}\right)(1-2\lambda_y+2\lambda_y^2)\right\} - N_I\right] \quad (3.6)$$

$$LSAV(\hat{X}_c^*) = E[\hat{X}_c^* - E(\hat{X}_c^*)]^2$$
$$= \frac{\sigma_v^2}{np}\left[1 + \frac{p\lambda_x(1-\lambda_y)}{\lambda_y(1-\lambda_x)} + \theta\left\{\frac{1}{\lambda_y} + 2\frac{(1-\lambda_x)}{(p-1)\lambda_x} + \frac{p\lambda_x(1-\lambda_y)}{\lambda_y(1-\lambda_x)}\right\}\right] \quad (3.7)$$

where

$$N_c = \left(\frac{1-\lambda_x}{\lambda_x}\right)[2d\gamma_{1v} - \theta\gamma_{2v}] \quad (3.8)$$

$$N_I = \lambda_y^{3/2}(1-\lambda_y)^{3/2}\left(\frac{\lambda_x}{1-\lambda_x}\right)^{1/2}\left[2d\gamma_{1u} - \theta(\lambda_x\lambda_y)^{1/2}\left(\frac{1-\lambda_y}{1-\lambda_x}\right)^{1/2}\gamma_{2u}\right]. \quad (3.9)$$

Proof See Appendix.

It is interesting to observe from the above expressions that the skewness and kurtosis of the distribution of measurement errors associated with the independent variable or working standards influence the LSAV of the calibration estimation $\hat{X}_c$, while the skewness and kurtosis of the distribution of errors associated with the measurements on the dependent variable influence the LSAV of $\hat{X}_I$. However, the LSAV of $\hat{X}_c^*$ remains the same for all kinds of error distributions. Thus, the efficiency properties of $\hat{X}_c$ and $\hat{X}_I$ may be altered by the departures from normality of errors but such is not the case with $\hat{X}_c^*$ and it is fully robust, at least asymptotically, with respect to departures from normality.

Now, if we assume that the measurement errors are normally distributed so that N_c and N_I given by (3.8) and (3.9), respectively are zero, it follows from (3.6) and (3.7) that the $\hat{X}_c^*$ with zero LSAB has always larger LSAV than the calibration estimator $\hat{X}_I$ having nonzero LSAB. However, $\hat{X}_c^*$ is not having only zero LSAB in comparison to $\hat{X}_c$ but it also has smaller LSAV than $\hat{X}_c$ when

$$\left[\frac{p\lambda_x(1-\lambda_y)}{\lambda_y(1-\lambda_x)} + \theta q\right] > 0 \quad (3.10)$$

where

$$q = \frac{1}{\lambda_y} + \frac{p\lambda_x(1-\lambda_y)}{\lambda_y(1-\lambda_x)} - 2\left(\frac{p}{p-1}\right)\left(\frac{\lambda_x}{1+\lambda_x}\right). \quad (3.11)$$

The reliability ratio λ_y associated with the dependent variable (the instrument being calibrated) will generally be smaller than the reliability ratio λ_x associated with the independent variables (working standards). In any case, λ_y will not be substantially larger than λ_x. In such situations, the quantity q will be positive for $p = 2$ and the condition (3.10) will be satisfied implying the superiority of $\hat{X}_c^*$ over $\hat{X}_c$. However, if $p > 2$, the quantity q is always positive and it then follows from (3.10) that $\hat{X}_c^*$ has better performance than $\hat{X}_c$.

A more exciting and challenging issue will be to extend the analysis for the multiple linear regression models. A question arises how to define the inverse calibration estimator in such a case and then to use it in the measurement error models. The finite sample properties and confidence interval estimation for the calibration estimators are also a challenge in itself as it is not so straightforward to derive the confidence intervals for regression coefficients in measurement error models.

4 An Example

Consider the dataset, given in Fuller (1987, p. 198), related to the yield (y) of corn and the soil nitrogen (x) collected at 25 sites on Marshall soil in Iowa (U.S.A.) presented in Table 1. The level of nitrogen is estimated by the chemical analysis of soil, and two determinations are made at each site.

From the given observations, σ_v^2 and s^2 are consistently estimated by

$$\hat{\sigma}_v^2 = \frac{1}{n(p-1)}\sum\sum(x_{ij} - \bar{x}_i)^2 = 54.80,$$

$$\hat{s}^2 = \frac{1}{np}\sum\sum(x_{ij} - \bar{\bar{x}})^2 - \hat{\sigma}_v^2 = 283.53,$$

so that a consistent estimator of λ_x is obtained as $\hat{\lambda}_x = \frac{\hat{s}^2}{\hat{s}^2 + \hat{\sigma}_v^2} = 0.838$.

Similarly, the consistent estimator of β and σ_u^2 is obtained as follows:

$$\hat{\beta} = \frac{p\sum(\bar{x}_i - \bar{\bar{x}})(y_i - \bar{y})}{\sum\sum(x_{ij} - \bar{\bar{x}})^2 - \left(\frac{p}{p-1}\right)\sum\sum(x_{ij} - \bar{x}_i)^2} = 0.476,$$

$$\hat{\sigma}_u^2 = \frac{1}{n}\sum(y_i - \bar{y})^2 - \hat{\beta}^2\hat{s}^2 = 50.40,$$

so that λ_y is consistently estimated by $\hat{\lambda}_y = \frac{\hat{\beta}^2\hat{s}^2}{\hat{\beta}^2\hat{s}^2 + \hat{\sigma}_u^2} = 0.56$.

Using the consistent estimates $\hat{\lambda}_x$ and $\hat{\lambda}_y$ in place of λ_x and λ_y in (3.1) and (3.2) respectively, the estimates of LSABs of the calibration estimates $\hat{X}_c$ and $\hat{X}_I$ are obtained as follows:

Table 1 Yield of corn and the soil nitrogen of 25 sites in Iowa, USA from Fuller (1987, p. 198)

Observation number	Soil nitrogen (x)		Corn yield (y)
	Determination 1	Determination 2	
1	71.00	70.00	106.00
2	78.00	66.00	119.00
3	76.00	77.00	87.00
4	59.00	58.00	100.00
5	97.00	87.00	105.00
6	53.00	69.00	98.00
7	76.00	63.00	98.00
8	43.00	45.00	97.00
9	86.00	81.00	99.00
10	44.00	58.00	88.00
11	89.00	71.00	105.00
12	46.00	66.00	91.00
13	66.00	53.00	90.00
14	62.00	54.00	94.00
15	76.00	69.00	95.00
16	59.00	57.00	83.00
17	61.00	76.00	94.00
18	70.00	69.00	101.00
19	34.00	47.00	78.00
20	93.00	87.00	115.00
21	59.00	62.00	80.00
22	48.00	40.00	93.00
23	64.00	48.00	91.00
24	95.00	103.00	111.00
25	100.00	97.00	118.00

$$\widehat{LSAB}(\hat{X}_c) = -1.431d$$
$$\widehat{LSAB}(\hat{X}_I) = 3.257d$$

which may give us an idea about the nature of asymptotic bias.

Similarly, the estimates of LSAVs from (3.5), (3.6) and (3.7) are given by

$$\widehat{LSAV}(\hat{X}_c) = 12.80 + 1.454d^2 - 0.387d\gamma_{1v} + (0.152 + 0.032d^2)\gamma_{2v}$$
$$\widehat{LSAV}(\hat{X}_I) = 2.743 + 0.196d^2 - 0.609\, d\gamma_{1u} + (0.270 + 0.665d^2)\gamma_{2u}$$
$$\widehat{LSAV}(\hat{X}_c^*) = 9.438 + 1.091d^2$$

where use has been made of the result $\theta = 0.786 + 0.193\, d^2$.

We thus observe that, under the normality of errors, $\hat{X}_c^*$ has not only zero LSAB in comparison to $\hat{X}_c$, but it has smaller LSAV too. However, $\hat{X}_c^*$ has larger LSAV than $\hat{X}_I$, and this is the price one has to pay for asymptotic unbiasedness.

Comparing the calibration estimator $\hat{X}_c$ and $\hat{X}_I$ having nonzero LSAB under the normality of errors, we find that $\hat{X}_c$ has a smaller magnitude of LSAB but larger LSAV than $\hat{X}_I$.

5 Conclusions

We have considered the problem of calibration when both the measurements on the dependent and independent variables are subjected to errors, and replicated observations are available on the independent variable. We have presented a modified version of the classical calibration estimator such that it has zero LSAB and in this sense, it is asymptotically unbiased. Interestingly enough, the proposed calibration estimator always has the same LSAV whether or not the distributions of the measurement errors are normal, and is thus robust, at least asymptotically with respect to departures from the normality of errors. This is, however, not the case with the classical and inverse calibration estimators whose LSAVs are influenced by the skewness and kurtosis of the error distributions.

When the distributions of the measurement errors are assumed to be normal, the calibration estimator with zero LSAB fails to perform better than the inverse calibration estimator with nonzero LSAB under the criterion of LSAV, and this is the price that one has to pay for the elimination of LSAB. When compared with the classical calibration estimator, the proposed calibration estimator has only zero LSAB but smaller LSAV too in most of the practical situations. It may be added that the relative performance of one calibration estimator over the other under the normality of errors may not remain the same when the measurement errors are not normally distributed. One can consider using the replicated observations on the dependent variable or when the replicated observations are available on both the variables to construct other calibration estimators.

Acknowledgements The author gratefully acknowledges the support from the MATRICS project from Science and Engineering Research Board (SERB), Department of Science and Technology, Government of India.

Appendix

Let us first introduce the following notation

$$
\begin{aligned}
A &= I_{np} - \frac{1}{np} e_{np} e'_{np} \\
B &= \frac{1}{p}(I_n \otimes e'_p) - \frac{1}{np} e_n e'_{np} \\
C &= I_n - \frac{1}{n} e_n e'_n \\
D &= \frac{1}{p}(I_n \otimes e_p e'_p) - \frac{1}{np} e_{np} e'_{np}
\end{aligned}
$$

where $\otimes$ denotes the Kronecker product operator of matrices and e denotes a column vector with all elements unity and its suffix indicating the number of elements in it.

The following properties of these matrices may be noted:

$$
\begin{aligned}
AD = D, \quad BD = D, \quad pB'B = D, \quad pBB' = C, \\
trA = (np - 1), \quad trC = trD = (n - 1), \\
Ae_{np} = 0, \quad Be_{np} = 0, \quad e'_n B = 0, \quad Ce_n = 0.
\end{aligned} \tag{5.1}
$$

Further, if Z denotes a $m \times 1$ random vector such that its elements are independently and identically distributed with mean 0, variance σ_z^2, third moment $\sigma_z^3 \gamma_{1z}$ and fourth moment $\sigma_z^4(\gamma_{2z} + 3)$, we have

$$
\begin{aligned}
E(Z'GZ) &= \sigma_z^2 trG \\
E(Z'GZ.Z) &= \sigma_z^3 \gamma_{1z}(I * G)e_m \\
E(Z'GZ.Z'MZ) &= \sigma_z^4[\gamma_{2z} trH(I * G) + (trG)(trH) + 2trGH]
\end{aligned} \tag{5.2}
$$

where G and H are $m \times m$ symmetric matrices with nonstochastic elements and $*$ denotes the Hadamard product operator of matrices.

If we define

$$
\begin{aligned}
f_{xx} &= \frac{2}{n^{1/2}} X'Bv + \frac{n^{1/2}}{p}\left(\frac{v'Av}{n} - p\sigma_v^2\right) \\
f_{xy} &= \frac{1}{n^{1/2}}\left[\frac{1}{\beta}(X'Cu + u'Bv) + X'Bv\right] \\
f_{yy} &= \frac{2}{\beta n^{1/2}} X'CY + \frac{n^{1/2}}{\beta^2}\left(\frac{u'Cu}{n} - \sigma_u^2\right) \\
f &= n^{1/2}\left[\frac{v'(A - D)v}{n(p - 1)} - \sigma_v^2\right]
\end{aligned}
$$

$$f_x = \frac{1}{pn^{1/2}} \sum_i \sum_j v_{ij}$$

$$f_y = \frac{1}{\beta n^{1/2}} \sum_i u_i,$$

it is observed that all these quantities are of order $O_p(1)$ where $u = (u_1, u_2, \ldots, u_n)$ and $v = (v_{11}, v_{12}, \ldots, v_{np})$ are column vectors of order $n \times 1$ and $np \times 1$ respectively.

Now, from (2.6), we can express

$$(\hat{X}_c - X) = \sigma_v d + \frac{f_x}{n^{1/2}} + \frac{s^2 + \sigma_v^2 + \frac{f_{xx}}{n^{1/2}}}{s^2 + \frac{f_{xy}}{n^{1/2}}} \left(\frac{U}{\beta} - \sigma_v d - \frac{f_y}{n^{1/2}} \right)$$
$$= \sigma_v d + \frac{f_x}{n^{1/2}} + \left(\frac{U}{\beta} - \sigma_v d - \frac{f_y}{n^{1/2}} \right) \left(\frac{1}{\lambda_x} + \frac{f_{xx}}{s^2 n^{1/2}} \right) \left(1 + \frac{f_{xy}}{s^2 n^{1/2}} \right)^{-1}$$

where s^2 and λ_x are defined in (3.1).

Expanding and retaining terms to order $O_p(n^{-1/2})$, we get

$$(\hat{X}_c - X) = \frac{1}{\lambda_x} \left[\frac{U}{\beta} - (1 - \lambda_x) \sigma_v d \right] + \frac{1}{n^{1/2}} \hat{\xi}_c + O_p\left(\frac{1}{n} \right)$$

where

$$\hat{\xi}_c = f_x - \frac{1}{\lambda_x} f_y + \frac{1}{s^2} \left(\frac{U}{\beta} - \sigma_v d \right) \left(f_{xx} - \frac{1}{\lambda_x} f_{xy} \right).$$

Similarly, from (2.7), we obtain

$$(\hat{X}_c^* - X) = \sigma_v d + \frac{f_x}{n^{1/2}} + \left(\frac{U}{\beta} - \sigma_v d - \frac{f_y}{n^{1/2}} \right) \left[1 + \frac{f_{xx} - f}{s^2 n^{1/2}} \left(1 + \frac{f_{xy}}{s^2 n^{1/2}} \right)^{-1} \right]$$
$$= \frac{U}{\beta} + \frac{1}{n^{1/2}} \hat{\xi}_c^* + O_p\left(\frac{1}{n} \right)$$

where

$$\hat{\xi}_c^* = f_x - f_y + \frac{1}{s^2} \left(\frac{U}{\beta} - \sigma_v d \right) (f_{xx} - f_{xy} - f).$$

Likewise, it is easy to see from (2.8) that

$$(\hat{X}_I - X) = \sigma_v d + \frac{f_x}{n^{1/2}} + \left(\frac{U}{\beta} - \sigma_v d - \frac{f_y}{n^{1/2}} \right)$$

$$\left(\frac{\beta^2 f_{xy}}{(\beta^2 s^2+\sigma_u^2)n^{1/2}}+\frac{\beta^2 s^2}{\beta^2 s^2+\sigma_u^2}\right)\left(1+\frac{\beta^2 f_{yy}}{(\beta^2 s^2+\sigma_u^2)n^{1/2}}\right)^{-1}$$
$$=\sigma_v d+\frac{f_x}{n^{1/2}}+\left(\frac{U}{\beta}-\sigma_v d-\frac{f_y}{n^{1/2}}\right)\left(\lambda_y+\frac{\lambda_y f_{xy}}{s^2 n^{1/2}}\right)\left(1-\frac{\lambda_y f_{yy}}{s^2 n^{1/2}}+\ldots\right)$$
$$=\frac{\lambda_y U}{\beta}+(1-\lambda_y)\sigma_v d+\frac{1}{n^{1/2}}\hat{\xi}_I+O_p(n^{-1})$$

where

$$\hat{\xi}_I=f_x-\lambda_y f_y+\frac{\lambda_y}{s^2}\left(\frac{U}{\beta}-\sigma_v d\right)(f_{xy}-\lambda_y f_{yy}).$$

Observing that

$$E(f_{xx})=-\frac{\sigma_v^2}{n^{1/2}p},\quad E(f_{yy})=-\frac{\sigma_u^2}{n^{1/2}\beta^2},$$
$$E(f_{xy})=E(f)=E(f_x)=E(f_y)=0,$$

the LSABs of $\hat{X}_c$, $\hat{X}_I$ and $\hat{X}_c^*$ to $O(n^{-1/2})$ are given by

$$E(\hat{X}_c-X)=\frac{1}{\lambda_x}\left[\frac{1}{\beta}E(U)-(1-\lambda_x)\sigma_v d\right]+\frac{1}{n^{1/2}}E(\hat{\xi}_c)$$
$$=-\left(\frac{1-\lambda_x}{\lambda_x}\right)\sigma_v d$$
$$E(\hat{X}_c^*-X)=\frac{1}{\beta}E(U)+\frac{1}{n^{1/2}}E(\hat{\xi}_c^*)$$
$$=0$$
$$E(\hat{X}_I-X)=\frac{\lambda_y}{\beta}E(U)-(1-\lambda_y)\sigma_v d+\frac{1}{n^{1/2}}E(\hat{\xi}_I)$$
$$=(1-\lambda_y)\sigma_v d$$

which are the results stated in Theorem 1.

By virtue of the distributional properties of u and v, the following results are obtained:

$$E(f_{xx}^2)=\frac{s^4(1-\lambda_x)}{p\lambda_x}\left[4+\left(\frac{1-\lambda_x}{\lambda_x}\right)(2-\gamma_{2v})\right]+O\left(\frac{1}{n}\right)$$
$$E(f_{xy}^2)=\frac{s^4(1-\lambda_x)}{p\lambda_x\lambda_y}\left[1+p\lambda_x\left(\frac{1-\lambda_y}{1-\lambda_x}\right)\right]+O\left(\frac{1}{n}\right)$$
$$E(f_{yy}^2)=\frac{s^4(1-\lambda_y)}{\lambda_y}\left[4+\left(\frac{1-\lambda_y}{\lambda_y}\right)(2+\gamma_{2v})\right]+O\left(\frac{1}{n}\right)$$

$$E(f^2) = \frac{s^4(1-\lambda_x)^2}{p\lambda_x^2}\left[2\left(\frac{p}{p-1}\right) + \gamma_{2v}\right] + O\left(\frac{1}{n}\right)$$

$$E(f_{xx}f_{xy}) = \frac{2s^4(1-\lambda_x)}{p\lambda_x} + O\left(\frac{1}{n}\right)$$

$$E(f_{xx}f) = \frac{s^4(1-\lambda_x)^2}{p\lambda_y^2}(2+\gamma_{2v}) + O\left(\frac{1}{n}\right)$$

$$E(f_{xy}f_{yy}) = \frac{2s^4(1-\lambda_y)}{\lambda_y}$$

$$E(f_x f_{xx}) = \frac{s^2\sigma_v\gamma_{1v}(1-\lambda_x)}{p\lambda_x} + O\left(\frac{1}{n}\right)$$

$$E(f_x f) = \frac{s^2\sigma_v\gamma_{1v}(1-\lambda_x)}{p\lambda_x}$$

$$E(f_y f_{yy}) = \frac{s^2\sigma_u\gamma_{1u}(1-\lambda_y)}{\beta\lambda_y} + O\left(\frac{1}{n}\right)$$

$$E(f_x^2) = \frac{\sigma_v^2}{p}$$

$$E(f_y^2) = \frac{\sigma_v^2\lambda_x(1-\lambda_y)}{\lambda_y(1-\lambda_x)}$$

and

$$E(f_{xy}f) = E(f_x f_{xy}) = E(f_x f_{yy}) = E(f_y f_{xx}) = E(f_y f_{xy}) = E(f_y f) = E(f_x f_y) = 0$$

where repeated use has been made of (5.1) and (5.2)

Utilizing these results and we obtain the LSAVs to order $O(n^{-1})$ as follows:

$$E[\hat{X}_c - E(\hat{X}_c)]^2 = \frac{1}{n}E(\hat{\xi}_c^2)$$

$$E[\hat{X}_c^* - E(\hat{X}_c^*)]^2 = \frac{1}{n}E(\hat{\xi}_c^{*2})$$

$$E[\hat{X}_I - E(\hat{X}_I)]^2 = \frac{1}{n}E(\hat{\xi}_I^2)$$

which gives the results mentioned in Theorem 2.

References

Aitchison, J. A. (1977). Calibration problem in statistical diagnosis: The system transfer problem. *Biometrika, 64*(3), 461–472.

Bartlett, J. W., & Keogh, R. H. (2018). Bayesian correction for covariate measurement error: A frequentist evaluation and comparison with regression calibration. *Statistical Methods in Medical Research, 27*(6), 1695–1708.

Berkson, J. (1969). Estimation of a linear function for a calibration line: Consideration of a recent proposal. *Technometrics, 11,* 649–660.

Blas, B., Bolfarine, H., & Lachos, V. H. (2013). Statistical analysis of controlled calibration model with replicates. *Journal of Statistical Computation and Simulation, 83*(5), 939–959.

Brown, P. J. (1982). Multivariate calibration. With discussion. *Journal of Royal Statistical Society, Series B, 44*(3), 287–321.

Brown, P. J. (1993). *Measurement error, regression and calibration*. Oxford Statistical Science Series. The Clarendon Press, Oxford University Press.

Brown, P. J., & Sundberg, R. (1987). Confidence and conflict in multivariate calibration. *Journal of Royal Statistical Society, Series B, 49*(1), 46–57.

Brown, P. J., & Sundberg, R. (1989). Prediction diagnostics and updating in multivariate calibration. *Biometrika, 76*(2), 349–361.

Cheng, C., & Van Ness, J. W. (1999). *Statistical regression with measurement error*. Arnold.

Dunsmore, I. R. (1968). A Bayesian approach to calibration. *Journal of Royal Statistical Society, Series B, 30,* 396–405.

Friedland, B. (1977). On the calibration problem. *IEEE Transactions on Automatic Control, AC-22*(6), 899–905.

Fuller, W. A. (1987). *Measurement error models*. Wily.

Gleser, L. J. (1992). The importance of assessing measurement reliability in multivariate regression. *Journal of American Statistical Association, 87*(419), 696–707.

Gleser, L. J. (1993). Estimators of slopes in linear errors-in-variables regression models when the predictors have known reliability matrix. *Statistic & Probability Letters, 17,* 113–121.

Gray, C. M., Carroll, R. J., Lentjes, M. A. H., & Keogh, R. H. (2019). Correcting for measurement error in fractional polynomial models using Bayesian modelling and regression calibration, with an application to alcohol and mortality. *Biometrical Journal, 61*(3), 558–573.

Gregory, A. W., & Smith, G. W. (1993). Statistical aspects of calibration in macroeconomics. In *Handbook of statistics*. Econometrics (Vol. 11, pp. 703–719)

Halperin, M. (1970). On inverse estimation in linear regression. *Technometrics, 12,* 727–736.

Han, Y., Liu, W., Bretz, F., Wan, F., & Yang, P. (2016). Statistical calibration and exact one-sided simultaneous tolerance intervals for polynomial regression. *Journal of Statistical Planning and Inference, 168,* 90–96.

Huang, S. Y. H. (2005). Regression calibration using response variables in linear models. *Statistica Sinica, 15*(3), 685–696.

Hunter, W. G., & Lamboy, W. F. (1981). A Bayesian analysis of the linear calibration problem. *Technometrics, 23*(4), 323–350.

Joseph, V. R., & Yan, H. (2015). Engineering-driven statistical adjustment and calibration. *Technometrics, 57*(2), 257–267.

Krutchkoff, R. G. (1967). Classical and inverse regression methods of calibration. *Technometrics, 9,* 425–439.

Krutchkoff, R. G. (1969). Classical and inverse regression methods of calibration in extrapolation. *Technometrics, 11,* 605–608.

Krutchkoff, R. G. (1971). The calibration problem and closeness. *Journal of Statistical Computation and Simulation, 1*(1), 87–95.

Lee, S., & Yum, B. (1989). Large-sample comparisons of calibration procedures when both measurements are subject to errors: The unreplicated case. *Communications in Statistics, Theory and Methods, 18,* 3821–3840.

Lwin, T., & Maritz, J. S. (1980). A note on the problem of statistical calibration. *Journal of Royal Statistical Society, Series C, 29*(2), 135–141.

Lwin, T., & Spiegelman, C. H. (1986). Calibration with working standards. *Journal of Royal Statistical Society, Series C, 35*(3), 256–261.

Minder, C. E., & Whitney, J. B. (1975). A likelihood analysis of the linear calibration problem. *Technometrics, 17*(4), 463–471.

Misquitta, P., & Ruymgaart, F. H. (2005). Some results on nonparametric calibration. *Communications in Statistics, Theory and Methods, 34*(7), 1605–1616.

Osborne, C. (1991). Statistical calibration: A review. *International Statistical Review, 59,* 309–336.

Pepper, M. P. G. (1973). A calibration of instruments with non-random errors. *Technometrics, 15,* 587–599.

Rao, C. R., Toutenburg, H., Shalabh, & Heumann, C. (2008). *Linear models and generalizations: Least squares and alternatives*. Springer.

Salter, J. M., & Williamson, D. (2016). A comparison of statistical emulation methodologies for multi-wave calibration of environmental models. *Environmetrics, 27*(8), 507–523.

Sansó, B., & Forest, C. (2009). Statistical calibration of climate system properties. *Journal of Royal Statistical Society, Series C, Applied Statistics, 58*(4), 485–503.

Scheffé, H. (1973). A statistical theory of calibration. *Annals of Statistics, 1,* 1–37.

Shalabh, & Toutenburg, H. (2006). Consequence of departure from normality on the properties of calibration estimators. *Journal of Statistical Planning and Inference, 136*(12), 4385–4396.

Skrondal, A., & Kuha, J. (2012). Improved regression calibration. *Psychometrika, 77*(4), 649–669.

Spiegelman, D. (2013). Regression calibration in air pollution epidemiology with exposure estimated by spatio-temporal modeling. *Environmetrics, 24*(8), 521–524.

Spiegelman, D., Logan, R., & Grove, D. (2011). Regression calibration with heteroscedastic error variance. *International Journal of Biostatistics, 7*(1), 1–34. Article 4

Srivastava, A. K., & Shalabh. (1997). Asymptotic efficiency properties of least squares estimation in ultrastructural model. *TEST, 6*(2), 419–431.

Strand, M., Sillau, S., Grunwald, G. K., & Rabinovitch, N. (2015). Regression calibration with instrumental variables for longitudinal models with interaction terms, and application to air pollution studies. *Environmetrics, 26*(6), 393–405.

Sun, Z., Kuczek, T., & Zhu, Y. (2014). Statistical calibration of qRT-PCR, microarray and RNA-Seq gene expression data with measurement error models. *Annals of Applied Statistics, 8*(2), 1022–1044.

Tallis, G. M. (1969). Note on a calibration problem. *Biometrika, 56,* 505–508.

Wang, C. Y., Hsu, L., Feng, Z. D., & Prentice, R. L. (1997). Regression calibration in failure time regression. *Biometrics, 53*(1), 131–145.

Williams, E. J. (1969). A note on regression methods in calibration. *Technometrics, 11,* 189–192.

Williford, W. O., Carter, M. C., & Field, J. E. (1979). A further look at the Bayesian approach to calibration. *Journal of Statistical Computation and Simulation, 9*(1), 47–67.

Yu, M., & Nan, B. (2010). Regression calibration in semiparametric accelerated failure time models. *Biometrics, 66*(2), 405–414.

Yum, B., & Lee, S. (1991). Calibration procedures when both measurements are subject to error: A comparative simulation study of the unreplicated case. *Computers and Industrial Engineering, 20,* 411–420.